Principles in Weed Management

SECOND EDITION

Principles in

Weed

Management

SECOND EDITION

R.J. Aldrich and R.J. Kremer

Richard J. Aldrich is professor emeritus of agronomy at the University of Missouri. He received his bachelor's degree from Michigan State College—now University—and his doctorate from Ohio State University. Dr. Aldrich was employed jointly by the USDA and Rutgers University, Michigan State University, the University of Missouri, the USDA, and jointly by the USDA and the University of Missouri. He was one of the first persons to receive a Ph.D. in weed science and was a pioneer in several aspects of the field early in his career. He was in agricultural research administration during the middle years of his career, then returned to weed science research and teaching. After retiring, he served five years as editor of *Weed Science*.

Robert J. Kremer is a USDA-ARS microbiologist and associate professor of soil science at the University of Missouri, Columbia. He has B.S. and M.S. degrees in agronomy from the University of Missouri and a Ph.D. in soil microbiology and biochemistry from Mississippi State University. He is the author of 60 refereed publications on soil–plant–microorganism interactions, biological control of weeds, allelopathy, and fates of pesticides in the environment. He has co-organized symposia on biological control and alternative weed management and has participated in a foreign exploration for biocontrol agents on weeds. He serves on the North Central Sustainable Agriculture Research and Education Committee and is a member of regional committees on bioeconomic modeling for weed management and biological control of weeds. He is also associate editor for *Weed Science, Communications in Soil Science and Plant Analysis,* and the *Journal of Plant Nutrition.*

© 1997 Iowa State University Press, Ames, Iowa 50014.
First edition ©1984 Wadsworth, Inc., Belmont, California 94002

♾ Printed on acid-free paper in the United States of America

Second edition, 1997

Library of Congress Cataloging-in-Publication Data

Aldrich, R. J. (Richard J.)
   Principles in weed management / R.J. Aldrich and R.J. Kremer.—
2nd ed.
      p.   cm.
   Previous ed.: Weed-crop ecology. N. Scituate, Mass.: Breton Publishers,
c1984.
   Includes bibliographical references and index.
   ISBN 0-8138-2023-5
    1. Weeds—Control.   2. Weeds.   3. Weeds—Ecology.
I. Kremer, R. J. (Robert J.)   II. Aldrich, R. J. (Richard J.). Weed-crop ecology.
III. Title.
SB611.A4   1997
632'.58—dc20   96-28049

Last digit is the print number: 9 8 7 6 5 4 3 2 1

CONTENTS

PREFACE

Selective chemical weed control may well have been the most important advance of the 20th century in agriculture. It has had a direct impact on nearly every aspect of plant production from trees to turf to crops and an indirect impact on such other activities as highway maintenance, industrial site maintenance, water recreation, and many more.

In the 21st century, weed management is poised to have an equally great impact on plant production. Any successful system of sustainable food production must effectively prevent losses from weeds, which will continue to be a threat every year for the foreseeable future. Preventing such losses represents a large part of crop production costs; the associated tillage also contributes to soil and water erosion, and herbicides to soil and water contamination. By utilizing the complementary roles of weed control and weed prevention, weed management has the potential to reduce weed numbers and costs for dealing with them over time while maintaining a high-quality food supply and reducing hazards to the environment.

For the full potential of weed management to be realized, prevention must have a place alongside control in our mind-set toward weeds. Discovery of the selective weed control properties of 2,4-D fueled a flurry of research and development in the chemical industry that led to a large number of additional selective herbicides. There was a concurrent rapid growth in demand from farmers and others for information on how best to use the new herbicides to control weeds in their particular situations. This left little time for weed scientists to seek answers to such basic questions as: Why do we have weeds? Why do we have the weeds we do? What is the nature of competition? What is the relationship between weeds and cultural practices? What are the possibilities for predicting future weed problems?

Enough information has now accumulated on these ecological questions to justify preparation of this book that articulates principles of weed control and weed prevention that can be drawn upon in developing effective weed management systems. There are several good texts on weed control, and some on weed biology/demography, but none that brings together principles in control and prevention. The focus in application of these principles is the individual plant production enterprise; this recognizes the dynamic relationships that exist between weeds and desired plants at the production-unit level that can best be observed and responded to by the owner/operator on the land.

This book is aimed at college students enrolled in an introductory course in weed management. A basic botany or plant science course (or a general biology course that covers the fundamental principles of plant science) is the only recommended prerequisite. Instructors desiring more specific coverage of weed control procedures might supplement this text with information from available weed control texts, weed science journals, the chemical industry, and especially Extension Services of the land-grant universities. Such publications contain more thorough, timely, and locally appropriate information than any general textbook.

For the student pursuing weed science, the book furnishes an understanding of ecological relationships between weeds and the environment provided by agricultural and other uses of land. Further, it provides a basic understanding of competition and of approaches for minimizing effects of such competition on desired plants. It also provides background for utilizing concepts identified to develop effective weed management systems. Weed science students with this knowledge will bring to the profession of weed science an appreciation of the companion roles of weed control and weed prevention in weed management. It is our hope that such knowledge will stimulate young weed scientists to accept the challenges of using weed management principles in the development of more sustainable agriculture systems.

For those with a general interest in agriculture or the plant sciences, the book tells why all plant production enterprises require a basic understanding of weed management. Further, it identifies the relationships between weeds (and weed science) and the other aspects (and disciplines) involved in plant production. Finally, it shows how the ecological concepts and relationships can be brought together and used in a total weed management approach.

Additionally, the book will help weed scientists and other scientists identify opportunities for cooperative effort. There are many fruitful areas for multi- and interdisciplinary efforts. One is in developing systems of sustainable food production that do not harm the environment. To be fully effective, such systems require input of scientists from such diverse disciplines as genetics, ecology, entomology, sociology, economics, chemistry, and botany.

Much material was drawn from the research of colleagues, many of whom generously provided photographs and drawings used as illustrations in the book. Many in the private sector also graciously provided material used in the illustrations. Comments and suggestions to an early draft of the manuscript by Diane Kintner, John Smith, and John Albright—weed science graduate students in the Agronomy Department of the University of Missouri-Columbia—helped identify critical omissions, clarify certain points, and develop a tone appropriate to a student audience. Lynn Stanley, USDA technician, provided able assistance in compiling and preparing material for the latest draft. Valuable council and advice on some aspects were also obtained from Dr. Laurel Anderson, Dr. O. Hale Fletchall (deceased), and Dr. Harold Kerr (deceased) in the Agronomy Department and from Dr. Elroy Peters and Dr. William Donald of the USDA. The publisher's review board, which included Kriton K. Hatzios of Virginia Polytechnic Institute and State University and Richard D. Ilnicki of Rutgers, The State University of New Jersey, offered many useful criticisms during the development of the original manuscript. We are sincerely grateful for all of these important contributions.

R.J. Aldrich
R.J. Kremer

Principles in Weed Management

SECOND EDITION

1

Introduction

CONCEPTS TO BE UNDERSTOOD

1. Weeds change in response to plant production practices, especially herbicides.
2. A preventive approach to weeds calls for a mind-set geared to the reproductive parts: seeds of annuals and biennials and seeds and perennating parts of perennials.
3. A weed must be defined in terms that relate it to its ecological heritage, establish it as a biological entity, and recognize its continuing potential to change.
4. Weediness connotes a condition related to numbers.

Have you ever wondered why we have weeds and why we have the ones we do? Why is dandelion (*Taraxacum officinale*) common to lawns everywhere? Why is chickweed (*Stellaria media*) a particular problem in fall-seeded alfalfa in the Northeast, giant foxtail (*Setaria faberi*) a particular problem in corn and soybeans in the Midwest, cocklebur (*Xanthium strumarium*) a plague to cotton producers from the early days of cotton production in the South, and kochia (*Kochia scoparia*) commonly a problem in sugarbeets in the West? A quick answer might be that these are weeds that herbicides do not effectively control. Only a little deliberation tells us this answer is inadequate. Available herbicides can control each of these weeds and, in fact, are providing good control on many hectares each year. Indeed, wherever they occur, there are very few weeds that cannot be satisfactorily controlled with one or more of the herbicides available today.

Since the widespread availability of selective herbicides has only occurred in the second half of this century, is it possible that there simply has not been enough time for these herbicides to have their full effect? Given time, will the storehouse of weed seeds, rhizomes, and other reproductive parts be used up, with a consequential disappearance of weeds?

Both the explanation and the projection must also be rejected. Weed control has been practiced from the very beginning of our managed production of food. Hand removal and tillage used prior to the modern herbicide era provided enough control on many farms for full crop yields to be realized. Weeds have persisted in spite of control. Why? Must it always be so?

As a student of weed science, you will find that answers to these questions can do more than simply satisfy your curiosity about puzzles of the plant world. The answers provide the necessary background to explore and understand competition of weeds with desired plants; to identify appropriate roles for allelopathy, biotic agents, cropping and tillage practices, and herbicides in preventing losses from weeds; and to better relate weeds and weed science to the many other areas of plant science studies.

SCOPE OF THE BOOK

Nature of Weeds and Competition

Chapter 2 of this book identifies the many ways weeds reduce crop yields and add to the cost of growing desired plants. Chapter 3 provides answers to why we have weeds, why we have those we do, and why they have persisted. The ecological concepts explored that apply to weed–crop relationships tell us why. Chapters 4 and 5 consider the remarkable capacity of weeds for survival, even under very adverse conditions. Seeds are the vehicle for survival of annual weeds. Thus, long-term control—prevention—must focus on the seed. Most perennial weeds reproduce by both seeds and vegetative parts. However, it is the protection offered by the vegetative reproductive parts during both climatic and crop competition stress that makes perennial species especially troublesome. Long-term control, or prevention, must focus on such reproductive parts. Factors affecting production and longevity of seed and perennating parts are identified and explained.

Chapter 6 looks specifically at factors that affect resumption of growth from seed and from perennating parts. Resumption of growth, either of seed or of vegetative parts, is a key phase for crop production in the life history of weeds since its timing and numbers in relation to the crop life cycle determine the severity of competition or whether there is competition at all. Chapter 7 examines the nature of competition between weeds and desired plants. What is competed for and when?

Weeds in Production Systems

With an understanding of the nature and competition of weeds as background, the remainder of the book explores ways of dealing with weeds in production systems. There is much evidence that metabolites of many weeds and desired plants have an adverse effect on the growth of plants. In Chapter 8, allelopathy, as this effect is known, is considered as a possible factor in the persistence of certain weeds, in the shift in composition of weed communities following changes of cultural practices, in the direct interference with production of a given crop, and as a potential tool for preventing losses from weeds.

It is a rare organism that has no natural enemies, if indeed such an organism exists. Weeds are no exception. Chapter 9 draws upon this fact to identify the role of biotic agents in preventing losses from weeds. Concepts involved in their successful use are listed and discussed.

Discovery of the auxin-type herbicides in the 1940s has led to the widespread use of herbicides common throughout agriculture today. Whereas most herbicides prior to the 1940s were inorganic chemicals, all those of the modern era are organic. The majority of the latter are systemic—that is, readily translocated in plants. Chapter 10 examines the current usage of systemic herbicides and introduces their broader potential role in weed management. Chapter 11 develops a general understanding of entry, transport, and gross effects as they relate to plant anatomy and to the environment. With this knowledge as background, Chapters 12 and 13 discuss the place of herbicides in each of three broad approaches to weed management: (1) preventing emergence of weeds with crops; (2) minimizing competition from weeds growing with crops; and (3) reducing the number of viable propagules in the soil. Both the physical environment and characteristics of the weeds themselves are discussed as they influence successful use of herbicides.

It has been said that the history of weeds is essentially the history of human beings. The significance of this statement for weed–crop ecology is the implication that weeds are associated with disturbed environments. Production agriculture involves repeated disturbance of the environment. Chapter 14 examines specific effects of cropping practices, tillage practices, and weed control practices on weed numbers and species. Understanding these effects is a necessary first step in predicting future weed problems and in designing more effective ways of dealing with them.

The preceding chapters provide a sequence of subject matter coverage that builds logically from the constraints of our current emphasis on control to the expanded opportunities of weed management. The concepts developed are brought together in Chapter 15. The need to base weed management on prevention is explained and the relationship between prevention and control in such an approach is discussed. In effect, Chapter 15 provides the necessary foundation for establishing weed management as the next higher level in a hierarchy of approaches to weeds.

A NEW MIND-SET TOWARD WEEDS

For weed management to contribute fully to a sustainable agriculture, there must be a change in our mind-set toward weeds. For the most part, weed control as practiced in farming today is geared toward dealing with the specific weed(s) in a given crop in a single year. Within this overall approach, a number of different methods for dealing with weeds have been developed in response to different weed problems faced. Preemergence and preplant-incorporated application of herbicides thus are based on the expectations that there will be a competing population of weeds—mostly annuals—with the crop. Machines have been adapted and new ones built to remove weeds growing with the crop. Sprayers have been designed to place chemicals selectively on the weeds and not on the crop. Similar practices have been followed in other plant-production enterprises.

Under this year-to-year approach, a new problem resulting from a change in weed composition or from a change in an individual species is dealt with only after the new problem is established. The result can be a need for drastic changes in weed control methods with associated costs to the farmer. Further, this approach implies that some measure of loss is to be accepted as a result of a change in the weed community before an effective control is developed. Finally, as we shall see, the year-to-year approach unnecessarily restricts the technology that can be used to minimize losses from weeds.

Changes in Weeds

There is much evidence to show that weeds do change in response to weed control and other production practices. Shifts in weed composition in the Corn Belt of the United States provide a striking example of such changes over a broad area. Following its introduction in the late 1940s, the herbicide 2,4-D[1] was widely used in corn throughout the Corn Belt. It selectively controlled many of the broadleaf annual weeds, causing a shift in the weed community by the late 1950s to one dominated by annual grasses. Herbicides (such as triazines) especially effective against annual grasses were developed and widely used in the 1960s. By the 1970s, species such as fall panicum (*Panicum dichotomiflorum*) and honeyvine milkweed (*Ampelamus albidus*) had become prevalent because by germinating and making their growth later in the season, they escaped the effects of the commonly used preemergence and early postemergence herbicides.

Changes have also occurred within weed species in response to herbicide use. There are now populations of many weeds resistant to herbicides that once provided effective control (Thill et al., 1991; Holt and LeBaron, 1990). This is because some plants within the species were resistant to a particular herbicide. Continual use of the herbicide led to selection of a resistant population. Resistance to the selecting herbicide may also impart resistance to other unrelated herbicides (Burnet et al., 1994; Holt and LeBaron, 1990).

---

[1]The common names of herbicides accepted by the Terminology Committee of the Weed Science Society of America are used in this text. A list of accepted common names and their chemical names are in Appendix 2.

Need for a Changing Approach

Weed science has much to gain by developing an understanding of the ecological relationships among weeds, desired plants, and production practices, even though the prediction of future weed problems and the development of corresponding prevention programs are still only goals. Current approaches to control can be improved if there is a better understanding of such relationships.

What is needed is recognition and acceptance of an appropriate companion role for prevention in dealing with weeds. A simple analogy identifies the change needed. Our historical approach, which has been to deal with weeds when and where we have them, was built upon the view that weed vegetative growth must be kept out or controlled if crop losses are to be avoided. In this view, weed reproductive parts—that is seeds, rhizomes, and so forth—are viewed as simply the vehicles for producing more plants. Conversely, the preventive view focuses on production of seed and perennating parts and on their germination. This view sees the plant as a vehicle for producing reproductive parts.

The former, or historical, view is adequate when competition from weeds is the only focus. It is inadequate for a perspective broadened to encompass changes in the weed community in response to changes in production practices. The first view recognizes that numbers of individual weeds determine the consequences of weeds for a particular crop. The latter, or preventive, view recognizes that it is the individual plant and its fitness for both the natural environment and the environment we provide that determines shifts in the weed community and changes within the individual weed species. The reason is that the individual—not the population—transmits the genetic message. Under a preventive mind-set, accumulation of knowledge about the individual weed's response to our imposed environment will ultimately be enough to make forecasting a reality. Rapid expansion of herbicide resistance adds a sense of urgency in adopting a preventive mind-set toward weeds.

DEFINITIONS

Weeds are constant associates of our cultivated plants. Anyone familiar with plants has a perception of what is a weedy and what is a clean crop, lawn, vegetable garden, or other managed use of plants. Yet, as we shall see, it is not easy to determine the degree of weediness that represents a threat to crop yield or to survival of a lawn. Neither is it easy to define a weed in terms that clearly distinguish it from desired plants and from other wild plants.

We will see, however, that weeds do have important characteristics different from cultivated plants. Also, out of 250,000 plant species in the world, fewer than 250 species (0.1%) account for the readily apparent weediness often seen. Thus, weeds are hardly synonymous with wild plants. Having in mind a clear definition of a weed and a grasp of weediness is important to a full appreciation and understanding of the concepts and principles this book encompasses.

Weed Defined

How should a weed be defined? Numerous definitions can be found in the literature. Such common definitions as "a plant out of place," "a plant interfering with the intended use of the land," and "a plant with negative value" fairly well describe a weed in a vegetation control context. That is, they are adequate when our focus is only on the problems posed by the growth of unwanted plants at a particular point in time. Further, these definitions embrace the key fact that a plant appropriately is a weed *only in reference to humans*. The problem comes when we try to use these definitions beyond a point in time and a particular place. Put another way, the definitions do not fit the dynamic aspects implied in a focus on ecological relationships. Rather, these definitions imply that a weed is fixed in time in a static state. As we shall see, weeds are continually changing. We will also see, contrary to the first definition, that we have weeds because our production practices provide a place for them.

The second definition poses a definitional problem for those plants that are present but become an interfering associate only when we change crops or production practices. Dandelion that infest lawns in subdivisions developed on land removed from agricultural use is one of many examples that could be cited. Under a strict interpretation of the second definition, a dandelion in a field used for food or feed production would not be called a weed because it is probably not interfering, but a dandelion in the lawn would. Calling one dandelion a weed and not the other is not based on any inherent difference between the plants, only on a difference in when and where they occur. Similarly, negative value can only apply to plants when they are present where they are not wanted. This definition accommodates cultivated plants that escape or volunteer to interfere with our intended use of the land. This loophole in itself suggests a weakness in the definition since it becomes confounded with the definition of a crop.

A more useful definition of a **weed** is *a plant that originated under a natural environment and, in response to imposed and natural environments, evolved and continues to do so as an interfering associate with our desired plants and activities.* Such a definition continues to relate to us as it should, but as contrasted with previous definitions, it provides both an origin and continuing change perspective. Recognizing that weeds are part of a dynamic, not static, ecosystem helps expand our thinking on how best to prevent losses from them. Excluding plants developed by us for our uses properly recognizes inherent differences between such plants and weeds with respect to seed dormancy, seed longevity, and length of the life cycle. This definition of a weed is used throughout the text.

Cultivated plants as weeds.   The new definition of a weed leaves the question of what to call escaped ornamental and volunteer cultivated plants. It is suggested that they be called exactly what they are—that is, *volunteer* corn, *volunteer* wheat, or some other *volunteer* crop growing where it is not wanted. Similarly, plants introduced as ornamentals that have escaped to interfere with our activities would be identified as *escapes*. For example, *escaped* Japanese knotweed (*Polygonum cuspidatum*), *escaped* prickly pear cactus (*Opuntia* spp.), or some other escaped ornamental.

Weediness Defined

Identifying as volunteer or escaped cultivated or ornamental plants that behave like weeds helps us grasp what is meant by weediness. **Weediness** is defined as *"the state or condition of a field, flower bed, lawn, and so forth in which there is an abundance of weeds"* (Winburne, 1962). Thus, weediness connotes a condition. By contrast, under the definition proposed in this book, a weed is a particular *biological entity*. It follows that it is appropriate to refer to soybeans as being weedy with corn and to a flower bed as being weedy with bluegrass.

Costs of weeds. Weediness also connotes "numbers." To suggest that a crop is weedy implies that it has many weeds—enough to reduce crop yield. Although weeds have many other negative aspects, it is this threat to yield, to the production of food, that accounts for most of the effort devoted to their control since the beginning of agriculture. Crop losses due to weeds are still very large. In the United States alone, as shown in Table 1-1, the estimated average annual loss exceeds $6 billion. Costs associated with weed control exceed $9 billion annually.

TABLE 1-1. Losses due to weeds in the United States.

Commodity	Loss $ (millions)
Field crops, fruit, and nuts	4,200
Pasture, hay, and range	1,521
Animal health	469
Total	6,190

Source: Data from Bridges, 1994.

Weeds represent a cost in many other respects. They directly affect human health through allergies and poisoning. Weeds adversely affect livestock production: Animals die from eating poisonous weeds, off-flavor is imparted to milk and other dairy products when weeds such as wild garlic (*Allium vineale*) are eaten, and hides and carcasses are physically damaged by weeds. Weeds may harbor insect and disease organisms that affect cultivated plants. Finally, cost for their control in fields and on such nonagricultural sites as along highways, around buildings, and in waterways is more than $1 billion a year (Bridges, 1994). Nearly all of these aspects imply the occurrence of excessive numbers. It is understandable, therefore, that the most commonly used definitions for a weed are also synonymous for weediness. This is one more reason why the traditional definitions are inadequate in a weed–crop ecology context because, as we shall see, the focus must frequently be on the individual, not the group, in order to understand properly the ecological forces involved.

Weed Management Defined

Simply stated, **weed management** is *an approach in which weed prevention and weed control have companion roles.* It implies a systems context in which all available tools are used to reduce the propagule seedbank, prevent weed emergence, and minimize competition from weeds growing with desired plants. Thus, weed management has both immediate and long-term objectives. This approach also implies a consideration of weeds in the broader context of their interactions with production practices. It follows that weed management requires knowledge of weeds themselves and of the ecological principles that involve them. The remainder of this book provides the necessary knowledge for such an expanded approach to weeds.

WEED NOMENCLATURE AND CLASSIFICATION

Scientific Classification

Binomial names.   Because weed species have ecological characteristics unique to each, it is essential that they be identified by a standard system. Further, the entire science of weeds and weed management is built upon the weed species present or anticipated. The standard of plant nomenclature in use throughout the world today gives the name in Latin and is based upon the 1753 publication *Species Plantarum* by Linnaeus in which a two-part, or *binomial*, naming system was used. The first part identifies the genus, or generic name, and the second part identifies the specific epithet.

For example, in the binomial name of giant foxtail (*Setaria faberi*), *Setaria* identifies the genus and *faberi* the specific epithet. Under the International Code of Botanical Nomenclature, only one genus can have the name *Setaria*, although the same specific epithet may be used for plants of different genera. Since reading is easier when common names are used, the generic name is used in this book only the first time the species is referred to in the body of the text. The binomial name of every species referred to by common name is listed in Table 1 of the Appendix.

Hierarchy of the Plant Kingdom.   Using lambsquarters (*Chenopodium album*) as an example, the major subdivisions of the taxonomic hierarchy of the plant kingdom include the following rank of taxa:

Kingdom—Plantae
    Division—Tracheophyta
        Subdivision—Spermatophytina
            Class—Angiospermae
                Order—Caryophyllales
                    Family—Chenopodiaceae
                        Genus—*Chenopodium*
                            Species—*Chenopodium album*

The ranks used most often in weed science are family, genus, and species. There are about 450 families of flowering plants in the world. A relatively few families contain the majority of the approximately 200 species that account for most of the losses in food production worldwide. The main contributing families and number of weed species in each are as follows: Poaceae (44), Cyperaceae (12), Asteraceae (32), Polygonaceae (8), Amaranthaceae (7), Brassicaceae (7), Leguminosae (6), Convolvulaceae (5), Euphorbiaceae (5), Chenopodiaceae (4), Malvaceae (4), and Solonaceae (3). The first three families contain 44% of the problem weeds, and the 12 combined, 68% (Holm, 1978). Some families and representative weed species of each are shown in Table 1-2.

TABLE 1-2. Some plant families and a representative weed species of each.

Family		Representative Weed	
Common Name	Latin Name	Common Name	Latin Name
Amaranth or pigweed	Amaranthaceae	Redroot pigweed	*Amaranthus retroflexus*
Buckwheat or smartweed	Polygonaceae	Pennsylvania smartweed	*Polygonum pennsylvanicum*
Chickweed or pink	Caryophyllaceae	Common chickweed	*Stellaria media*
Composite	Compositae	Common ragweed	*Ambrosia artemisiifolia*
Goosefoot	Chenopodiaceae	Lambsquarters	*Chenopodium album*
Grass	Gramineae	Quackgrass	*Elytrigia repens*
Mallow	Malvaceae	Velvetleaf	*Abutilon theophrasti*
Milkweed	Asclepiadaceae	Common milkweed	*Asclepias syriaca*
Morning-glory	Convolvulaceae	Field bindweed	*Convolvulus arvensis*
Mustard	Cruciferae	Wild mustard	*Brassica kaber*
Nightshade	Solonaceae	Jimsonweed	*Datura stramonium*
Parsley	Umbelliferae	Wild carrot	*Daucus carota*
Plantain	Plantaginaceae	Buckhorn plantain	*Planatago lanceolata*
Purslane	Portulacaceae	Common purslane	*Portulaca oleraceae*
Sedge	Cyperaceae	Yellow nutsedge	*Cyperus esculenetus*
Spurge	Euphorbiaceae	Leafy spurge	*Euphorbia esula*

Common Systems of Classification

Grasslike and broadleaf weeds.   The seed-producing plants (Spermatophytina) have two classes: Angiospermae (covered seed) and Gymnospermae (naked seed). Most weeds are in the Angiospermae. The Angiospermae is further separated into two subclasses based upon the number of cotyledons (seed leaves). *Monocotyledoneae* are those weeds whose embryo has one cotyledon. *Dicotyledoneae* have two cotyledons. The name *grassy* or *grasslike* is commonly applied to the monocotyledons, and the name *broadleaf*, to the dicotyledons. The common names are descriptors of both leaf shape and growth form. The monocotyledons commonly have a nonbranched growth form with leaves that are usually narrow or linear in shape with parallel veins. The dicotyledons commonly have a branched growth form with leaves that are usually broader and net-veined. As we shall see later, these differences in leaves and growth form may have far-reaching implications for a weed's competitive ability and for its control or management.

Annual, biennial, and perennial weeds.   Weeds are also commonly classified according to their life cycles. *Annuals* are species that can complete their life cycle in 1 year or less from seed germination to seed production. There may be two kinds: winter annuals and summer annuals. *Summer annuals* germinate in the spring, produce seed, and die later that same year. *Winter annuals* germinate in the fall of one year, overwinter, then resume growth, produce seed, and die the next year. Annuals represent by far the largest number of weed species that compete with annual row crops. The reason is that completion of their life cycle is interfered with less in the production of annual crops than is that of biennials or perennials. This statement is further emphasized by the fact that summer annuals, such as redroot pigweed, giant foxtail, and lambsquarters, are most commonly associated with spring-seeded crops such as soybeans, corn, and spring wheat. Winter annuals, such as chickweed and pepperweed (*Lepidium* spp.), are most common with fall-seeded crops, such as legumes and winter wheat.

Biennials may complete their life cycle in 2 years. They germinate one year, overwinter, then resume growth the next year and may produce seed, after which they usually die. There are only a few biennial weeds. Among the common ones are wild carrot (*Daucus carota*), mullein (*Verbascum thapsus*), and bull thistle (*Cirsium vulgare*). Because more than 1 year is required to complete the life cycle, such species are not a common problem with annual row crops but may be with perennial crops used for hay and pasture. In nature, many individual plants of so-called biennial species may take more than 2 years to complete the cycle. For example, if flowering and seed production of wild carrot are prevented in the second year, it has been shown that the plants continue vegetative growth into subsequent years, continuing possibly for 4 to 5 years (Holt, 1972). The terminating feature is seed production, not years from germination.

Perennials, as the name implies, are species that may live more than 2 years. A distinguishing feature is that many such species can reproduce from vegetative parts, as well as from seed, in the same and subsequent years. At any given time, new plants may be arising both as a result of regeneration from the perennating parts and from seed. It is the perennating characteristic that makes some perennials so difficult to prevent from interfering with crop production. Vegetative reproduction is discussed in detail in Chapter 5. Note here, however, that such reproduction may occur from roots (dandelion), tubers (yellow nutsedge *Cyperus esculentus*), rhizomes (quackgrass

Elytrigia repens), stolons (bermudagrass *Cynodon dactylon*), and bulbs (wild garlic).

Some care must be exercised in classifying weeds according to life cycle. Life cycle itself is not fixed; rather, it is greatly influenced by the environment. As mentioned earlier, the life cycle of the biennial wild carrot can be extended beyond 2 years. Also, plants of some biennials may behave as annuals. Plants of some species commonly classified as annuals can be made to live for more than 1 year by providing conditions that encourage vegetative growth and discourage flowering. Younger (1961) observed that large crabgrass (*Digitaria sanguinalis*), classified as a summer annual, behaved otherwise in bermudagrass turf in southern California. Some plants in the turf, which was regularly mowed to 2.5 cm (1 in.), overwintered, and if allowed to do so, flowered and produced seed the next spring. At the other extreme, plants of some species commonly classified as perennials can be forced to behave as annuals. This situation is illustrated by the fact that bermudagrass, widely used as a perennial forage species in the southeastern United States, fails to become a perennial if overgrazed. In fact, the objective of cultural practices used to check perennial weeds is to prevent the production of perennating parts, which prevention, in effect, causes them to behave as annuals. In summary, it is helpful to view classification by life cycle as applying to the entire population of a species. In this way, general behavior is emphasized rather than behavior of individuals. Plasticity of life cycle of individual specimens should be borne in mind, however.

FRAMEWORK OF THE BOOK

Units of Measurement

The metric system is used throughout the body of the text. Where data in references cited are presented in the English system, the English unit is shown in parentheses after the metric unit. This method is used to avoid the confusion that would otherwise occur in switching back and forth from metric to English units, depending upon the system used in the data cited. Conversion to metric units of data published in English units results in some awkward numbers. For example, a depth of 3 in. reported in a tillage study becomes 7.62 cm. To improve readability, only one decimal place is used in such transformed figures. In this case, the depth as given in the text is 7.6 cm. To correctly reproduce figures and tables taken from other publications, the system used in the other publication is retained. Transformation in the text to metric units with English units in parentheses minimizes the need for conversion.

Throughout the text negative exponents are used with the abbreviations to indicate type of measurement, e.g., kg ha^{-1} for kilograms per hectare, g m^{-2} for grams per square meter, and plants m^{-1} of row for plants per meter of row.

Literature Treatment and Key Points

A large body of literature has accumulated on weed science. This book is not intended as a review of that literature. Only that literature needed to establish the concepts and principles covered is cited. Because there is limited literature on some important concepts and principles, some citations necessarily go back many years. However, these citations should not be construed as a review of the particular literature. Similarly, some recent publications on particular subjects are not cited simply because they neither add new concepts nor modify concepts already established from earlier publications.

A list of important concepts to be understood is given at the beginning of each chapter. This list will help you recognize and retain as a reference key points in each chapter. It is important to grasp the concepts in each chapter as you go along because each succeeding chapter, to some extent, builds upon concepts and principles of the preceding chapter(s).

REFERENCES

Bridges, D.C. 1994. Impact of weeds on human endeavors. Weed Technol. 8:392-395.

Burnet, Michael W.M., A.R. Barr, and S.B. Powles. 1994. Chloroacetamide resistance in rigid ryegrass (*Lolium rigidum*). Weed Sci. 42:153-157.

Holm, L.G. 1978. Some characteristics of weed problems in two worlds. Proc. West. Soc. Weed Sci. 31:3-12.

Holt, B.R. 1972. Effect of arrival time on recruitment, mortality, and reproduction in successional plant populations. Ecology 53:668-673.

Holt, J.S., and H.M. LeBaron. 1990. Significance and distribution of herbicide resistance. Weed Technol. 4:141-149.

Thill, D.C., C.A. Mallory-Smith, L.L. Saari, J.C. Cotterman, M.M. Primiani, and J.L. Saladini. 1991. Sulfonylurea herbicide resistant weeds: discovery, distribution, biology, mechanism, and management. In J.C. Caseley, G.W. Cussans, and R.K. Atkin, eds., Herbicide resistance in weeds and crops, pp. 115-128. Oxford: Butterworth-Heinemann.

Winburne, J.N. 1962. A dictionary of agricultural and allied terminology. E. Lansing: Michigan State University Press.

Younger, V.B. 1961. Winter survival of *Digitaria sanguinalis* in subtropical climates. Weeds 9:654-655.

—2—

Competitiveness of Weeds

CONCEPTS TO BE UNDERSTOOD

1. Weeds may cut crop yields by reducing harvestable yield and/or by reducing yield harvested.
2. Weeds must be prevented from growing with the crop during the first few weeks or must be removed after only a few weeks cohabitation to avoid loss of crop yield.
3. Weeds that emerge after about one-third of the life cycle of annual crops usually do not reduce harvestable crop yield.
4. The effect of weed numbers on crop yield tends to be curvilinear.
5. Broadleaf weeds tend to cause greater losses in crop yields than grassy weeds.
6. Most weeds, especially annuals, are very intolerant of competition.
7. Crops, and varieties within crops, differ in their competitiveness toward weeds.
8. Annual weeds commonly have a shorter life cycle than the crop with which they are competing.

A general understanding of losses caused by weeds is important background to an understanding of the complicated nature of competition and ultimately to successful approaches for preventing or minimizing such losses. Answers to such questions as, When must weeds be controlled to avoid losses in crop yield? and How many weeds can a crop tolerate without suffering losses in yield? have been, and continue to be, central to weed management. Thus, it is not surprising that a sizable literature has accumulated on weed–crop competition. A review of such literature to mid-1978 by Zimdahl (1980) includes 586 citations. A voluminous literature on the subject continues to accumulate. No attempt is made to review this literature. The concepts and general trends and their relationships to established ecological concepts and principles, where such can be established, are covered here.

DIRECT AND INDIRECT LOSSES FROM WEEDS

In general terms, losses may be subdivided into two categories: direct and indirect. *Direct losses* are those that reduce the quantity or cash return of the crop produced. Included are reductions in yield, both that produced and that harvested, and contamination resulting in dockage for the crop when it is sold. *Indirect losses* are those that represent a cost to society at large or to the producer and property owner but in themselves do not represent a reduction in cash return. Included are the harboring of insects, diseases, and pests of other crops and desired plants; creation of safety hazards, such as limited visibility at highway intersections; creation of health hazards, such as those posed by common ragweed pollen for hay fever sufferers and by poison ivy (*Toxicodendron radicans*) for those sensitive to its toxin; reduction in property values; the increased cost of right-of-way maintenance; and the increased cost of crop production since many of the practices involved in producing crops are primarily for weed control or weed prevention.

Reductions in Crop Yield

The losses of main concern to weed–crop ecology are those related to yield. Thus, direct losses are our concern as we turn to a consideration of competition. Weeds may reduce crop yields in two ways: (1) by reducing the amount of harvestable product (grain, stover, forage, and so on) produced by the crop, and (2) by reducing the amount of crop actually harvested.

Reductions in harvestable crop.   As we shall see, the extent of reduction in crop yield caused by weeds varies greatly depending upon the crop, the weed, and the growing conditions. A severe infestation present for the entire growing season may result in complete loss of some crops. Because weeds cut yields, means for controlling them have been an integral part of agriculture for all of recorded history. Great strides have been made in weed control, especially since the 1940s. The impetus for control, of course, has been the availability of selective new herbicides. The net effect has been to very much reduce crop losses from weeds. Even so, as we saw in the previous chapter, annual losses from weeds still exceed $6 billion.

Reductions in harvested crop.   Weeds can cause sizable machine losses when grain crops are combined. The losses can be a result of: (1) the additional bulk provided by weeds interfering with threshing and separating of grain, and (2) weeds interfering with actual cutting and movement of the grain into the combine. Some representative losses of this nature are shown in Table 2-1.

The major part of the loss occurred at the header. Weeds, particularly redroot pigweed, resulted in more beans remaining in pods left on the stubble and more beans on stalks cut but not delivered into the combine. Threshing and separating losses were about the same for both weeds and approximately 3 times the losses for weed-free soybeans.

TABLE 2-1. Soybean harvest losses caused by redroot pigweed and giant foxtail.

	Percent Loss		
Weed	Header Losses*	Threshing and Separating Losses	Total Losses
Redroot pigweed	5.35	0.73	6.08
Giant foxtail	1.55	0.81	2.36

*Includes losses from (a) beans free of pods or in pods free of stalks, (b) beans in pods remaining attached to the stubble, (c) beans remaining in pods attached to stalks that were not cut, and (d) beans remaining in pods attached to stalks that were cut but not delivered into the harvester.
Source: Data from Nave and Wax, 1971.

Weeds may also reduce harvested yield and lower the quality of fiber crops such as cotton. Figure 2-1 shows that trash in mechanically harvested cotton and cotton left in the field increased as the number of tall morning-glory (*Ipomoea purpurea*) plants in the cotton row increased. As few as 4 weeds in a 15-m row more than doubled the amount of trash.

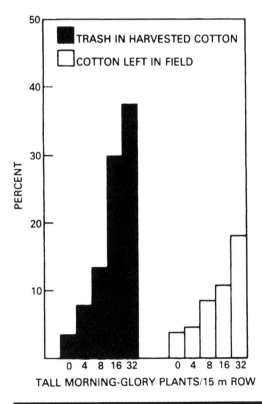

FIGURE 2-1. Reduction in the quality of mechanically harvested cotton and increase in the amount missed by the picker caused by weeds.

Source: Data from Crowley and Buchanan, 1978.

Some weeds, when present in the harvested crop, reduce the price received by the producer. For example, in Missouri, if the number of wild garlic bulblets is 2 or more per 1000 g of wheat (about 65 per bushel), the wheat is graded "garlicky" and docked in price. Harvest losses and dockage are direct losses from weeds and should be kept in mind as possible additional losses to those caused by competition.

FACTORS IN CROP YIELD LOSS

The extent of crop yield loss is closely tied to the number of competing weeds and their weight. That is, there is some number or weight above which loss or damage occurs and below which it does not. These are not the only factors involved since, as we shall see, *when* the weed is present in relation to the life cycle of the crop also influences importantly the degree of competition. Nonetheless, numbers and weight are the basic elements of competition, with the time of presence of the weed serving in what might be termed a modifying role. A sparse weed stand cannot cut crop yield, no matter how long it is present. As we shall see, a dense weed stand cannot remain long without reducing the crop yield.

Weed Numbers

The fact that crop losses increase as weed numbers increase is common knowledge. What is surprising is how few weeds may cut yields. For example, as few as 1 kochia plant per 3 m (10 ft) of row cut sugarbeet yields 26% (Weatherspoon and Schweizer, 1971); 1 wild mustard plant per 30 cm (1 ft) of row cut soybean yields 30% (Berglund and Nalewaja, 1969); and a barnyardgrass (*Echinochloa crus-galli*) plant per 0.1 m² (1 ft²) cut rice yields 57% (Smith, 1968). These examples, plus results from a large number of other studies, show that relatively few weeds may reduce crop yield.

Because of physical aspects of the environment, precise density–crop yield relationships cannot be projected for field conditions. The general relationship is the solid line in Figure 2-2. This relationship is curvilinear, not linear, and simply means that one weed on the sparse-density end has a greater effect on yield than one weed on the high-density end. This effect is due to plasticity in plant form on the part of both the weed and the crop plant. As the number of weeds on a given area increases, the size of each plant decreases (as expressed in numbers of branches or tillers, numbers of leaves and size, and size of root system), thus reducing the individual plant's competitiveness. The number of weeds tolerated by the crop is commonly called *threshold* value.

To the producer the important question is where on the density-yield curve the cost of weed control equals the increased return in yield from the control action. This point is commonly called the *economic threshold* (Cousens, 1987). Efforts to eliminate weeds represent a cost to producers, so they must decide how much they can afford to

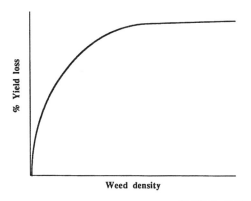

FIGURE 2-2. Yield loss relationship to weed density.

Source: According to Cousens rectangular hyperbola, 1985a.

spend for control or prevention. Using the economic threshold to make weed management decisions will be covered in Chapter 15. At this point, we simply need to recognize that the effect of an individual weed plant on crop yield varies greatly depending upon time of emergence, growing conditions, crop cultivar, and other factors.

Modifying effect of time of emergence. When the weed emerges relative to when the crop emerges has a dramatic effect on competitiveness of the weed. This is shown by the effect of lambsquarters on sugarbeet yield losses in Table 2-2. Loss was only half as much with the same density of lambsquarters that emerged 10 days after sugarbeets as it was if the lambsquarters emerged at the same time as sugarbeets. Nearly twice as dense a stand of lambsquarters that emerged 21 days after sugarbeets reduced yield only 7%.

TABLE 2-2.   Effect of time of lambsquarters emergence on yield of sugarbeets.

Weed Density (plants m^{-2})	Weed Emergence (days after crop)	Yield Loss (%)
5.5	0	79
5.5	10	37
9.1	21	7

Source: Data from Kropf and Lotz, 1992.

Period Weeds Are Present,

Period tolerated. How soon must weeds be removed from the crop? Logic tells us that there is a time very early in the crop's growing period when the presence of small weeds will not cut crop yield. Duration of this tolerant period tells us when weeds must be removed to avoid crop yield losses. Most agronomic annual row crops germinate quickly and have seedlings that grow rapidly. Plant breeding programs indeed have tended to select for these attributes. Thus, competition from weeds is usually not expressed in reduced stands of the crop.

Some species of grasses and legumes used as forages are not this aggressive. Birdsfoot trefoil is one legume that establishes slowly. With these crops weeds may establish well in advance and have a very pronounced effect upon the stand of the crop, as shown in Table 2-3. Thus, for crops that establish slowly, competition from weeds may occur at the very outset of growth.

TABLE 2-3.   Stand of birdsfoot trefoil as affected by weeds.

Weeding Treatment	Trefoil Plants (no. 1.4m^{-2})
No weed removal	7.3
All weeds removed	35.6

Source: Data from Kerr and Klingman, 1960.

The more usual situation is for competition to affect growth, rather than stand, and be reflected in reduced crop yield. Results from a large number of studies, summarized in Table 2-4 (column A), show that most crops can tolerate the presence of weeds for a relatively short time only, depending upon the weed and the crop. For example, mixed annual weeds left longer than 3 weeks in corn resulted in a measurable reduction in crop yield. As can be seen, crops vary greatly in the period of weed presence tolerated, from as little as 3 weeks to as much as 22 weeks. Also, for any given crop,

TABLE 2-4.   Range of weed–crop relationships with time: (A) duration of weed presence tolerated without yield loss and (B) weed-free period required to prevent crop yield reduction.

Crop	Weed	Location	Weeks after Seeding/Emergence* (A)	(B)
Bean	Mixed annuals	Washington, USA	8	5
Beet, red	Mixed annuals	England	4	2-4
Cabbage	Mixed annuals	England	3-4	2
Corn	Mixed annuals	Mexico	3	5
Corn	Giant foxtail	Illinois, USA	6	3
Cotton	Mixed annuals	Alabama, USA	8	6
Cotton	Mixed annuals	Mexico	9	4
Lettuce	Mixed annuals Sicklepod	England	3	3
Peanut	Florida beggarweed	Alabama, USA	4	8
Potato	Redroot pigweed, lambsquarters	Lebanon	6	9
Sorghum	Mixed annuals	Nebraska, USA	4	3
Soybeans	Giant foxtail	Illinois, USA	8-9	3
Sugarbeets	Barnyardgrass	Washington, USA	12	10
Sugarbeets	Kochia	Colorado, USA	4	6
Wheat, winter	Downy brome	Oregon, USA	22	
		Nebraska, USA		2

*Some studies counted weeks from planting and others weeks from crop emergence. Studies reported include only those where one or the other was used for both (A) and (B).

Source: Adapted from Zimdahl, 1980.

the period tolerated depends upon the weed. Without exception, however, crops tolerated weeds for a time early in the growing season. The practical importance of this fact to the producer is that weeds must be controlled or prevented during the early part of the crop's growing period if losses in yield are to be avoided.

Weed-free period required.   Selected results of studies to determine the weed-free period required, shown in Table 2-4 (column B), indicate that once well established, most crops effectively compete with weeds. Here, too, the period required is relatively short and, for many crops, is only the first few weeks. As is true for the period tolerated, the weed-free period required varies widely for the different crops and within a single crop. Without information on specific competitive effects, we can only speculate that the explanation is to be found in differences in plant form interacting with growth factors.

Figure 2-3, which shows the two effects of weed duration in sugarbeets, is representative of a large number of weed–crop relationships. Once competition from weeds begins, crop yields drop sharply from continued weed presence. Similarly, the curve depicting the effect of weed-free period shows that crops quickly become competitive. The shapes (slopes) of the two curves are different, however. The slope showing yields for weed-free duration is skewed to the left compared with that for yields for weed duration, indicating that the required weed-free period is shorter than the presence-tolerated period. In a very general way, the weed-free period measures the relative competitiveness of the crop, and the presence-tolerated period measures the competitiveness of the weed in the crop. As we shall see, most annual weeds are relatively poor competitors compared to most crops, which explains the general relationship between the length of the weed-tolerant and weed-free periods. The difference in length of these two periods has important implications for the choice of weed management approaches and is expanded upon in Chapter 15.

Life Cycle Differences

Annual weeds commonly have a shorter life cycle than the crop with which they are competing. This fact is pointed out in Figure 2-4, which compares the time needed to develop maximum leaf area (as measured by leaf area index, LAI) for soybeans and velvetleaf (*Abutilon theophrasti*). LAI is the ratio of the surface area of leaves to a given area of ground. Velvetleaf reached its maximum LAI at 8 weeks and soybeans at 10 weeks after emergence. This is a factor in competitiveness of weeds toward crops.

The short life cycle of weeds may explain why early competition from weeds cuts yields (Li, 1960). In his studies, he estimated that weeds made 15% to 18% of their total growth during the first 2 to 3 weeks after emergence compared with less than 1% growth for corn during that time. Note, however, that competition for light continues after the weed has stopped increasing in height and leaf area. That is, the weed leaf in place when growth stops will continue to shade the crop leaf. The significance of short life cycles of many weeds for weed management is that efforts to check growth may need to be followed for only a relatively short time.

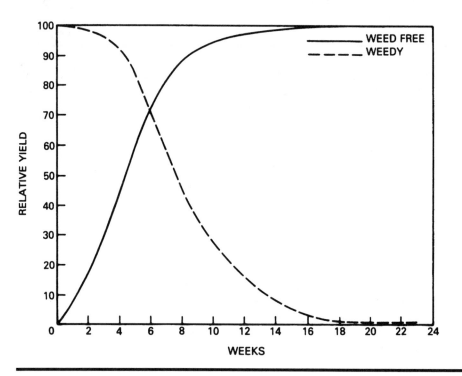

FIGURE 2-3. Effects of weed-free period and weedy period on yields of sugar beets.

Source: Data from Weatherspoon and Schweizer, 1969.

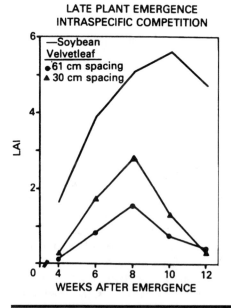

FIGURE 2-4. Faster development than crops of many annual weeds partly accounting for weeds' competitiveness.

Source: Oliver, 1979. Reproduced with permission of the Weed Science Society of America.

Grasses vs. Broadleaf Weeds

Broadleaf weeds tend to cause greater reductions in crop yields than grass or grasslike weeds. Figure 2-5 shows this relationship in peas for numbers of weeds. In terms of reduction in pea yields, 2 mustard plants per 0.1 m² (1 ft²) were equal in effect to 4 foxtail plants. Staniforth (1965) observed approximately a 2:1 relationship based on weed weights. In his study, soybean yields were reduced 0.47 kg ha[-1] for each kilogram dry matter of velvetleaf, and 0.22 kg ha[-1] for each kilogram dry matter of foxtail. Cocklebur was found to be about twice as damaging to soybean yields as the perennial weed johnsongrass (*Sorghum halepense*); depending upon the soybean variety, yields were reduced from 27% to 42% by johnsongrass and from 63% to 75% by cocklebur (McWhorter and Hartwig, 1972).

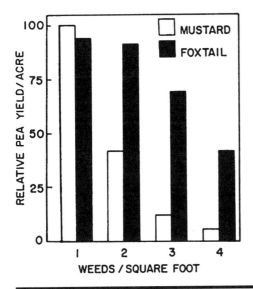

FIGURE 2-5. Relative effect of a grass (foxtail) and a broadleaf (mustard) weed on yield of peas. Weeds were present the first 10 weeks following pea planting.

Source: Data from Nelson and Nylund, 1962.

Since most weedy grasses are of the C_4 type,[1] the greater effect from broadleaf species is likely explained by their more spreading growth form and more horizontal leaves that make them relatively more competitive for light. The lesser competitiveness of grasses must not be confused with their relative seriousness as weeds or difficulty to control. Of the world's 10 most serious weeds, 8 are grasses or are grasslike (Holm et al., 1977).

---

[1]The terms C_3 and C_4 are commonly used to denote the CO_2-fixing characteristics of plants, although other characteristics are also associated. C_4 plants fix CO_2 into sugars about 50% faster than C_3 plants.

Weed Weight

As was true for numbers, relatively small weights of weeds can reduce crop yields. For example, as shown in Table 2-5, it was observed in Iowa that 1 kg ha^{-1} reduced soybean yields 0.16 to 0.65 kg ha^{-1}, depending upon the weed and year. Kilogram for kilogram, velvetleaf was about twice as damaging as the foxtails. Also, kilogram for kilogram, both foxtail and velvetleaf were about twice as damaging to soybean yields in 1963 as in the previous 2 years. These differences between species and years emphasize the complexities of competition. At the same time, the relative constancy for weight effects in a given species and under a given set of growing conditions suggests that it may be possible to develop mathematical constants for the weight–yield relationship.

TABLE 2-5.   Reduction in soybean yields for given weights of three weeds.

Weed Species	Year	Weed Yield (kg ha^{-1})	Soybean Yield (kg ha^{-1}) Weed-free	Weedy	Reductions Due to Weeds kg ha^{-1}	kg kg^{-1} weeds
Yellow foxtail	1961	4936	2288	1460	828	0.17
	1962	2486	2618	2221	397	0.16
Green foxtail	1963	3304	2571	1454	1117	0.34
Average for foxtail		3584	2490	1710	780	0.22
	1961	1422	2430	1932	498	0.35
Velvetleaf	1962	1792	2140	1481	659	0.37
	1963	1568	2571	1548	1023	0.65
Average for velvetleaf		1590	2383	1642	741	0.47

Source: Staniforth, 1965. Reproduced with permission of the Weed Science Society of America.

Weed weight reflects growth factors captured by the weed. Therefore, it is more closely correlated to crop yield loss than is weed number. Nevertheless, only limited attempts have been made to identify the economic threshold for weight. This is because the effect of weed weight is affected by some of the same things that cause variability in the affect of number, although the effect on the weed weight relationship is less. Further, it is not as easy to obtain weed weight as it is weed number.

PREDICTING CROP YIELD LOSS

A way of predicting crop loss for a given weed infestation is a logical expectation in view of what we have just learned about competitiveness. Such information is important to crop producers in at least two ways. It is needed to identify the economic threshold for their particular weed–crop situation. It can also allow them to use the minimum amount of herbicide for the weed problem, thereby minimizing herbicide entry into the environment. The weed scientist needs such information to target research at the most pressing problem.

Early attempts to predict loss resulted in fairly satisfactory mathematical descriptions of an observed effect of weed numbers on crop yield but predictive value was limited. In 1972 an *index of competition* was suggested to estimate yield losses of barley, wheat, and flax for given densities of wild oat (*Avena fatua*) (Dew, 1972). However, the equation was based on assumptions that restrict usefulness. One assumption was that weed and crop emerge at the same time. Many weeds germinate over an extended period, even if the flush tends to be concentrated in time. A second assumption was that the crop and the weed had equal access to water, nutrients, and light. As we shall see in Chapter 7, plants may differ greatly in their ability to compete for such growth factors. The effect of kochia on sugarbeet yields was predicted within 5% of the actual yield by use of a mathematical equation based upon the number of kochia plants per meter of row (Schweizer, 1973). However, which of the three polynomial regression equations derived—linear, quadratic, or cubic—best represented the density–yield relationship depended upon the actual density of kochia. No one equation perfectly represented the relationship over the entire density range because of the curvilinear shape (Figure 2-2). A number of regression models have been suggested since then to explain effects of competition (Cousens, 1985a; Spitters, 1983; Spitters et al., 1989). It is now generally accepted that the quantitative relationship between weed numbers and crop yield is best described by a quadratic hyperbola (Cousens, 1985b). Models having practical value for predicting crop yield loss have been developed (Coble, 1985; Wilkerson et al., 1991), but there is no infallible way of predicting yield loss based on weed densities. Useful models will be presented in Chapter 15, where we consider production practices.

A major problem in attempts to predict crop yield loss based only on weed numbers is that time of weed emergence, as we saw earlier, greatly affects competitiveness of weeds, as do other factors. In practical terms, all such attempts at prediction had the common weakness of not being tied to factors that explained competition. It is not weed numbers that determine crop yield loss but what the numbers mean for capture of resources needed by the crop. This fact has moved recent efforts to model crop loss toward inclusion of growing conditions and competition for growth factors (Spitters and Aerts, 1983; Kropf and Spitters, 1991; Spitters, 1989; Weaver et al., 1992). These approaches will be examined more closely in Chapter 15, where we consider production practices.

A *weed-loss survey system* designed for estimating weed losses in soybeans provides an interesting example of a practical way of incorporating several factors involved in competition. The sequence of photographs shown in Figure 2-6 is the basis for estimating yield losses. A percentage yield loss assigned each level of infestation is based upon actual losses recorded in experiments throughout the soybean production region for each infestation. The percentage losses assigned each level of infestation from A to E are, respectively, less than 5%, 5% to 10%, 10% to 20%, 20% to 35%, and greater than 35%. When the photographs are matched with actual field infestations, the loss falls within the level for the photograph. It is interesting that the survey system is apparently fairly accurate, even though it is applied to different weeds, different soils, and different climatic regions. The explanation is found in the fact that the visual appearance of a weed infestation captures several aspects of weeds important to determining competition, including their numbers, their size, the species present, when they became established, and how long they have been present. This survey system serves as an excellent reminder that competition frequently is the net result of several interacting factors.

| A. Weed free, no losses (less than 5%) | B. Slightly weedy, 5% to 10% loss | C. Moderately weedy, 10% to 20% loss | D. Heavy weeds, 20% to 35% loss | E. Disaster, 35% to 100% loss |

FIGURE 2-6. Weed-loss survey in soybeans.

Source: Reproduced with permission of Elanco Products Co.

WEED'S-EYE VIEW OF COMPETITION

Crop–Weed Competition

To this point, we have been looking at the effects of weeds on crops. To understand the relationship between crops and weeds for long-term weed management approaches, we need to examine competition from the weed's standpoint. There is much evidence that weeds, especially annuals, are very intolerant of competition. Giant foxtail made very little growth in either corn or soybeans if the crop was given a 3-week head start (Knake and Slife, 1965). In fact, as shown in Figure 2-7, foxtail seeded into soybeans 3 weeks after the soybeans were planted made practically no growth. In similar studies involving soybean competition for barnyardgrass, even a 1-week head start caused more than a 60% reduction in growth of the weed (Maun, 1977). Additionally, even though the barnyardgrass managed some growth regardless of how much head start the soybeans had, it produced no seed if soybeans had as much as a 2-week head start (Figure 2-8). Barnyardgrass weight of the whole plant and of the caryopsis was determined 15 weeks after soybean emergence and compared with the barnyardgrass weight in pots without soybeans. Clearly, the sensitivity of weeds toward competition suggests good opportunities for success in managing production practices to minimize losses from them.

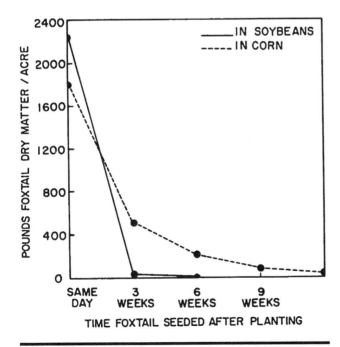

FIGURE 2-7. Effect of competition from corn and soybeans on growth of giant foxtail.

Source: Data from Knake and Slife, 1965.

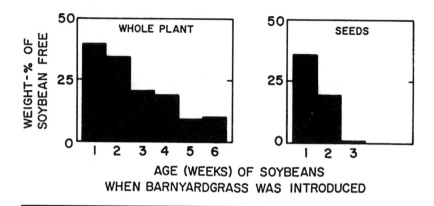

FIGURE 2-8. Effect of competition from soybeans on vegetative growth and seed production of barnyardgrass. Three 1-week-old barnyardgrass plants were transplanted into pots containing a soybean plant of the age shown.

Source: Data from Maun, 1977.

Further, the results certainly strongly imply that light is frequently the factor involved, an implication examined in greater detail in Chapter 7. At this point, however, it is worth noting that soybeans were relatively more competitive than corn toward foxtail if the crop had a head start, whereas the opposite was true if the crops and foxtail emerged together (Knake and Slife, 1965). Corn yields were cut about 13%, but soybeans 29% if the foxtail emerged with crops and remained only 3 weeks.

Weed–Weed Competition

Weed-to-weed competition may also have important implications for weed management. Table 2-6 shows redroot pigweed to be much more competitive toward lambs-

TABLE 2-6. Plant population and shoot dry weights of redroot pigweed (RR) and common lambsquarters (LQ) in a mixed culture in the greenhouse.

Treatment*		Species	Plants (no. m⁻²)	Dry Weight of Shoots (g m⁻²)
'0	'20			
RR	LQ	RR	2307	398
		LQ	0	0
LQ	RR	RR	0	0
		LQ	1985	255
1 RR:1LQ	0	RR	1862	320
		LQ	457	5
1 RR:10 LQ	0	RR	437	230
		LQ	2240	57

Note: Plants grown for 36 days under average temperatures of 29°C day and 24°C night.
* '0 = initial sowing; '20 = sown 20 days later.
Source: Chu et al., 1978. Reproduced from Crop Science 18:308-310, 1978, by permission of the Crop Science Society of America.

quarters than the reverse. Even when seeded at 10 times the rate of pigweed, lambs-quarters had much less growth at 36 days. However, if either weed is given a time advantage over the other, it completely suppresses the late-planted species.

The explanation for differential competitiveness lies in the germination and growth of the two species in response to temperature. As can be seen in Table 2-7, pigweed germinates much more rapidly than lambsquarters under the higher temperatures but much more slowly under the lower temperatures. Additionally, from Table 2-8 it can be seen that pigweed grows more rapidly under higher temperature, but under the lower temperature regimes, there is little difference between the two species. Pigweed and lambsquarters are common associates in the weed community in horticultural and agronomic row-crop agriculture in many parts of the world. Which species dominates the weed problem can be expected to vary from year to year, depending upon temperature during the early period of their association. Temperature, in turn, could be affected by time of planting. In a relatively dry spring in the Midwest, early planting, when temperatures are relatively cool, can be expected to favor lambsquarters. Wet weather, necessitating late planting when temperatures are warmer, gives a decided edge to redroot pigweed. This relationship between lambsquarters and redroot pigweed provides another reminder that data on growing conditions are needed for valid comparisons of weed populations among different experiments.

TABLE 2-7.   Effect of temperature on the germination of redroot pigweed and common lambsquarters.

Temperature (day/night °C)	Species	Germination Rate Index*
24°/18°	Redroot pigweed	14.8
24°/18°	Lambsquarters	6.6
13°/7°	Redroot pigweed	0.4
13°/7°	Lambsquarters	2.6

*Germination determined daily for 30 days incubation and an index computed as follows

$$\text{Germination rate index} = \sum_{n=1}^{n=30} \frac{\text{number germinated since } n-1}{n}$$

where n = days of incubation.
Source: Chu et al., 1978. Reproduced from Crop Science 18:308-310, 1978, by permission of the Crop Science Society of America.

TABLE 2-8.   Effect of temperature on the growth of redroot pigweed and common lambsquarters at 11 weeks.

Temperature (day/night °C)	Species	Shoot Dry Weight (g plant⁻¹)	Leaf Area (dm² plant⁻¹)
29°/24°	Redroot pigweed	48.5	86.2
	Lambsquarters	11.4	22.0
24°/18°	Redroot pigweed	12.3	31.8
	Lambsquarters	11.8	20.5
18°/13°	Redroot pigweed	6.4	13.7
	Lambsquarters	6.3	10.5

Source: Chu et al., 1978. Reproduced from Crop Science 18 :308-310, 1978, by permission of the Crop Science Society of America.

The usual situation with annual weeds in row-crop agriculture is for the makeup of the weed community to change somewhat from year to year. This situation is observed even under monoculture and even though the composition of the seedbank in the soil may be relatively constant from year to year. The relationship between lambsquarters and redroot pigweed in response to environmental conditions may help explain why.

Weeds interacting with other weeds may have long-term effects on the weed community as well. Common milkweed (*Asclepias syriaca*) has been increasing steadily over the past many years. A survey of 13 states in the north-central region of the United States indicated that 10.5 million hectares (26 million acres) were infested in 1976 (Evetts, 1977). Every state indicated that this weed had increased in the last 5 years. The explanation, no doubt, is found in the fact that many annual weeds, both grasses and broadleafs, have been rather well controlled with herbicides in the several years preceding 1976. As seen in Tables 2-9A and 2-9B, common milkweed is very intolerant of competition from both green foxtail (*Setaria viridis*) and redroot pigweed, two weeds common to row-crop agriculture in this region but effectively controlled with available herbicides. The experiment was arranged to measure the independent effects of competition above (light) and below ground (soil) and of the two combined (full). The pronounced reduction in reproduction from roots is especially dramatic. Sensitivity of milkweed to competition from sorghum supports crop-to-weed competition discussed in the previous section.

Many problem weeds are extremely intolerant of competition. As more and more specific data are obtained on the nature and extensiveness of such competition, possibilities for manipulating production practices to give the crop the slight edge it needs should become evident.

COMBINED IMPACT OF PEST COMPETITION

Before leaving competition, we need to consider the potential combined impact on the crop of competition from a weed and from an insect, disease, or both. It is not uncommon for a crop to be subjected to stress from at least two classes of pest. In fact, this situation may be the usual one. Very little information is available to document the combined effect of weeds, insects, and diseases on crop yield and the relative contribution of each. A key question from the weed–crop ecology perspective is, Are such effects synergistic or merely additive? If the relationships are synergistic, a crude generalized conclusion would be that relatively less competition could be tolerated from a weed than if the effects were only additive.

A hypothetical weed–insect–crop situation illustrates this conclusion. An anticipated 95% control of a given weed and a given insect would be adequate if their effects were additive (5% loss from each) and the cost of 100% control of each was greater than the 10% yield saved. On the other hand, if the presence of the weed and insect together had a synergistic effect, causing the 5% population of each to inflict a 10% loss, the resulting 20% loss might justify attempts to further reduce the pest infestation.

TABLE 2-9.   Competition of two weeds and sorghum toward common milkweed measured 35 days after planting in the greenhouse.

A. Effects on Shoot Weight

Competing Plant	Shoot Weight for Types of Competition (cg)			
	None	Light	Soil	Full
Green foxtail	41	22	18	11
Redroot pigweed	41	27	25	15
Sorghum	33	23	18	15

B. Effects on Reproduction Percentage

Competing Plant	Percentage Reproduction from Roots			
	None	Light	Soil	Full
Green foxtail	78	9	56	8
Redroot pigweed	81	3	52	11
Sorghum	73	46	56	40

C. Competition Boxes Used for the Studies*

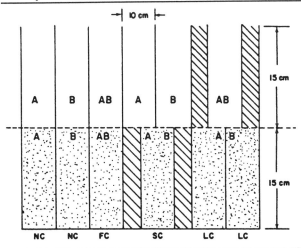

*Diagram of competition boxes with aerial portions above and soil compartments below. No competition (NC) contains two soil and two aerial compartments. Full competition (FC) contains one soil and one aerial compartment. Soil competition (SC) contains one soil and two aerial compartments. Light competition (LC) contains two soil and one aerial compartments. One plant species is represented by (A) and the second by (B).

Source: Adapted from Evetts and Burnside, 1975. Reproduced with permission of the Weed Science Society of America.

REFERENCES

Berglund D.R., and J.D. Nalewaja. 1969. Wild mustard competition in soybeans. In Proc. NCWCC., p. 83. Sioux Falls, SD.

Chu, C., P.M. Ludford, J.L. Ozbun, and R.D. Sweet. 1978. Effects of temperature and competition on the establishment and growth of redroot pigweed and common lambsquarters. Crop Sci. 18:308-310.

Coble, H.D. 1985. Development and implementation of economic thresholds for soybeans. In R.E. Frisbes and P.L. Adkisson, eds., Integrated pest management of major agricultural systems, pp. 295-307. Texas A&M University, College Station, TX.

Cousens, R. 1985a. An empirical model relating crop yield to weed and crop density and a statistical comparison with other models. J. Agric. Sci. 105:513-521.

———— 1985b. A simple model relating yield loss to weed density. Ann. Appl. Biol. 107:239-252.

————. 1987. Theory and reality of weed control thresholds. Plant Prot. Quart. 2:13-26.

Crowley, R.H., and G.A. Buchanan. 1978. Competition of four morningglory (*Ipomoea* spp) species with cotton (*Gossypium hirsutum*). Weed Sci. 26:484-488.

Dew, D.A. 1972. An index of competition for estimating crop loss due to weeds. Can. J. Plant Sci. 52:921-927.

Evetts, L.L. 1977. Common milkweed—the problem. In Proc. NCWCC, pp. 96-99. St. Louis, MO.

Evetts, L.L., and O.C. Burnside. 1975. Effect of early competition on growth of milkweed. Weed Sci. 23:1-3.

Holm, L.G., D.L. Plucknett, J.V. Pancho, and J.P. Herberger. 1977. The world's worst weeds. Honolulu: University Press of Hawaii.

Kerr, H.D., and D.L. Klingman. 1960. Weed control in establishing birdsfoot trefoil. Weeds 8:157-167.

Knake, E.L., and F.W. Slife. 1965. Giant foxtail seeded at various times in corn and soybeans. Weeds 13:331-334.

Kropf, M.J., and C.J.T. Spitters. 1991. A simple model of crop loss by weed competition from early observations on relative leaf area of the weeds. Weed Res. 31:97-105.

Kropf, M.J., and L.A.P. Lotz. 1992. Optimization of weed management systems: The role of ecological models of interplant competition. Weed Technol. 6:462-470.

Li, M.Y. 1960. An evaluation of the critical period and the effects of weed competition on oats and corn. Ph.D. dissertation. Rutgers University, New Brunswick, NJ.

Maun, M.A. 1977. Suppressing effect of soybeans on barnyardgrass. Can. J. Plant Sci. 57:485-490.

McWhorter, C.G., and E.E. Hartwig. 1972. Competition of johnsongrass and cocklebur with six soybean varieties. Weed Sci. 20:56-59.

Nave, W.R., and L.M. Wax. 1971. Effect of weeds on soybean yield and harvesting efficiency. Weed Sci. 19:533-535.

Nelson, D.C., and R. E. Nylund. 1962. Competition between peas grown for processing and weeds. Weeds 10:224-229.

Oliver, L.R. 1979. Influence of soybean (*Glycine max*) planting date on velvetleaf (*Abutilon theophrasti*) competition. Weed Sci. 27:184-188.

Schweizer, E.E. 1973. Formula for predicting sugarbeet root losses based on kochia densities. Weed Sci. 21:565-567.

Smith, R.J., Jr. 1968. Weed competition in rice. Weed Sci. 16:252-254.

Spitters, C.J.T. 1983. An alternative approach to the analysis of mixed cropping experiments. I. Estimation of competition effects. Neth. J. Agric. Sci. 31:1-11.

Spitters, C.J.T. 1989. Weeds: population dynamics, germination, and competition. In R. Ragginge, S.A. Ward, and H.H. van Loar, eds., Simulation and systems management in crop protection, pp. 182-216. Simulation Monographs. Pudoc, Wageningen.

Spitters, C.J.T., and R. Aerts. 1983. Simulation of competition for light and water in crop weed associations. Aspects App. Biol. 4:467-484.

Spitters, C.J.T., M.J. Kropf, and W. deGroot. 1989. Competition between maize and *Echinochloa crus-galli* analyzed by a hyperbolic regression model. Ann. Appl. Biol 115:541-551.

Staniforth, D.W. 1965. Competitive effects of three foxtail species on soybeans. Weeds 13:191-193.

Weatherspoon, D.M., and E.E. Schweizer. 1969. Competition between kochia and sugarbeets. Weed Sci. 17:464-467.

———. 1971. Competition between sugarbeets and five densities of kochia. Weed Sci. 19:125-128.

Weaver, S.E., M.J. Kropf, and R.M.W. Groeneveld. 1992. Use of ecophysiological models for crop–weed interference: the critical period of weed interference. Weed Sci. 40:302-307.

Wilkerson, G.G., S.A. Modena, and H.D. Coble. 1991. HERB: Decision model for postemergence weed control in soybean. Agron. J. 83:413-417.

Zimdahl, R.L. 1980. Weed–crop competition: A review. Corvallis: International Plant Protection Center, Oregon State University.

3

Ecological Relationships and Concepts

CONCEPTS TO BE UNDERSTOOD

1. Weeds are part of a dynamic ecosystem that also includes desired plants, the natural environment, and human beings.
2. The microenvironment varies greatly at the level of the germinating and growing weed.
3. Individual plants within a weed species differ in growth form.
4. Weeds, within a species and within a mixture of species, change in response to plant management practices.
5. Most problem annual weeds have a survival strategy based upon production of a large number of seeds.
6. Most problem perennial weeds have a survival strategy based upon strong competitive ability.

Ecology by definition is concerned with the relationship between organisms and the environment. Understanding the relationship between structure (growth form, species makeup, and other growth characteristics) and function (role in the system) is the ultimate goal of a study of any level of a plant system. Understanding the relationship between the weeds associated with a given set of production practices and the function of that ecosystem is the goal of *weed–crop ecology.*

RELATIONSHIPS BETWEEN DESIRED PLANTS AND WEEDS

Historical Perspective

Weeds are part of a dynamic system. However, our approach to dealing with them, especially in recent times, has failed to adequately recognize this fact. This failure is largely a consequence of rapid advances in herbicide technology. The discovery of 2,4-D as an effective herbicide to selectively control many broadleaf weeds in grain crops ushered in the modern era of weed control. The discovery and development of additional selective herbicides progressed rapidly, and there are now well over 100. Prior to this widespread availability of herbicides, prevention was involved in many of the practices used to deal with weeds. The availability of a herbicide to selectively remove nearly any weed from any crop has reduced the pressure to gear tillage and crop choice, crop sequence, and crop spacing to weed control. Herbicide use served to mask the importance of prevention and the need to understand weed–crop ecological relationships. Growing concern for possible hazards with pesticides and for sustainability of an agriculture heavily dependent on external inputs raises serious questions about nearly total reliance on herbicides for weed control.

The need to deal with weeds has influenced farm machinery development and use from the beginning. The first such machine, the plow, was recognized as an effective implement for interrupting growth or destroying weeds by cutting them off and turning them under. The design of most tillage equipment that followed the plow took weeds into account. Some, such as the field cultivator, the rotary hoe, and the row cultivator, were designed specifically for weed control. The primary reason for planting crops such as cotton, corn, and soybeans in rows was so that machines could be used to remove weeds during the growing season. The choice of what crop to plant and in what sequence in the field was influenced by the need for weed control during production of that crop and the crops to follow. Perennial forage crops grown for hay, especially alfalfa, were used to check some problem weeds. In fact, the long-standing practice of rotating crops itself was an important way to minimize losses from weeds prior to the modern herbicide era.

Ecological Perspective

A number of events have served to emphasize the need for more and better information about the ecological relationships involved in weed management. Shifts in weed composition that have occurred where weeds were differentially controlled with herbicides is one. Weed problems associated with crop monoculture is another. Multiple cropping and intercropping in themselves require knowledge of ecological concepts. These challenges cannot be satisfactorily met without an understanding of reproduction of weeds, dormancy of reproductive parts, and competition as these phenomena relate to the environment imposed by our production practices. The move to develop such an understanding begins with the system of which weeds are a part.

The ecological relationships are depicted schematically in Figure 3-1. This schematic representation shows that there are four key components: (1) the weeds, (2) the desired plant, (3) the natural environment, and (4) human beings. Our more traditional viewpoint and approach have been to focus mainly on the weed and the desired plant in this four-part system. In this chapter and those that follow, we will see that

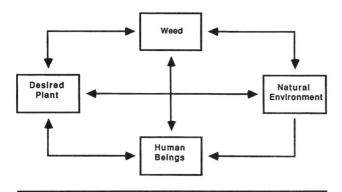

FIGURE 3-1. Weed-desired plant ecosystem. Each part has a
potential impact on all other parts.

weeds also influence the environment. Thus, the four components create a dynamic
system called the *weed–desired plant ecosystem.*

Hierarchy of Biosystems

Developing weed science around the ecosystem presented schematically in Figure 3-1
logically necessitates its being founded upon ecological concepts and principles. In
ecological terms, there is a hierarchy of biosystems that begins at the lowest level with
genetic systems and ends with the largest systems, ecosystems. These systems, composed of both living (biotic) and non-living (abiotic) components, are shown in Figure
3-2. Our primary concern in weed management, as with ecology in general, is species

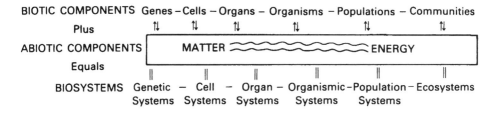

FIGURE 3-2. Levels of organization spectrum.

Source: Odum, 1971. From Fundamentals of ecology, 3rd ed., by Eugene P. Odum. Copyright © 1971 by
W.B. Saunders Co. Copyright 1953 and 1959 by W.B. Saunders Co. Reprinted with permission of Holt,
Rinehart & Winston, CBS College Publishing.

(organisms), populations, communities, and their corresponding systems in this spectrum of levels of organization. Note the two-way relationship between the biotic and abiotic components. This two-way relationship serves as a reminder that the biosystem is a product of the interactions of these two components. Note also that one level is not independent of another. No sharp line can be drawn separating them. In terms of weed science, this fact simply recognizes that an individual weed cannot survive—at least not for long—independent of its population any more than a population can survive without the individual as the means for the production of seed or other reproductive parts. However, each level has unique characteristics that hold significance for weed science. Therefore, it is important to examine some of the characteristics of individual organisms, populations, and communities.

Weed's-Eye View

To understand the ecological relationships discussed, it is helpful to have a weed's-eye view—that is, to see the several factors to be discussed from the perspective of the weed. How does a weed view intensive versus minimum tillage, corn versus soybeans, and Delsoy 5500 soybeans versus Linford soybeans, for example? Only through an awareness of the weed's-eye view can we fully appreciate the extreme heterogeneity of the environment and of the individual plant growth form interacting with that environment.

Environmental heterogeneity. The environment where we are apt to find a particular weed is indeed extremely heterogeneous when viewed by the plant growing in it. The diversity extends both laterally and vertically. This concept may run counter to our initial view of the environment; for example, an environment provided for cotton and weeds in a newly planted cotton field. After all, the field was uniformly tilled and fertilizer applied as uniformly as possible. Depending upon the size of the field, each hectare received the same amount of rainfall as each neighboring hectare. Thus, from our view, there is environmental uniformity in the lateral dimension.

The individual cotton seed or seedling or weed seed or seedling experiences quite a different situation. On a microscale level, minor differences in elevation across the field create differences in temperature and in wetness and dryness, depending upon the direction of slope. The final tillage operation itself creates a series of ridges and furrows that are deep valleys and hills to a weed seed. The relatively small ridges, 7.5 cm, shown for a newly planted wheat field in Figure 3-3 are fairly representative of millions of hectares of crops. Figure 3-4, showing a ridge of about 25 cm for cotton, indicates an extreme, but one nonetheless used for many hectares of cotton and some other

FIGURE 3-3. Relatively small ridges in a newly planted wheat field.

FIGURE 3-4. Large ridges in a ridge-planted cotton field.

field and vegetable crops. A height difference between the top of a ridge and the bottom of a furrow of only 5 cm represents a height factor of 250 to 1 for our smallest weed seeds. Thus, even though the growth factors of water and nutrients the plant must get from the soil may be relatively evenly distributed laterally, the conditions under which they are presented to the germinating seed and developing seedling may vary greatly.

In the vertical dimension, there may be diversity in both the growth factors and the conditions under which they are available to the plant. In the soil, water and nutrients typically are unevenly distributed throughout the soil profile, at least on the scale pertinent to a germinating weed seed. These differences are further influenced by microsite differences in other environmental conditions, such as temperature.

The magnitude of differences at the weed site are shown in studies on medusahead (*Taeniatherum asperum*) (Evans and Young, 1970). In gross terms, the temperature range at the surface of bare soil was about twice that of soil with plant residue on it (Figure 3-5). In a more general study of plant residue effects on environmental conditions, measurable differences were found in temperature for as little as 20% cover of soil with residue (Van Doren and Allmaras, 1978). Thus, there is at least circumstantial evidence suggesting that weed seeds at the soil surface under a disked-down corn stalk or other plant residue are exposed to quite a temperature difference from neighboring weed seeds germinating at or just beneath the surface of bare soil.

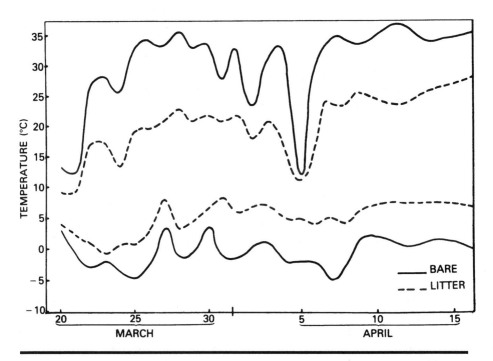

FIGURE 3-5. Influence of soil cover (litter) on maximum and minimum soil temperatures at the soil surface.

Source: Evans and Young, 1970. Reproduced with permission of the Weed Science Society of America.

The moisture environment may also be influenced by litter, as shown in Figure 3-6. Even at midday, the relative humidity did not fall below 60% where there was plant litter, whereas it was below this level for much of the daylight period over bare soil. As we shall see later, because seeds of many weeds germinate at or slightly below the soil surface, such minor temperature and moisture differences may well have a significant effect on germination.

Plants obtain light and carbon dioxide (CO_2) through their aboveground portion. Here, too, from the developing plants' perspective, considerable variation may exist, depending upon the amount and type of vegetation present. Representative variations

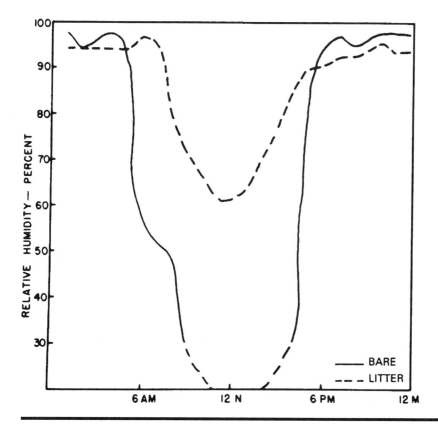

FIGURE 3-6. Influence of soil cover on relative humidity at 0 cm to 3 cm above the soil surface. Daily patterns are the average of readings in March and April. Values below 30% are extrapolated.

Source: Evans and Young, 1970. Reproduced with permission of the Weed Science Society of America.

of these two factors are shown in Figures 3-7 and 3-8. In Figure 3-7, light intensity drops rapidly with depth in the canopy. Also the *leaf area index* (LAI)—that is, the area of leaf blades relative to the soil surface—shows that clover provides more than twice as much leaf area for light interception as does the grass. Although not as dramatic as light, CO_2 content may vary as much as 15 ppm, depending upon the level within the canopy (Figure 3-8).

Clearly the *abiotic environment* is extremely heterogeneous at the level being sampled by an individual weed seed or seedling. Because the microenvironment is so varied, weeds with quite different requirements may occupy a given area, which helps explain why we usually have a mixture of weeds.

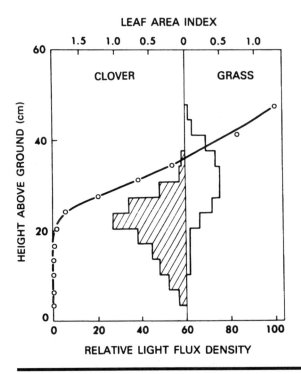

FIGURE 3-7. Light extinction with height in a mixed clover-grass sward.

Source: Trenbath, 1976. Reproduced from Multiple Cropping, ASA Special Publication no. 27, 1976, with permission of the American Society of Agronomy, Crop Science Society of America, and Soil Science Society of America.

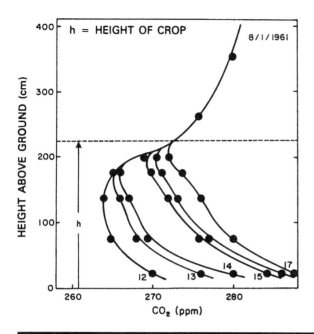

FIGURE 3-8. CO_2 profiles within and above a maize crop for the period 1200-1700 hours.

Source: Wright and Lemon, 1966. Reproduced from Agron. J. 58:265-268, with permission of the American Society of Agronomy.

Differences in growth form.   Growth form varies considerably from weed species to weed species and from plant to plant within a species. In Figure 3-9, we see that the aboveground portions vary from single-stemmed to multiple-branched and from upright to prostrate to climbing. The form of root growth may be equally as varied, from a single taproot to a network of root fibers (Figure 3-10). These differences in aboveground and belowground growth forms impart different capabilities to the weed to sample the varied abiotic factors of the environment. Differences in the environment and in the individual weed's ability to sample the environment further explain why we usually have a mixture of weeds. That is to say, in almost any situation that can be envisioned, there will be differences in environment from the weed's perspective that can be differentially sampled by different individuals or different species.

Weed's ecological niche.   If we add to the differential sampling ability of individual weeds their effect on the environment, we have defined the weed's *ecological niche*. The reader is referred to Odum's *Fundamentals of Ecology* (1971) for a detailed treatment of ecological niche. Here we simply need to recognize that weeds do change the environment. Their germination, growth, and death change moisture, temperature, nu-

A. Upright, branched, broad leaves of lambs-quarters

B. Upright, nonbranched, narrow leaves of giant foxtail

C. Prostrate growth of carpetweed

D. Climbing growth of wild buckwheat

FIGURE 3-9. Different aboveground growth forms of weeds.

 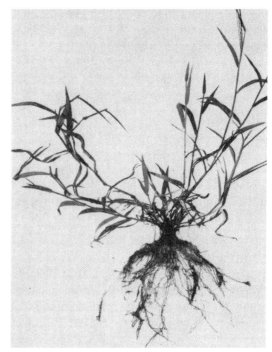

A. Taproot of giant ragweed   B. Fibrous roots of weeds

FIGURE 3-10. Different belowground growth forms of weeds.

trient, and ultimately organic matter of the soil. They are active not passive participants in the ecosystem. Thus, removal of a weed by whatever means creates a void in the ecosystem. The void is soon filled by another weed unless the environment of which the niche is a part is itself changed by a change in the production system. Recognition of this ecological principle is fundamental to a management approach to weeds.

At this point, it may be helpful to remind ourselves that the net effect and response is at the individual plant level. A mustard plant, or any other individual weed plant, may differ greatly from a neighboring mustard plant, depending upon microscale differences in the environments available to the two plants. Further, growth within an individual plant varies with location on the plant. A leaf at the top of a plant has more light available to it than one in a lower portion of the plant. The CO_2 level may well be different for the two leaves. Temperatures can vary within the canopy. These facts have particular significance for weed science in understanding and interpreting results of research on individual species.

We have seen that environments are heterogeneous and that a weed's ability to sample different environments depends upon its growth form, which also varies greatly. For this reason, we commonly have a mixture of weeds. This mixture is not static. It changes in response to imposed changes in the environment. To understand why and how the mixture of weed species changes, it is first necessary to understand how new species arise.

SPECIES

A *species* is the natural biological unit, or organism, (Figure 3-2) tied together by the sharing of a common gene pool (Merrell, 1962). In simple terms, *speciation* is the product of natural selection and genetic mutation, resulting in a new gene pool to be shared. This definition tells us *what* happens. To understand *why* it happens, it is necessary to examine the processes in terms of the forces exerted by the environment.

Limiting Factors

The concept of *limiting factors* is a good place to begin. Organisms are controlled in nature by (1) the quantity and variability of materials for which there is a minimum requirement and physical factors that are critical and (2) the limits of tolerance of the organisms themselves to these and other components of the environment (Odum, 1971). Thus, organisms are controlled by both too little and too much of the factors needed for growth and the conditions under which they are available. The weeds' response to limiting factors is the beginning of speciation. As we have already learned, a weed is an actor in the game of evolution, not just a passive observer. Its very presence modifies the physical environment and it itself changes. These dynamic aspects of a plant's presence serve to lessen the effects of limiting factors. In ecological terms, such lessening of effects is known as *factor compensation*. Compensation for temperature and light or other factors outside the optimum range may involve fixation in new genetic combinations or simply a wide range of growth responses (plasticity).

Allopatric and Sympatric Speciation

Keeping in mind the weed's-eye view of the environment, it can be seen that factor compensation may be operable within a relatively restricted geographic area. Therefore, examination of time and space aspects are especially important to a full understanding of speciation in weeds. Ecology uses the term *sympatry* to describe speciation in a very local situation and *allopatry* to describe that situation where segments of a species—that is, individual plants or groups of plants—are widely separated geographically. (An example of allopatry is the introduction of weeds into North America from other continents.) In time, compensation for conditions of the new environment becomes fixed in the genetic makeup to the point that interbreeding can no longer occur. In a sense, allopatry is speciation in response to influences outside the plant.

Sympatry, by contrast, may be thought of as the result of changes from within. That is, isolation of the gene pool occurs as a result of polyploidy, hybridization, self-fertilization, and asexual reproduction, all of which can occur in a very local situation. Clearing the forests and plowing the prairies provided a rich setting for *interspecific hybridization*—the crossing of two species—by bringing together closely related species previously well isolated. Such natural hybridization is common among weeds, more so than for other groups of plants. Also common among weeds is *introgressive hybridization,* the gradual infiltration of the germplasm of one species into that of another by hybridization and repeated backcrossing. In both allopatry and sympatry, speciation is the result of interruption in gene movement within the common pool illustrated in Figure 3-2.

An obvious time difference is implied between allopatry and sympatry. Allopatric speciation is assumed to be a relatively long-term process, whereas sympatry may be a short-term process. For example, duplication of chromosome sets in polyploidy brings about immediate genetic isolation.

It is important to emphasize that weeds have had and continue to have thrust upon them conditions that encourage speciation. They have been moved from continent to continent and from place to place within agricultural regions (allopatric influences). Continual disturbance of the land and ever-changing agricultural practices provided repeated opportunities for hybridization and for selection in response to all genetic phenomena (sympatric influences). Thus, we expect a given weed species to be different today from what it was in the past and to be different in the future from what it is today. There is much evidence in the weed research literature to show that this indeed is so.

Ecotypes

Many studies have shown that the same weed species differ from one location to another. *Ecotype* is the term used to describe such locally adapted populations. The presence of ecotypes is well documented for a wide variety of weed species, including johnsongrass (*Sorghum halepense*) (McWhorter, 1971; Burt, 1974; Wedderspoon and Burt, 1974; McWhorter and Jordan, 1976); Canada thistle (*Cirsium arvense*) (Hodgson, 1964); common ragweed (Dickerson and Sweet, 1971); yellow nutsedge (Yip, 1978); purslane (*Portuclaca oleracea*) (Gorske et al., 1979); annual bluegrass (*Poa annua*) (Warwick and Briggs, 1978); creeping buttercup (*Ranunculus repens*) (Soane and Watkinson, 1979); and medusahead (Young et al., 1970).

Differences in growth form of ecotypes may be relatively large. In a study of johnsongrass ecotypes, 55 morphologically distinct ecotypes from Mississippi and from selected other states were grown at Stoneville, Mississippi (McWhorter, 1971). The data for growth form in Table 3-1 show about a twofold range for leaf length (31 cm to 59 cm), leaf blade width (1.7 cm to 3.4 cm), width of clumps (99 cm to 156 cm), and plant height (128 cm to 212 cm). The range in density of culms is even greater (65 to 226 culms m^{-2}).

TABLE 3-1.   Average seasonal growth of johnsongrass ecotypes in 1964 and 1965 at Stoneville, Mississippi.

Ecotype Source	Leaf Length (cm)	Leaf Blade Width (cm)	Width of Clumps (cm)	Plant Height (cm)	Culm Density (culms m^{-2})
Arkansas	41	2.1	118	154	226
Arizona	37	1.8	123	174	182
California	40	1.9	120	160	114
Georgia	43	1.8	117	146	118
Illinois	37	3.0	133	171	81
Louisiana-BR	39	1.8	155	165	88
Louisiana-H	39	2.5	122	192	101
Missouri-1	47	2.4	140	178	65
Missouri-2	40	2.0	130	170	113
North Carolina	40	1.9	153	160	107
Texas	44	1.9	140	163	140
Washington	39	2.0	119	142	66
Mississippi S1	36	1.7	132	172	112
Mississippi S2	40	3.1	112	162	105
Mississippi S3	46	3.4	152	189	111
Mississippi S4	42	1.9	139	188	100
Mississippi S5	35	1.9	129	170	117
Mississippi S6	31	1.8	116	175	163
Mississippi S7	44	2.0	124	168	111
Mississippi S8	59	2.7	109	212	87
Mississippi S9	42	2.2	119	146	110
Mississippi S10	38	2.1	130	140	123
Mississippi S11	57	2.4	155	177	108
Mississippi S12	46	2.2	121	182	90
Mississippi S13	44	1.9	124	180	96
Mississippi S14	43	1.9	128	185	85
Mississippi S15	48	2.3	123	178	73
Mississippi S16	41	1.9	124	166	96
Mississippi S17	43	1.9	135	170	110
Mississippi S18	41	1.9	156	162	82
Mississippi S19	53	2.3	146	192	114
Mississippi S20	46	2.2	128	177	140
Mississippi S21	40	2.1	126	163	81
Mississippi S22	39	1.9	128	158	104
Mississippi S23	45	2.6	132	141	91
Mississippi S24	42	2.0	99	131	73
Mississippi S25	35	2.0	112	128	151

Source: McWhorter, 1971. Reproduced with permission of the Weed Science Society of America.

Ecotypes may occur within relatively small geographic areas. For example, ecotypes of johnsongrass from northern Maryland had nearly completed flowering (96%) 7 weeks after planting, compared with only 46% flowering of the ecotypes from southeastern Maryland (Burt, 1974). The linear distance between the two regions is about 100 km, and the latitudinal distance is approximately 2°. Variations for ecotypes from within the state of Mississippi were greater than the range in these measurements for ecotypes from outside Mississippi for all measurements except culms m^{-2} (Table 3-2).

From these studies, it is not possible to identify the specific processes involved in the development of a particular ecotype. They do provide the basis for some interesting speculation. Local variations support the established ecological principle for

TABLE 3-2.   Range in measurements of growth characteristics for johnsongrass ecotypes from within and outside Mississippi when grown in Stoneville, Mississippi.

	Range				
Ecotype Source	Leaf Length (cm)	Leaf Width (cm)	Clump Width (cm)	Height (cm)	Culm Density (culms m^{-2})
Within Mississippi	31–59	1.7–3.4	99–156	128–212	73–165
Outside Mississippi	37–47	1.8–3.0	117–155	142–192	65–226

Source: Adapted from McWhorter, 1971.

species in close association to diverge. Caution is necessary in doing more than simply speculating that this is the explanation in these instances since the studies were not established to test the principle and, of course, the sample size and time are probably too small. Irrespective of the specific mechanism involved, it is significant that differences occurred in a comparatively small geographic area, serving to emphasize the need for care in interpreting results of studies on individual weed species by different researchers in different locations. The researchers may not, in fact, be working on precisely the same species. This possibility is reinforced by a study of population differences in annual bluegrass (Warwick and Briggs, 1978). They found two very different types of annual bluegrass in bowling greens in England and in the associated flower beds along the greens. The ecotype in the green was mainly of prostrate growth form. The flower bed contained an ecotype that was upright in growth form. The difference was shown to be genetically controlled.

The development of ecotypes has far-reaching implications for weeds and their management or control. As we have seen, ecotypes vary markedly in growth form. For example, johnsongrass ecotypes within Mississippi had approximately a twofold range in leaf length, leaf width, clump width, height, and culms per unit area. As we shall see in Chapter 7, these differences can be expected to influence the ability of the particular ecotype to compete for growth factors. Later (Chapter 11), we shall see that these differences influence the extent of herbicide contact and penetration. Finally, although internal biochemistry is not identified as differing in these studies, there may be differences as pronounced as those in growth form. Such differences can be expected to influence the total response to specific herbicides. Together, the external and internal differences of ecotypes can be expected to influence the entire range of interrelationships of the ecotype in the ecosystem depicted in Figure 3-1.

Role of Apomixis

Asexual reproduction in plants is called *apomixis*. There are two kinds: agamospermy—reproduction by means of seeds that contain embryos without the benefit of meiosis and fertilization of the female egg cell, and vegetative—reproduction by propagules that develop in place of flowers (vivipary), by apical and axillary buds on modified shoots, and by adventitious buds on roots, stems, or leaves. Agamospermy apparently is not widespread among weeds but there are some examples (dandelion, dallisgrass (*Paspalum dilatatum*), guineagrass (*Panicum maximum*), and rhodesgrass (*Chloris gayana*). Vegetative reproduction is widespread and is characteristic of perennial weeds. This form of asexual reproduction will be examined in depth in Chapters 5 and 6. At this point, we simply need to recognize that vegetative reproduction provides genetic stability conducive to rapid buildup of an adapted genotype. Further, many perennial weeds combine sexual reproduction with apomixis. This no doubt allows more rapid adaptation to changing environments than in a species depending on only one or the other. This is so since the sexual process produces new genotypes that the asexual process can exploit via clones.

Ploidy Relationships

The inherent genetic makeup of a species, as expressed by ploidy level (multiple of the basic chromosome number) and by life cycle (annual, biennial, perennial), also influences the extent and rate of natural selection in response to changes in the environment. Apomixis in weeds is closely associated with polyploidy (Gustafson, 1948; Warburg, 1960). Mulligan (1960) made a detailed analysis of polyploidy in Canadian weeds that provides genetic background information significant for speciation. The study included 151 common weeds widespread in Canada. As shown in the top graph of Figure 3-11, there was a difference in the distribution in *diploids*—2 times the basic chromosome number—and *polyploids*—more than 2 times the basic chromosome number—for different habitats.

Diploids were more common to grain fields and polyploids were more common to row crops, pastures, and hayfields. More polyploids in hayfields and pastures was explained on the basis that such crops have more perennial than annual weeds, as shown in the bottom graph of Figure 3-11. Nearly 80% of the weeds in pastures are perennials, compared with less than 20% in grain fields and row crops. Polyploidy is more common with perennials than with annual species. Because the same five perennial species occurred in grain fields and row crops, the differences observed for these two habitats are presumed to lie with the annual species. The higher incidence of polyploidy in row-crop than in grain-field weeds was ascribed by Mulligan to the selection pressure imposed by different agricultural practices.

In grain fields, selection is presumed to be for weeds with their maximum germination in the spring and with a life cycle and growth habit similar to the grain. Thus, it would be advantageous to be homogeneous for these characteristics. Since diploids are usually more uniform than polyploids, they might be expected at high frequency in grain fields where strong selection pressure for homogeneity occurs. In annual row crops, cultivation provides selection pressure for weeds with variable dormancy, germination periodicity, and photoperiod, thus favoring the less uniform nature of polyploids. Thus, it would appear that the type of agriculture within a given area does influence speciation through the relationship with ploidy.

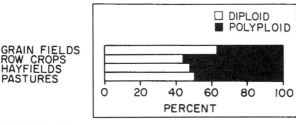

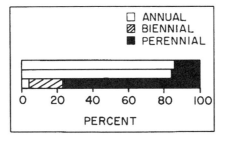

A. Ploidy level

B. Life cycle

FIGURE 3-11. Different ploidy levels and life cycles of weeds
for different crops.
Source: Data from Mulligan, 1960.

In summary, compensation in growth form and modification in genetic makeup on
the part of individual plants within a species in response to alterations in the environment cause new species to arise.

POPULATION

Population, the accumulation of organisms of a single species, is the level of ecological organization around which weed control programs and recommendations are built.
Because density—that is, the number of plants on a given area—is the attribute of a
weed population that figures most prominently in the development of control efforts,
some of the ecological concepts associated with it are examined here. However, a population has many characteristics other than density that are unique to it, including age
distribution, growth form, adaptiveness, persistence, reproduction fitness, birth and
death rates, and dispersion. Thus, an individual weed germinates but does not have a
germination rate as does a population, a weed has age but does not have an age ratio,
and so forth. Refer to Odum (1971) for a detailed review of the several characteristics
unique to a population. The emphasis in this book is on density.

Inherent Variability of Weeds

Before moving to a discussion of density, it is helpful to consider briefly the cross-fertilization heritage of weeds. Some knowledge of this evolutionary background of weeds will contribute to our understanding of the significance of density aspects, particularly survival strategy. The key question is: How do we explain the fact that many weeds survive under a range of environments? The answer lies in their evolutionary background. Most weeds evolved, and continue to do so, under hostile environments. They have had to survive unfavorable conditions imposed by both the natural environment—for example, drought and low temperature—as well as the manipulated environment—for example, disruption from tillage and competition with crops.

The products of these ecological forces are termed *general-purpose genotypes* (Baker, 1965, 1974). These are genotypes that allow a wide range of growth responses (phenotypic plasticity) from an individual while still maintaining the ability to evolve new forms through genetic recombinations. As discussed later in the chapter, such attributes are particularly important to annual weeds, commonly the early colonizers in ecological succession. Most weeds, especially annuals, are normally self-pollinated; however, the potential exists for some cross-pollination. Self-pollination assures the seed production of weeds invading an area even if plants are widely scattered or even if only a single individual is present. It also allows the rapid accumulation of individuals as well suited to that environment as the immigrant parent. Vegetative reproduction in perennial species also accomplishes this purpose; the risk, of course, is the development over time of a very uniform population poorly suited to a different environment. Cross-pollination provides variability in offspring that improves the chances for surviving such changes in the environment.

The ability to cross-pollinate has also contributed to variability in weed species through the crossing of segments of populations introduced at different times and from different sources. The movement of weed seed, coincidental to human migration and to trade, introduced new species and genotypes to new areas over a long period of time. Such movement provided repeated opportunities for introgressive and interspecific hybridization, described earlier.

Density Concepts and Relationships

As previously mentioned, weed density is an important determinant of the need for weed control. Several aspects of population density influence weed–desired plant ecology. Among them are the survival strategy of weeds, the effects of physical and biological factors on density, the relationship between density levels and carrying capacity, and the effects of agriculture itself.

K- and r-survival strategies.   Weeds have two broad strategies with respect to survival. In general terms, one strategy is based on numbers, and the other is based on exploitive ability. These strategies are founded on the two parameters, r and K, that together identify population behavior in response to disturbance. The *parameter r* identifies the potential rate of increase of a population for a given set of environmental conditions and is a measure of the rate at which a population will increase where there is no shortage of resources or constraints on population growth. The *parameter* K identifies an upper limit beyond which a population cannot go. This upper limit is determined by available resources and constraints of the population itself. For a species, the r-survival strategy, or *r-strategy,* is one that relies on the production of a large number of seeds (or vegetative reproduction units) and high dispersibility. The K-survival strategy, or *K-strategy,* relies upon fewer reproduction units, relatively low dispersibility, and strong exploitive ability. Further, in r-strategy species, a larger portion of resources is allocated to seed production.

The r-strategy is representative of many of the annual weeds with which we deal in agriculture. It is an important factor to keep in mind in our efforts to develop a predictive and preventive posture toward such weeds, which are discussed in greater detail in later chapters. That is the r-strategy comes into play in competitive ability, germination, dormancy, and longevity of weed seeds. The larger-seeded annual weeds and perennial weeds are characteristic of the K-strategy. Such species are usually not first colonizers but enter after there has been some amelioration of the environment.

As already pointed out, many perennials produce seed as well as perennating parts. Such species may exhibit either the r- or the K-strategy for both seed production and the production of perennating parts. For example, in a study of five perennial Compositae, it was found that Canada thistle was the most r-strategic in vegetative reproduction but quite K-strategic in seed production (Bostock and Benton, 1979). Dandelion, on the other hand, was strongly r-strategic in seed production and intermediate in vegetative reproduction. Thus, the strategy for both seed production and vegetative reproduction must be considered in designing control and prevention approaches.

Survival vs. competition. In considering r- and K-strategies, care must be taken to distinguish survival strategy from competition of weeds toward desired plants. Otherwise, the r-strategy ascribed to survival of many annual weeds and the K-strategy of many perennial weeds may seem contradictory. It is common knowledge that annual weeds can be very damaging to crop yield, frequently more so than perennials. This is so largely because their r-survival strategy assures a large number of potential individuals on every hectare of a field of a given crop. As we shall see when we examine competition in Chapter 7, individual weed plants may well be poor competitors with the crop plants under equal conditions. By contrast, many perennial weeds are restricted to patches within a field. The individual plants of such weeds commonly are more competitive for growth factors available to the patch area than individual plants of either the crop or of most annual weeds. For the entire field, however, collective losses from the perennial weed plants may well be much less than collective losses from the annual weed plants.

Physical and biological factors.   The upper limit beyond which density cannot go is determined by both physical and biological factors. Further, the imposition of these factors may be both independent of and dependent upon density of the population. Physical factors such as light, water, and nutrients tend to exert their influence independent of the density of the population. That is, the amount of these factors is outside the control of the population itself. For example, the amount of light available to the plants in 1 m² is the same whether there is 1 plant or 100 plants in that area. Biological factors, on the other hand, such as growth form, seed production, and competitive ability are all influenced in their effect by the density of the population. It must be kept in mind, however, that these effects are not mutually exclusive. Although the population cannot govern the amount of a physical factor available to it, it can modify its influence. For example, as we shall see later, stem extension, leaf area, and leaf arrangement influence a plant's ability to compete for light. Conversely, a physical factor can influence the growth form and other characteristics of the plant.

The broad overall principles of density relationships can be applied to weed–crop situations to help explain observed differences in weed population density for different crops. The general relationship between number of species and their density is shown graphically in Figure 3-12. The solid line is representative of undisturbed plant communities. Such communities have a few species in large numbers and many species in small numbers. The dotted line is representative of communities in disturbed environments. Such communities have a variety of species with none predominating.

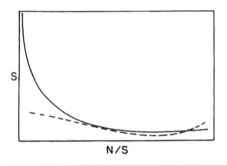

FIGURE 3-12. General density relationship between number of species (S) and number of individuals per species (N/S).

Source: Odum, 1971. From Fundamentals of ecology, 3rd ed., by Eugene P. Odum. Copyright © 1971 by W.B. Saunders Co., Copyright 1953 and 1959 by W.B. Saunders Co. Reprinted with permission of Holt, Rinehart & Winston, CBS College Publishing.

The implication for weed–crop ecology is that we would expect the relationship between species number and individuals per species to be like that represented by the dotted line. That is, tillage imposes a physical stress that can be expected to flatten the relationship curve. Such a flattening, in fact, has happened in Germany (Koch and Hess, 1980). On the experimental farms of Hohenheim University, intensive farming is credited for the reduction by more than half in the number of weed species from 1860 to 1980. Since the greatest loss of species occurred prior to 1940, herbicides were not an important factor. Further, in comparing the weed flora in an area in southwestern Germany, the number fell from 124 species in 1948-1949 to 61 species in 1975-1978, when cereals were grown. Similar reductions in species numbers occurred in Ohio with no-tillage of continuous corn (Table 3-3). There were about one-half as many species

TABLE 3-3.   **Weed species in seedbanks following long-term continuous tillage.**

Tillage	Average Number of Species in Soil Series		
	Wooster	Crosley	Hoytville
Conventional	4.6	4.0	3.0
No-tillage	10.1	7.0	5.7

Source: Data from Cardina et. al., 1991.

in the seedbank after 27 years of conventional tillage compared to no-tillage in all three soils.

The density-dependent relationships established in these ecological principles support the following generalizations: (1) density of the population tends to be controlled by physical factors in low-diversity, physically stressed agroecosystems representative of row-crop agriculture and (2) biological factors are the controlling influence in higher-diversity agroecosystems, such as permanent pasture.

Density levels and carrying capacity.   In considering density relationships, it is necessary to distinguish between densities that can be maintained (supported over time) and densities that exceed this level as a consequence of a large flush of seed germination, which may be caused by any number of factors. A term commonly used in agriculture to describe the density level that can be maintained over time is *carrying capacity*. Carrying capacity can be exceeded both by us in our agricultural production practices and by the weed populations. Overshooting the carrying capacity is followed by fluctuations before ultimately settling down to a density level that can be maintained. In weed–crop situations, the development of ecotypes and of plasticity is likely to cause weed numbers to stabilize at a somewhat lower level than the carrying capacity.

Overshooting carrying capacity has implications for weeds in still another aspect. There is evidence from limited studies that the mortality rate is high during the seedling period and then constant at a lower rate for the rest of the weed's life cycle. The flush of weed seed germination, especially that associated with row-crop agriculture, provides a large number of individuals from which natural selection can choose those best fitted to the particular set of cultural practices imposed by the agricultural production system. Rapid development of herbicide resistance, referred to in the previous chapter, is an example of this phenomenon.

Effects of agriculture on density.   Although there is a fixed upper limit to density for any given weed, as we have seen, wide fluctuations are common both above and below this limit before the population settles back to this level. The maturity of the agroecosystem and the cultural practices followed in agriculture affect the density relationships, especially the fluctuation above and below the carrying capacity. The most violent fluctuations tend to be associated with the most simple agroecosystems, which are those of row-crop agriculture. In this regard, it should be emphasized that systems of production that have been used for long periods of time, even though the practice itself is unsettling, can nevertheless be assumed to have a stabilizing aspect. For example, the common practice of plowing, used throughout agriculture for many years, although disruptive, can be said to have a stabilizing aspect. The recent move to reduced or minimum tillage represents a departure from this long-used practice. Because of this change, we have seen wide fluctuations in associated weeds.

Individual Plant as a Population

Before leaving a discussion of population, it is helpful to consider the concept that an individual plant is also a population in a certain sense (Harper, 1977). An individual plant actually has an age structure in that the first-formed leaves are older than more recently formed leaves, and the plant can respond to stress usually by altering the number of its parts. Thus, Harper suggests viewing populations as having two structures. One structure is described by the number of individuals (N) that come from original zygotes (genets). The other structure is a unit (n) of the genet—for example, a leaf with its bud, a tiller, a rhizome, and so forth. N and n combined identify the number of modular units and describe behavior of a population occurring through changes in N or n or some combination of the two. These structures have special significance for weed science. Weed numbers (N) are often used in measuring effectiveness of weed control practices. Although N provides a valid measure of herbicide efficacy, it may provide relatively little information about weed competition because it fails to take n into account. Also, as we shall see later, N especially may provide little information about seed production because it fails to identify the number of potential reproductive units (n structure).

COMMUNITY

The *community,* or mixture of weeds, has special significance for weed science because it represents the organizational level at which change in response to agricultural practices most commonly occurs. Further, a community of weeds is the most usual situation faced in agricultural use of land. There are instances of a single weed being the predominant problem in a given crop situation. For example, milkweed in wheat in Missouri at harvest time is often the only weed problem in many fields. However, these instances are exceptions.

The characteristic of a community to change in response to changes in production practices is of special concern to weed science. The foundation in ecological principles and theory for these changes needs to be understood to better predict future weed problems and prepare to deal with them. Two mechanisms can bring about a change in a community of weed species: (1) a change can occur within a species (such as ecotype development, mutation, and so on) and (2) one species can be replaced by another. As we have seen, changes do occur within an individual weed species as a result of the processes of allopatry and sympatry. The growing list of ecotypes for weeds attests to this fact. Nevertheless, the replacement of one species by another is the much more obvious mechanism and probably the more important one in the short run. A primary reason is found in the fact that the practice of agriculture provides repeated opportunities for selection of those species best suited to a particular practice. This opportunity is amplified by the observed common tendencies for populations to overshoot the carrying capacity. Shifts in weeds referred to earlier for corn in response to different herbicides provide an excellent example of species replacement in response to production practices.

Two ecological principles are at work in mixtures of weed species to influence the makeup of the community: (1) the competitive exclusion principle and (2) the factor compensation principle. The *competitive exclusion principle* is the process that expresses the ecological concept of closely related or otherwise similar species to separate ecologically. This process clearly operates essentially to minimize the number of different species within a community. The *factor compensation principle* is exemplified by coexistence of different species.

Coexistence of Different Weed Species

The fact that a single species has not won out in the struggle for adaptation to conditions imposed by particular agricultural practices indicates the importance of coexistence as a type of change. Four mechanisms may allow two species to persist together: (1) different nutritional requirements, (2) different causes of mortality, (3) differing sensitivity to toxins, and (4) different time demands on growth factors (Harper, 1977). The most common example of the first mechanism is that of legumes persisting with nonleguminous species because of their different requirements for nitrogen. Selective grazing by livestock of the forage species, leaving ungrazed such weeds as Canada thistle, is an example of coexistence because of different mortality. The differential responses of plants to allelochemicals (plant chemicals produced and introduced into the environment that inhibit other plants), which is discussed in greater depth in Chapter 8, is an example of differing sensitivity to toxins.

Different Time Demands

Differences among populations in the time of their demand on environmental factors is much the most common source of coexistence. A long-term crop rotation study established on the University of Missouri campus at Columbia, Missouri, provides an excellent example of coexistence. Timothy (*Phleum pratense*) has been grown continuously for about 50 years on one of these plots. In this plot, little barley (*Hordeum pusillum*) has become almost the sole weed species. It appears to be living as a winter annual. Thus, it germinates in the fall after the timothy has ceased active growth, overwinters, then completes its growth and seed production in the spring prior to the early flush of timothy growth. In effect then, these two species coexist by making their main demands upon moisture, light, and plant nutrients at different times of the year, thus avoiding competition. Such differences in time demands are common in the weed–desired plant world. This phenomenon has particular implications for a system of control geared to the entire rotation and for preventive approaches because it implies that the weed may originate (from seed or perennating parts) at a different time or in a different crop from the one in which it is a problem.

Edge Effect

Another ecological concept, known as the edge effect, has significance in explaining diversity of weed populations in weed–desired plant situations. The *edge effect* in ecology is the tendency for greater diversity of species at junctures of communities. The junction zone, or *ecotone*, contains some species common to both adjoining communities. The practice of agriculture in itself presents many edges. In the United States, for example, settlement involved "islands" of farms surrounded by forest or prairie from which the farm was taken. This settlement created two broad types of ecotone: (1) the forest–agricultural use ecotone and (2) the prairie–agricultural use ecotone. In more recent years, placement of land in the so-called soil bank in the 1950s and removal of much of that same land from the soil bank in the 1970s created ecotones in broad agricultural belts within the United States and even on individual farms. Many additional edge effects can be visualized as the result of the type of farming, including the relationship between farmed land and roadsides, no-tillage and conventional tillage, ditch banks, and others. Thus, the edge effect is no doubt a major source of diversity in weed–crop communities, especially in the United States.

ECOSYSTEM

The ecosystem of weeds and desired plants, also called the *agroecosystem,* is the plant community and nonliving environment functioning together. This system is the overall focus of this book. Focus on this ecosystem recognizes we cannot effectively deal with weeds unless we understand their relationship with desired plants and the environment, including humans. Further, it recognizes that the weed community is the reflection of the history of crops and other desired plants grown, tillage performed, and the environment. For weed management to be fully effective, we must have an understanding of both the currently interacting forces and the effects of past treatments.

Ecological Succession

Ecological succession is an orderly change in species resulting from modification of the physical environment by the community. In nature, it culminates in a stabilized ecosystem dominated by a few species. *Primary succession* is succession that begins on newly exposed rock, bare sand, or lava flow. *Secondary succession* is succession on an area from which a community has been removed. Secondary succession is the type in mind for the discussion that follows because some principles involved in secondary ecological succession provide important information for understanding the historical interactions of the weeds with which we deal.

A common thread in succession, of course, is the replacement of species. It should be emphasized that replacement is the result of action by the community, that is, by the living part of the system. Water, nutrients, and other physical factors of the environment may affect the rate of change and the ultimate balance or climax attained; but it is the living portion that modifies and changes the environment, thus triggering succession. Modification of the environment, which is the driving force for succession, has its foundation in the aspect of the ecological niche concept that dictates that each species has its particular requirements, with no two species being able to occupy exactly the same niche, at least not for long. Thus, in a natural ecosystem, the presence of populations results in new niches being formed that are occupied by different populations, which in turn results in new niches being formed. These same forces are at work in agroecosystems.

A commonly used method for evaluating the modifying effects of the community during succession is the study of abandoned fields. The abandoned wagon roads made by the pioneers as they crossed the Great Plains in the movement west provided an early opportunity for identifying succession in this region of the United States. The pattern observed involved four stages: (1) annual weed stage, (2) short-lived grass stage, (3) early perennial grass stage, and (4) climax grass stage. Although the species differ with geography, the pattern holds everywhere in the Great Plains. The species common to the four stages for Oklahoma and southern Kansas are shown in Table 3-4.

TABLE 3-4. **Dominant species for each of the four stages of plant succession in old fields in central Oklahoma.**

Stage 1 Pioneer Weed (2-5 years)	Stage 2 Annual Grass (3-10 years)	Stage 3 Perennial Bunch Grass (10-20 years or more)	Stage 4 True Prairie
Common sunflower (*Helianthus annuus*) Western ragweed (*Conyza canadensis*) Lambsquarters Johnsongrass Crabgrass Plus several others	Triple-awned grass (*Aristida oligantha*)	Little bluestem (*Andropogon scoparius*)	Little bluestem Big bluestem (*Andropogon gerardi*) Switchgrass (*Panicum*) *virgatum*) Indiangrass (*Sorghastrum nutans*)

Dominants. Dominance by a few species is a common characteristic of plant communities. Species that strongly affect the environment of all other species are termed *dominants*. For weeds, of the hundreds that might be present in a weed–crop situation, relatively few occur; and of these, a very few species are usually dominant in a given crop situation.

One of the established principles relating to removal of a dominant has special implications for weed ecology: Removal of a dominant is more disruptive of the environment than is removal of a nondominant. By intent, weed control focuses on the dominants in view of their overriding influence on crop yields. Thus, control itself creates a diversity of environments to be filled by other weeds unless it is filled by a desired plant. Any time effective control of a dominant weed is realized, we should be alert to the probability of its being replaced.

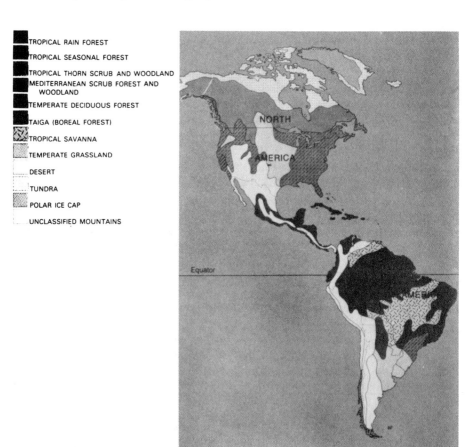

TROPICAL RAIN FOREST

TROPICAL SEASONAL FOREST

TROPICAL THORN SCRUB AND WOODLAND

MEDITERRANEAN SCRUB FOREST AND WOODLAND

TEMPERATE DECIDUOUS FOREST

TAIGA (BOREAL FOREST)

TROPICAL SAVANNA

TEMPERATE GRASSLAND

DESERT

TUNDRA

POLAR ICE CAP

UNCLASSIFIED MOUNTAINS

FIGURE 3-13. Major biomes of the world.

Source: Starr and Taggart, 1981. From Biology, the unity and diversity of life, 2nd ed., by Cecie Starr and Ralph Taggart. © 1981 by Wadsworth, Inc. Reprinted with permission of Wadsworth Publishing Company, Belmont, CA.

Biomes. The thrust, or function, of succession from the pioneer stage on is to try to achieve stability. If succession goes to completion, the vegetation present, known as the *climate climax,* will be representative of broad climate zones. These broad, world-wide areas, known as *biomes,* are shown in Figure 3-13. The names in the key signify the dominant species and general characteristics for the climax and represent the end toward which succession is reaching in each broad area. Within each of these large areas, there are climaxes called *edaphic climaxes* that are modified by local conditions (soil type, nutrient availability). Note that annual weeds are the first stage in succession for all of the climatic and edaphic climaxes. These annuals are r-strategy species that rely upon large numbers and high dispersibility for survival. In this regard, it is significant for weed science that seeds of such species dominate the reservoir of seeds in the soil, even under permanent pasture or mature agroecosystems.

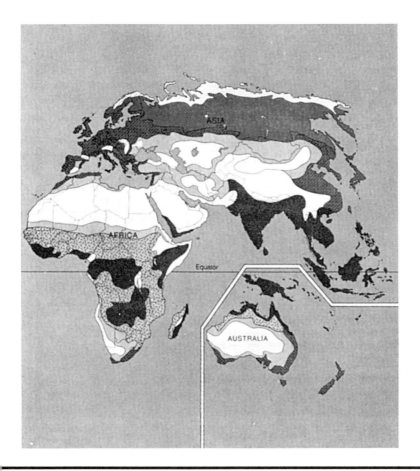

The striving for stability implies a progressive lessening of the impacts of disturbances. This situation is represented diagrammatically in Figure 3-14 for a two-species community. Competition is shown by a minus (−) sign, neutral effects by a zero (0), and synergism (helpful effects) by a plus (+) sign. The interactions that occur range from those in which each species is hurt (- and -) by association with one another to those in which each is helped (+ and +). Disturbance (agriculture) tends to create unstable, competing relationships. The challenge facing weed science in this regard is to do what it can to cause agroecosystems to behave more like natural ecosystems—that is, to move the weed–crop relationship in the direction of reduced competition and greater stability. Reduced tillage (facilitated by adequate chemical weed control) can have a stabilizing influence.

Two key phrases with respect to succession have implication for weed science: *orderly change* and *predicted end.* As we learn more about the changes in weed composition associated with particular agricultural practices, it seems reasonable to expect that we can achieve fairly accurate predictions of what our weed problem will be in the future. The tremendous contributions such predictions from weed science can make to agriculture are many. For weed science itself, intelligent predictions are an important first step toward effective preventive weed control. The status of predictive technology will be examined in Chapter 15.

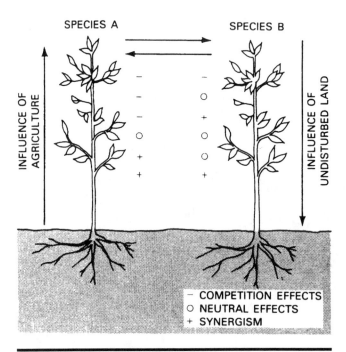

FIGURE 3-14. Diagrammatic representation showing possible interactions between two species and the general influence of land use on such interaction.

REFERENCES

Baker, H.G. 1965. Characteristics and modes of origin of weeds. In H.G. Baker and G.L. Stebbins, eds., The genetics of colonizing species, pp. 147-172. New York: Academic Press.
———— 1974. The evolution of weeds. Annu. Rev. Ecol. Syst. 5:1-24.
Bostock, S.J., and R.A. Benton. 1979. The reproduction strategies of five perennial Compositae. J. Ecol. 67:91-107.
Burt, G.W. 1974. Adaptation of johnsongrass. Weed Sci. 22:59-63.
Cardina, J., E. Regnier, and K. Harrison. 1991. Long-term tillage effects on seedbanks in three Ohio soils. Weed Sci. 39:186-194.
Dickerson, C.T., Jr., and R.D. Sweet. 1971. Common ragweed ecotypes. Weed Sci. 19:64-66.
Evans, R.A., and J.A. Young. 1970. Plant litter and establishment of alien annual weed species in rangeland communities. Weed Sci. 18:697-703.
Gorske, S.F., A.M. Rhodes, and H.J. Hopen. 1979. A numerical taxonomic study of *Portulaca oleracea*. Weed Sci. 27:96-102.
Gustafson, A. 1948. Polyploidy, life form and vegetative reproduction. Hereditas 34:1.
Harper, J.L. 1977. Population biology of plants. New York: Academic Press.
Hodgson, J.M. 1964. Variations in ecotypes of Canada thistle. Weeds 12:167-171.
Koch, W., and M. Hess. 1980. Weeds in wheat. In Wheat, technical monograph, pp. 33-40. Basle, Switzerland: CIBA-GEIGY.
McWhorter, C.G. 1971. Growth and development of johnsongrass ecotypes. Weed Sci. 19:141-147.
McWhorter, C.G., and T.N. Jordan. 1976. Comparative morphological development of six johnsongrass ecotypes. Weed Sci. 24:270-275.
Merrell, D.J. 1962. Evolution and genetics. New York: Holt, Rinehart and Winston.
Mulligan, G.A. 1960. Polyploidy in Canadian weeds. Can. J. Gen. Cyt. 2:150-161.
Odum, E.P. 1971. Fundamentals of ecology, 3rd ed. Philadelphia: Saunders.
Soane, I.D., and G.R. Watkinson. 1979. Clonal variation in populations of *Ranunculus repens*. New Phyto. 82:557-573.
Starr, C., and R. Taggart. 1981. Biology, the unity and diversity of life, 2nd ed. Belmont, CA: Wadsworth.
Trenbath, R.R. 1976. Plant interactions in mixed crop communities. In Multiple cropping, ASA special publication no. 27, pp. 129-169. Madison, WI: American Society of Agronomy, Crop Science Society of America, and Soil Science Society of America.
Van Doren, D.M., and R.R. Allmaras. 1978. Crop residue management systems. In Effect of residue management practices on the soil, physical environment, microclimate, and plant growth, ASA special publication no. 31, pp. 49-83. Madison, WI: American Society of Agronomy, Crop Science Society of America, and Soil Science Society of America.
Warburg, E.F. 1960. Some taxonomic problems in weed species. In J.L. Harper, ed., The biology of weeds, p. 11. Oxford: Blackwell Scientific.
Warwick, S.E., and D. Briggs. 1978. The genecology of lawn weeds, 1. Population differentiation in *Poa annua* in a mosaic environment of bowling green lawns and flower beds. New Phyto. 81:711-723.
Wedderspoon, I.M., and G.W. Burt. 1974. Growth and development of three johnsongrass selections. Weed Sci. 22:319-322.

Wright, J.L., and E.R. Lemon. 1966. Photosynthesis under field conditions, IX. Vertical distribution of photosynthesis within a corn canopy. Agron. J. 58:265-268.

Yip, C.P. 1978. Yellow nutsedge ecotypes, their characteristics and responses to environment and herbicides. Diss. Abstr. Int. B 39:1562-1563.

Young, J.A., R.A. Evans, and B.L. Kay. 1970. Phenology of reproduction of medusahead. Weed Sci. 18:451-454.

4

Reproduction from Seed

CONCEPTS TO BE UNDERSTOOD

1. Most problem annual weeds have the capacity to produce a large number of seeds per plant.

2. Invasion of weed seeds from outside is usually of only minor importance as a source of additions to the seedbank on a given area.

3. Many weeds, especially annuals, have a relatively short period from emergence to flowering and quickly reach the point of having viable seed.

4. Most problem annual weeds can adjust seed production per plant to compensate for losses in plant numbers over a relatively wide range.

5. Many annual weeds can produce some seed even under very adverse conditions.

6. Light is the growth factor to which seed production is most sensitive in many weeds: Production of seed is commonly retarded and may even be prevented completely by degrees of shading that have much less effect on vegetative growth.

7. Seeds of many weeds suffer sizable losses in viability if left on the soil surface for extended periods of time.

8. The number of seeds in the seedbank is commonly on the order of several million for the plow layer.

9. The seeds of some weeds may remain viable for many years in the soil, but for many, germination the year following production accounts for a majority of total germination for the life of the seed in the seedbank.

10. Tillage reduces the longevity of weed seeds in soil.

11. For most weed seeds, longevity is extended by incorporation in the soil and the relationship approaches linearity with depth in the top 15 cm.

12. Long-term tillage reduces species diversity in the seedbank.

The weed seed, especially of annual species, holds the key to success for both a control and a prevention approach to weeds. The number of seeds that germinate and survive control efforts largely determines crop loss for a given year. The number of seeds both in the soil (seedbank) and returned to the soil in a new crop determines whether the species will survive to pose a future threat. For this reason, as discussed in Chapter 1, it is important to adopt a mind-set that views the plant as a vehicle for producing seeds. Only with such a mind-set can there be appropriate focus on the seed commensurate with its key role. Its role is larger than simply serving as the vehicle for multiplication, however, important as that role is.

Four additional roles are played by the seed in a weed's life cycle: (1) dispersal, (2) protection during conditions unfavorable for germination and development (dormancy), (3) a temporary source of food for the embryo, and (4) a source for transfer of new genetic combinations. All five roles come into play in the natural selection of individuals best suited to specific conditions. In the remainder of this chapter, we examine multiplication, dispersal, and longevity. Dormancy and its particular effect on longevity are examined in Chapter 6.

WEED SEED CHARACTERISTICS

The seed is an interesting package indeed. By definition, a *seed* is a fertilized, mature ovule having an embryonic plant, stored food material (rarely missing), and a protective coat or coats. Thus, it contains all that is necessary to transmit the genetic material provided by the parents and sustain, at least temporarily, the new seedling that carries this genetic information.

External Characteristics

The seed varies tremendously in its external characteristics, as can be seen in Figures 1-30 in the appendix at the end of the chapter. It may vary in size from a 10 kg coconut to a seed of false pimpernel (*Lindernia anagallidea*) so small that approximately 300 million are required to weigh 1 kg. Its shape may vary from round to trapezoidal and from spherical to flat. It comes in all colors of the rainbow. Its surface may be smooth, rough, coated with mucilage, or covered with a variety of appendages important to widespread dispersal.

Internal Characteristics

Internal characteristics of seeds are equally variable. Differences include embryo characteristics, quantity of food reserve stored, and chemical composition. One obvious difference in embryo characteristics is in the number of cotyledons. Monocot embryos in the seed have one cotyledon, and dicot embryos, two cotyledons. The quantity of stored food in a coconut may be a billion times greater than that in an orchid seed. Seeds of corn are high in starch, and soybean and cotton seeds are high in protein and in oil.

Seed Numbers

Weeds vary greatly in their potential seed production capacity. Table 4-1 shows the relative potential seed production capacity per plant and per gram for a number of common weeds. The actual production per plant will vary greatly from the potential indicated depending upon environmental conditions under which the plant is grown. For example, seed production in the poppy (*Papaver rhoeas*) varied from 1 capsule containing only 4 seeds up to 400 capsules, each containing 2,000 seeds (Bleasdale, 1960). The significance of the figures in Table 4-1 lies in showing the large production *potential* of many of our weed species. For many species, a single plant has the potential to produce a competitive infestation if all the seeds are evenly distributed over an area and all germinated in a given season.

TABLE 4-1. Seed production capacities of selected weeds.

	Number of Seeds	
Common Name	Per Plant	Per Gram
Barnyardgrass	7,160[1,2]	714
Buckwheat, wild	11,900	143
Charlock	2,700	526
Dock, curly	29,500	714
Dodder, field	16,000[2]	1,299
Kochia	14,600	1,176
Lambsquarters	72,450	1,428
Medic, black	2,350	833
Mullein	223,200	11,111
Mustard, black	13,400[3]	588
Nutsedge, yellow	2,420[1]	5,263
Oat, wild	250[1]	57
Pigweed, redroot	117,400[1]	2,632
Plantain, broadleaf	36,150	5,000
Primrose, evening	118,500	3,030
Purslane	52,300	7,692
Ragweed, common	3,380[1]	253
Sandbur	1,110[1]	148
Shepherdspurse	38,500[1,2]	10,000
Smartweed, Pennsylvania	3,140	278
Spurge, leafy	140[3]	286
Stinkgrass	82,100[1,2]	14,286
Sunflower, common	7,200[1,2]	152
Thistle, Canada	680[1,2]	637

[1] Calculated immature seeds also present.
[2] Many seeds shattered.
[3] Yield of one main stem.
Source: Data from Stevens, 1932.

On a population basis, weeds have the ability in the production of seed to compensate greatly for loss in numbers. For example, it was observed with corn cockle (*Agrostemma githago*) that a 90% reduction in stand reduced the seed production of that population by only 10% (Bleasdale, 1960). Very large seed production potential and the ability to produce some seed even under very adverse conditions are two important difficulties facing a weed prevention approach. The weed numbers may be below the level to compete with desired plants but still sufficient to produce enough seed to assure a competing population in subsequent years. As shown in Figure 4-1, this situation simply reflects the allocation of more of the resources of such plants to production of seed than is true of perennial weed species and expresses the r- and K-survival strategies discussed in Chapter 3. The figure also shows that plants that produce only one seed per carpel (monocarpy) allocate proportionately more resources to seed production than do those that produce two or more seeds per carpel (polycarpy). A *carpel* is an individual segment of the ovary.

Relative Freedom from Pests

Seeds of most annual weeds apparently have fewer insect and disease pests than seeds of other species. As new colonizers, such species would not have been as exposed to such pests as later, more permanent residents in succession. For these annual species to have insect and disease pests, an alternate host would probably have been necessary. It can be speculated that the yearly occurrence of some of these species in agriculture has provided opportunity for the selection of pests that indeed have such alternate host traits. However, the tendency for pioneer species to have fewer native diseases and insect pests suggests that biological control may be more difficult with them than with perennial species. This matter is expanded upon in Chapter 9 when we consider biological control.

Entry of Weed Seeds

The significance of additions to the *seedbank*—that is, the reservoir of viable seeds in the soil—from outside a given site must be understood for the importance of weed seed production and longevity to be fully appreciated. Of concern here is the addition of numbers sufficient to significantly increase the size of the seedbank. Thus, we are not concerned at this point with the spread of weeds from place to place over relatively great distances; this topic is covered in Chapter 14.

In the past, contamination of crop seed with weed seed was a major source of additions to the seedbank. With the general use of high-quality seed today, this problem is not major, although it may be for selected weed species. For example, eastern black nightshade (*Solanum ptycanthum*) is increasing as a problem in soybeans, in part because the fruit is comparable in shape and size to the soybean seed and therefore can-

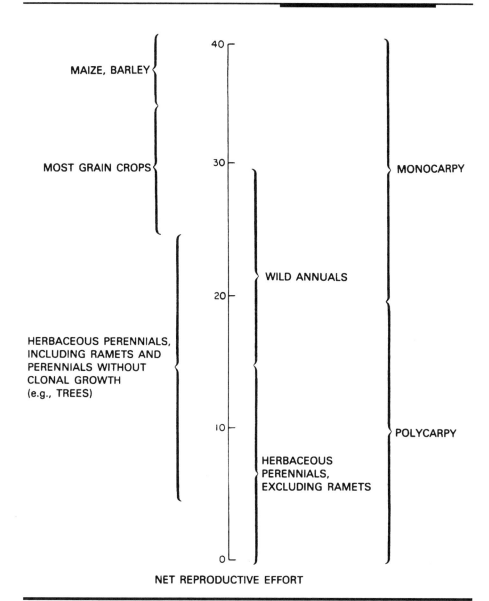

FIGURE 4-1. Proportions of annual net assimilation involved in allocation to reproduction in different groups of flowering plants.

Source: From Ogden, as reported by Harper, 1977. Reproduced courtesy of J.L. Harper, © copyright 1977.

not easily be separated during seed cleaning. Nevertheless, ample supplies of clean seed are usually available for all crops and thus should not be an important source of additions to the seedbank.

Birds and other animals, water, and wind are vehicles for the movement of weed seeds into an area. Weeds and weed seeds have a number of unique characteristics that aid such movement. Some of these characteristics are shown in Figure 4-2.

Animals.   Many weed fruit or seed readily attach to animals and can, in that way, be moved into a clean area (Figures 4-2C and 4-2D). Many weed seeds can survive passage through the digestive tract of birds and other animals and thus be the source of establishment in new sites. Nevertheless, although such movement may be important in the spread of species from a given source, and, therefore, in reinfestation, it is not an important source of additions to the seedbank. That is, individual, scattered plants may arise but not in sufficient numbers to be a factor in competition. Thus, wind and water are left as the vehicles that add significant numbers of weed seeds to the seedbank.

Water.   Irrigation water can move large numbers of weed seeds to the land. In studies of irrigation laterals of the Yakima and Columbia rivers in Washington, seed of 127, 84, and 77 species were identified in the water in three separate years of sampling (Kelley and Bruns, 1975). The total number of weed seeds found was such that if evenly distributed in average rates of irrigation, it would contribute 94,500, 10,400, and 14,000 seeds ha^{-1} for the 3 years, respectively. From what we learned about numbers and competition in Chapter 2, if all the seeds germinated at one time, the resulting weed stand of about 9.5, 1.0, and 1.4 weeds m^{-2}, respectively, could be competitive, depending upon the weed and the crop. Water from the North Platte River Project was also found to deliver potentially competitive numbers of weed seeds to cropland (Wilson, 1980). On nonirrigated land, movement with runoff water can be visualized as resulting in the concentration of seeds of certain weeds on new sites. For example, many seeds, such as common milkweed in Figure 4-2B, readily float on water.

Even though these examples indicate that significant numbers of weed seeds may move into an area with water, these circumstances are rather unusual. Further, these numbers are much less than the numbers already present in most agricultural soils. From the perspective of agriculture as a whole, therefore, influx with water must be considered as a minor source of addition to the seedbank.

Wind.   Some seeds, such as dandelion, have appendages that aid movement with wind. The umbrella-like structure shown for dandelion in Figure 4-2A is called a *pappus*. A pappus is characteristic of seeds of the Compositae family. The pappus of dandelion acts like a parachute, thus allowing it and the attached seed to literally float with the wind. Pappi vary in their efficiency to aid dispersal, depending mainly on their diameter compared to the diameter of the seed. Although such structures clearly can result in individual seeds being moved some distance, the important question in terms of the seedbank for a given area is: How many seeds enter in this way? Studies, such as

A. Parachute-like pappus of dandelion seed that floats with the wind

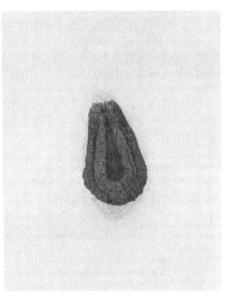

B. Flat and light common milkweed seed that readily floats on water

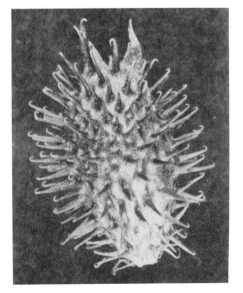

C. Common cocklebur seed with appendages that attach to animals

D. Beggarticks seed with appendages that attach to animals

FIGURE 4-2. Some unique characteristics that aid dispersal of weed seeds.

the one shown in Figure 4-3, clearly indicate that dispersal falls off rapidly with distance from the source. For the weed shown, tansy ragwort (*Senecio jacobaea*), 60% of all the seeds were at the base of the plants, and only about 1.4% were beyond 4.6 m. Regardless of wind direction, the preponderance of seeds was close to the source. Where movement was with the wind, the number of seeds moved to about 35 m was approximately 10 m^{-2}. If all survived to become a part of the viable seedbank, they would represent a significant addition. Beyond this distance, additions would be relatively insignificant.

Relative importance of seed entry.   In spite of evidence for movement of significant numbers of weed seed by water and by wind, the fact remains that invasion from outside is a relatively minor source of additions to the seedbank under most agricultural

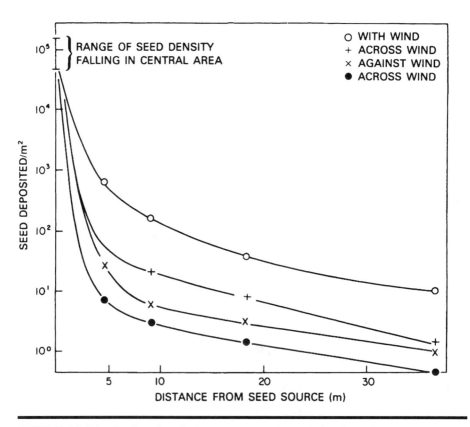

FIGURE 4-3. Distribution of seed of tansy ragwort (*Senecio jacobaea*) from a dense population about 20 m², measured in relation to the direction of the prevailing wind. Note that curves are eye-fitted.

Source: Harper, 1977. Reproduced courtesy of J.L. Harper, © copyright 1977.

conditions. This fact is well illustrated by the vastly different community of weeds in the different treatments in the long-standing crop production treatments in Sanborn Field at Columbia, Missouri. For example, the weeds on the plot that has been in continuous wheat since 1888 are almost exclusively shepherdspurse (*Capsella bursa-pastoris*), pepperweed, and pennycress (*Thlaspi arvense*), even though plots of other cropping systems only a few meters away are infested with a host of other species. Similarly, the soil in Broadbalk Field at the Rothamsted Station in England contained very different flora for the differences in manure applications on wheat, indicating invasion from neighboring plots was relatively unimportant (Brenchley and Warington, 1930).

Thus, except for cropland immediately adjacent to fence rows, ditch banks, roadways, and similar sources, invasion from outside does not add significant numbers to the seedbank. This factor is very significant for prevention in weed management. It means that efforts can be concentrated on seed production of those weeds occupying the area and on those seeds already present. Present approaches to weed management usually underemphasize the value of prevention, partly because the significance of infestation from outside is overemphasized. This statement, of course, does not mean that such additions from outside are unimportant because they may result in reinfestation and in new species becoming established that if not controlled, could become a problem. As we shall see in the next chapter, outside additions may be particularly significant for perennial species that spread relatively slowly by their vegetative parts.

FACTORS AFFECTING WEED SEED PRODUCTION

At the outset of this section it is important to emphasize two characteristics common to many annual weeds that further complicate preventive approaches. These characteristics are: (1) the relatively short period from emergence to first production of seed (juvenile period) and (2) the ability to produce viable seeds even when the flowering part is severed or the parent plant is destroyed before the seeds reach maturity.

Juvenile Period

Many annual weeds produce seed comparatively quickly. Such precociousness, where it occurs, is a major reason these species persist with cultivated plants. Annual bluegrass, shepherdspurse, and groundsel (*Senecio* spp.) are examples of weeds with a very short juvenile period. These weeds may produce viable seed in about 6 weeks from the time their seed is planted. Such weeds, where they occur in row-crop agriculture, may produce viable seed before the crop with which they are growing is harvested. Therefore, escaped plants from a weed control or management effort successful in reducing or keeping numbers below a competitive level could nevertheless contribute significant new additions of seed to the seedbank in the soil.

Common chickweed is an example of many weeds that may produce seed before a perennial crop, such as alfalfa, is harvested. Common chickweed in alfalfa germinates in the fall with a fall seeding or in an established stand. In either case, it flowers in the late winter or early spring and produces seed before the first cutting of alfalfa is harvested. Common ragweed following wheat harvest in the central United States is one example of many weeds that produce seed following crop harvest. In Missouri, winter wheat is usually harvested in June, leaving about 4 months of growing season. This time is more than ample for the many precocious weeds, such as ragweed, to produce a seed crop.

Precociousness may also allow a weed to produce seed in spite of management practices that control or prevent seed production of most weeds. Annual bluegrass in golf course turf is a good example. During ideal growing conditions, seedheads can appear within hours after mowing and may develop to the point that some viable seeds are present by the next mowing.

Maturity

As is true with crop seeds, viability of weed seeds increases with maturity. However, with many weeds—for example, wild oats—even very immature seeds may be viable. The data in Table 4-2 show that by the time the parent plants had begun to head, some seeds were able to produce a new plant even if the plants were cut. Clearly, viability increases with maturity, to the point where in the late milk to early dough stages, more than 50% emergence occurred. In view of the very large seed production potential of many annual weeds, even a rather low viability percentage on destroyed plants could contribute significant additions to the seedbank. It is therefore important in a preventive weed management program to know the effect of maturity on seed viability for the species involved. Any practice used to prevent seed production needs to be timed to precede the stage after which significant numbers of seeds may be produced.

TABLE 4-2. **Percent emergence, 2 weeks after planting, of wild oat seedlings from seeds harvested from mowed parent plants.**

Stage of Parent Plants When Mowed*	Percent Emergence
Late joint	0.0
Early boot	0.0
Boot	0.0
Early head	0.3
Early milk	18.0
Late milk to early dough	53.7

*Simulated by removing the panicles and storing dry at room temperature until planted in the greenhouse.

Source: Adapted from Andersen and Helgeson, 1958.

Nutrients and Moisture

It is a common characteristic of many weeds to produce some seeds under very poor conditions. This characteristic is understandable in view of their evolutionary background. That is, many of them as early colonizers faced circumstances of low nutrient and moisture availability and low organic matter. This situation provided for the selection of individuals that could tolerate such conditions.

There is circumstantial evidence that weed seed production is affected less than desired plants by supplies of nutrients and water, although there are no discrete data to support or reject this supposition. The reduction in corn cockle seed production of only 10% for a stand reduction of 90% referred to earlier is one such piece of evidence. There are data for additional weeds that also show production of seed for a given area to be relatively constant over a wide range of plant densities (Palmblad, 1968; Harper, 1961). Although most desired plants also exhibit plasticity of seed production and density, the extent is appreciably less than for the weeds cited. Relevance to moisture and nutrients is simply that population densities used were so high that some competition for these growth factors could have been expected. For example, in one study (Harper, 1961), there were 200 plants per each 15-cm pot. The relatively shorter life cycles of many annual weeds than the life cycles of crops with which they are growing is another piece of such evidence.

Thus, under conditions often encountered in agriculture, weeds may well obtain enough nutrients and water to assure production of a full seed crop before competition from the crop occurs. Further, because their life cycles are relatively shorter, they may experience fewer instances of moisture stress than the crop may face over its longer growing season.

Environmental differences under which weeds and desired plants developed provide additional theoretical evidence. Desired plants have been bred and selected under conditions of more or less adequate nutrient supply and sometimes even under supplemental irrigation. Thus, most desired plants have been purposely selected for the ability to perform best under good conditions. For their part, as we know, weeds, especially the early colonizing annuals, evolved under conditions of inadequate supply of nutrients and moisture. Because their survival strategy is based on numbers, a premium is placed on the ability to perform (produce seed) well under such adverse conditions.

Light

The evolutionary background provides reason to believe that light might have a greater effect on seed production of weeds than of desired plants. The reasoning is that as early colonizers, such species basically establish themselves in open areas without selection pressure for light. Although there is some work on the effect of shading on the growth of weeds, very little data exist concerning the effect on seed production. What data are available indicate that seed production is very much reduced by shade.

Data on the effects of shading on seed production of giant foxtail were obtained in connection with a competition study (Knake, 1972). Different levels of shading using a plastic shade cloth were initiated when foxtail plants were 30 cm tall (leaves extended). Light shade caused the foxtail to be taller (Figure 4-4A) and to have more tillers, or stems (Figure 4-4B), at the last measurement date. The latter effect probably kept the plants vegetative longer since those under no shade actually produced more tillers, but tillering occurred earlier in the test period. Shade of 70% or more reduced both height and number of tillers, with 98% shade ultimately being lethal. The effect on numbers of seedheads per plant was relatively greater than on growth (Figure 4-4C compared with 4-4A and 4-4B). Seedhead production was cut about 50% by 60% shade, whereas height was actually increased and stems per plant decreased only about one-third by this level of shading. Under 80% shade, the number of seedheads was only about one-fifth of the unshaded.

Shade also greatly reduced the number of seeds in each head and, thus, total seed weight, as can be seen in Figure 4-5. Under 80% shade, seed numbers were reduced about 50% and the weight of seed by about 70%. In this study, 98% shade was lethal to the foxtail plant, so, of course, no seed was produced.

To relate these findings under controlled conditions to what might occur under field conditions, light measurements were made in 50 cornfields selected at random in Illinois between July 8 and August 6, 1964. Table 4-3 shows the percent shade that occurred under corn canopies of different populations. Even the lowest population provided shade well above the 80% level and approaching the level found to be lethal to the foxtail plant itself.

Work with itchgrass (*Rottboellia exalta*) has shown that this weed is capable of producing seed under relatively dense shade. In corn that provided 92% shade, itchgrass produced 35 seeds per plant (Fisher and Burrill, 1981). Itchgrass, a native of India, is a weed of warm-season crops. Since giant foxtail is apparently a young, and itchgrass an old, species to agriculture, the contrasting response between it and foxtail with respect to shading suggests an interesting speculation. If given crop production and weed control programs are maintained long enough, will species such as foxtail, young in ecological terms, adapt their seed production to heavy shading? What we have learned about speciation suggests that they could.

TABLE 4-3.   Percent shade provided by different stands of corn under field conditions in Illinois between July 8 and August 6, 1964.

Number of Fields	Range in Population between Fields		Population Mean (plants ha⁻¹)	Mean Shade (%)
	From	To		
8	50,400	68,000	56,300	97.3
8	44,500	47,700	46,000	96.2
15	40,000	43,700	42,300	95.9
12	35,800	39,000	37,600	95.8
7	30,400	34,300	32,900	92.4

Source: Knake, 1972. Reproduced with permission of the Weed Science Society of America.

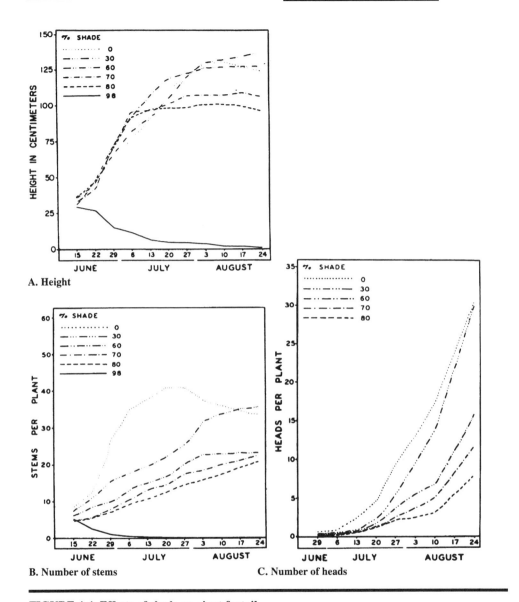

FIGURE 4-4. Effects of shade on giant foxtail.

Source: Knake, 1972. Reproduced with permission of the Weed Science Society of America.

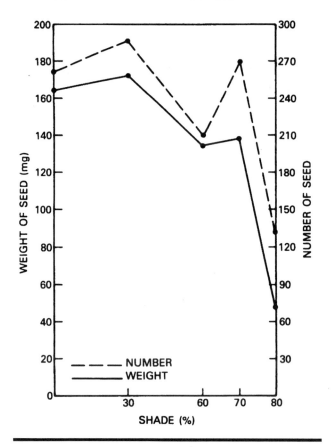

FIGURE 4-5. Effects of shade on seed production of giant fox-tail.

Source: Data from Knake, 1972.

SEEDBANK

At any given time, the soil contains viable weed seeds produced in several previous years. As we will see later, the number may be very large. The seeds that may, under favorable conditions, germinate and emerge to interfere with our use of the land constitute the seedbank. The seedbank consists of seed of different ages, some of which are dormant (development arrested), with some being exposed to favorable and some to unfavorable conditions for germination. Seeds several meters deep may be a part of the seedbank because they can be brought to the surface by excavating, by burrowing animals, or by other means. Ordinarily, however, those seeds of the seedbank in the tilled surface soil layer are of primary concern in weed management since they largely determine the number and species of weeds present to interfere with our use of land each year.

Safe Sites

After a weed seed is produced, many things can happen to it to prevent it from becoming a part of the seedbank and a potential source for production of new seeds. The concept of safe site for germination and seedling establishment developed by Harper (1977) can be adapted to seed survival. A *safe site* for seed survival, thus, is one that provides: (1) protection against specific hazards before reaching the soil (predation, harvesting, and so forth); (2) a place for physical residence on the soil surface; (3) protection against effects of predators in the environment after reaching the soil; and (4) conditions unfavorable for germination.

A large part of a weed seed crop can be lost before it ever reaches the soil. While still on the parent plant, there is an ever-present threat from birds, other animals, insects, and disease. Harvesting of the crop may physically remove the weed seeds from the area. With wild oat, for example, only 40 seeds of each 1,000 produced in cereal grain actually reached the soil (Sagar and Mortimer, 1976). The remainder were in the straw or in harvested grain.

Once the seed reaches the soil, establishing residence is dependent upon a matchup of seed size and shape with like size and shape of a site on the soil surface. Otherwise, the seed can physically move off the soil with wind or water unless it has a mucilaginous surface, such as that of Mediterranean sage (*Salvia aethiopis*) shown in Figure 4-6. The magnitude of such movement was measured for selected species by Mortimer and reported by Harper (1977). The movement of seeds sown on soil prepared to give different degrees of surface roughness was followed. Seeds of oat-grass (*Arrhenatherum elatius*) moved as far as 37 cm per day. This distance could be significant in dispersal but would be of only minor importance in terms of the seedbank on the macroscale of our interest in agriculture.

Seeds that find a physical place to reside face additional threats of both a physical and biological nature. They may be preyed upon, they may be moved by birds or by burrowing animals, or their viability may be substantially reduced by exposure to the elements. In the latter context, viability was rapidly lost by wild oat seeds resting on the soil surface (Sagar and Mortimer, 1976). A summary of their data shown in Figure 4-7, shows that only 10% of the seeds sown in September were still viable the following February where they overwintered on the soil surface. In a related study of a natural infestation, the number of viable seed dropped from 294 m^{-2} (246 yd^{-2}) on September 3 to 44 m^{-2} (37 yd^{-2}) on November 29. Other work (Papay and Thompson, 1979; Chepil, 1946a) has also shown that some protection, such as litter or physical burial, is necessary to avoid marked loss of viability in those seeds that reach soil.

Thus, sizable losses can occur in the weed seed crop before it has a chance to become a part of the seedbank in soil. Some of the hazards faced may be exploitable in

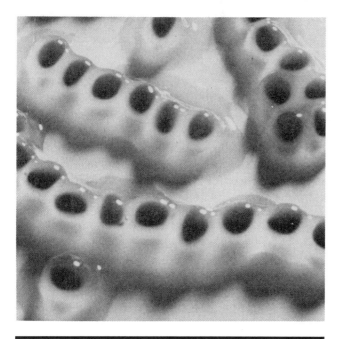

FIGURE 4-6. Mucilage surrounding seeds of Mediterranean sage to help hold them in place on the soil surface.

Source: Young and Evans, 1973. Reproduced with permission of the Weed Science Society of America.

designing weed management programs to reduce renewal of the seedbank. Such possibilities are examined in Chapter 15.

Size of the Seedbank

Many studies have shown that the size of a seedbank is very large. Some of the earliest work was done at the Rothamsted Station in England. Table 4-4 shows the millions of seeds per hectare in the top 15.2 cm (6 in.) layer of soil for several species. Although seedbanks often contain seeds of many species, those of cultivated soil are commonly dominated by one or two species (Forcella et al., 1992). The large number of seeds indicates the magnitude of the task facing a clean or preventive approach.

Longevity of Weed Seeds in the Seedbank

Longevity of weed seeds in the soil has long been a matter of continuing concern to weed science because of its obvious implication for potential weed numbers with de-

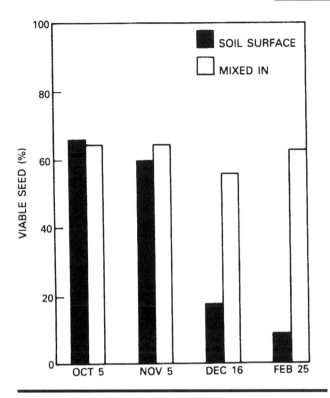

FIGURE 4-7. Viability of wild oat seed greatly extended by mixing in soil.

Source: Data from Sagar and Mortimer, 1976.

TABLE 4-4. Average number of viable weed seeds by species in the upper 15.2 cm (6 in.) layer of cultivated soil in England.

Weed Species	Millions of Seeds per Hectare
Scarlet pimpernel	0.20
Parsley-piert	1.09
Mouse-ear cress	0.82
Shepherdspurse	1.73
Lambsquarters	0.27
Fumitory	1.33
Wild chamomile	0.44
Poppy	0.72
Annual bluegrass	16.06
Prostrate knotweed	1.33
Common groundsel	2.96
Common chickweed	3.21
Birdseye speedwell	1.68
Total	32.38

Source: Data from Roberts, 1968.

sired plants. There have been anecdotal reports of weed seeds remaining viable for thousands of years.

Studies initiated by Beal in 1879 and Duvel in 1902 of retention of viability in weed seeds buried in soil in containers provided the first firm data on weed seed longevity. In Beal's study, 3 species of the 23 started still showed some viability after 100 years (Kivilaan and Bandurski, 1981). One, common mullein (*Verbascum thapsus*), had 2% germination, which was the first germination in 80 years. Another, dwarf mallow (*Malva rotundifolia*), also had 2% germination and was the first in 65 years. The third, moth mullein (*Verbascum blattaria*) had 42% germination, which represented a more or less steady decline over the years. In Duvel's study, more than one-third of 107 species buried showed some viability after 39 years. These studies provide valuable data on potential longevity. Of course, they are not representative of what happens under field conditions because the seeds were somewhat protected from the soil elements and from tillage.

More-recent work, in which seeds were buried in plastic screen bags, provides conditions more like those that would be faced by newly arrived seeds to a seedbank. In these studies that were initiated in 1972 and 1973, weeds common to the Stoneville, Mississippi, area varied greatly in their retention of viability (Table 4-5). Redvine (*Brunnichia cirrhosa*), chickweed, and barnyardgrass had lost all viability after 5.5 years. At the other extreme, johnsongrass still had nearly 50% viability and spurred anoda (*Anoda cristata*), velvetleaf, and purple moonflower (*Ipomoea turbinata*) about

TABLE 4-5.   Mean percentage of original population of weed seeds still viable after burial for up to 5.5 years.

Species	Mean Percent Viability after Burial for Indicated Years			
	0	1.5	3.5	5.5
Spurred anoda	99	89	71	30
Velvetleaf	100	87	65	36
Prickly sida	100	45	21	<1
Purple moonflower	100	84	65	33
Pitted morning-glory	100	36	23	13
Hemp sesbania	100	77	60	18
Sicklepod	100	13	16	6
Florida beggarweed	88	9	11	5
Common cocklebur	99	27	10	<1
Redroot pigweed	96	24	2	1
Common purslane	99	21	2	<1
Common evening primrose	95	36	12	4
Prostrate spurge	77	20	11	3
Redvine	66	5	2	0
Chickweed	100	1	<1	0
Johnsongrass	86	75	74	48
Goosegrass	97	39	15	4
Large crabgrass	97	48	4	<1
Texas panicum	89	39	15	5
Barnyardgrass	96	1	<1	0

Note: Seeds were buried in 1972 and 1973 at depths of 8 cm, 23 cm, and 38 cm.
Source: Adapted from Egley and Chandler, 1983.

one-third of their initial viability after 5.5 years. Of course, such data still does not tell us what the situation would be if we were to begin with a population containing both new and old seeds in the seedbank (the situation faced when a preventive weed management program is initiated). Nor does it tell us what the viability would be if the soil were tilled each year in association with crop production. Finally, it does not tell us how many seeds will have *germinated* after specified periods.

Calculated longevity.   Drawdown of the seedbank can be mathematically calculated if assumptions are made on germination and seedbank additions. Such calculations provide an indication of the time needed to accomplish a given seedbank reduction under a planned propagule reduction program. Age distribution of the seedbank is immaterial to the calculations. Resulting curves (Figure 4-8) show that it would take more than 30 years to deplete the soil of all propagules if 75% of the seeds present germinated each year and as few as 0.5% escaped control and produced new propagules. If, on the other hand, none of the 75% that germinated were allowed to produce new propagules, all of those in the soil would be depleted in 14 years. If it were possible to induce 98% of the propagules to germinate each year and no weeds produced seed, all of the propagules would be eliminated from the soil in 6 years.

When the calculated longevity of Figure 4-8 and actual germination of Table 4-5 are considered together, it can be seen that even if no weeds produce seed, natural loss of viability requires several years to reduce the seedbank of all weeds to the point where it does not pose a threat to crop production. However, they also show that much can be gained by preventing seed production. For some species, such as redvine and barnyardgrass, preventing additions to the seedbank for 2 years would effectively eliminate them as threats to crops. Further, preventing seed production simplifies the weed control task over time by narrowing the weed base. For example, if soil was uniformly infested with the 20 species shown in Table 4-5, only 6 would remain as problems after 5 years if seed production was prevented during that time. Finally, preventing seed production, coupled with the hastening of seed germination, can very much shorten the time needed to reduce weed numbers to a noncompeting level. Against this background, we examine the influence of tillage in the remainder of this chapter. Hastening germination is considered in Chapter 12.

Effects of tillage.   Much evidence exists that tillage hastens depletion of the seedbank. The situation most usually faced is where some tillage occurs each year. Figure 4-9, which compares reduction of velvetleaf under tillage and no tillage, is indicative of the magnitude of the effect of tillage. In this case, reduction was more than twice as rapid under tillage. Roberts (1970) suggested that the annual reduction from tillage is exponential for a given species and falls within the range of 30% to 60% for most weeds. For some, it is more rapid. Wilson (1978) found the annual rate of reduction of wild oat seeds to be about 80%. Standifer (1979) eliminated all viable barnyardgrass seeds with 2 years of continuous cultivation, which is about a 90% annual rate of reduction.

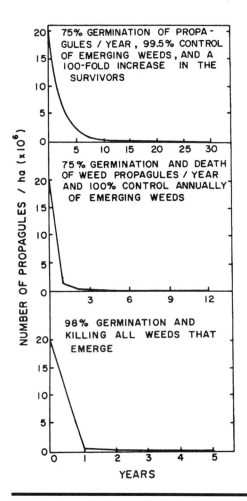

FIGURE 4-8. Calculated time needed to eliminate viable weed seeds from soil under different levels of germination and control.

Source: Ennis, 1977. Reproduced courtesy of University of Tokyo Press.

There is some evidence that the amount (frequency) of tillage in any 1 year has some effect on rate of reduction of the seedbank. However, differences are less, if they occur, between levels of tillage than between no tillage and tillage. For example, 12.1, 16.5, and 25.7 million viable seeds were found in soil after 5 years of 4, 2, and 0 cultivations per year (Roberts and Dawkins, 1967). Others found nearly identical reductions for 1 and 2 plowings per year (Lueschen and Andersen, 1980). Less effect for amount of tillage may not be too surprising in view of the fact that many weed seeds require light for germination (one tillage might be enough to meet this need), and many

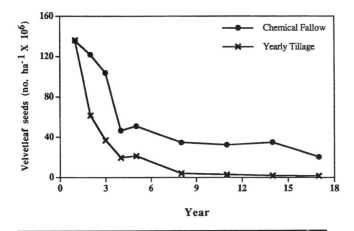

FIGURE 4-9. Reduction of velvetleaf seed in soil under tillage
and no tillage with seed production prevented for both conditions.

Source: Redrawn from Lueschen et al., 1993.

also have a fixed period of maximum germination. These aspects are considered in
greater detail in Chapter 6.

The extent of first-year germination is important to weed management. Table 4-6
shows that for many species, most of the germination that is going to occur did so by
this time under cultivation. Weed seeds were planted in three different soil types, then
alternately fallowed or cropped to cereals. Tillage was done only to a depth of 7.6 cm
(3 in.) so as not to bring up other weed seeds. Since the number of viable seeds was
known at the outset, emergence each year allowed the accumulated percentage to be
calculated. Table 4-6 shows the accumulated emergence and first-year portion at the
end of 5 years for the clay soil. There were some differences in germination and
longevity between clay and the other two soils, but the results in clay serve to illustrate
the depletion relationships. The first-year emergence shows that for many species, most
of the germination that is going to occur does so by this time under cultivation. In fact
for the 54 species, 30 had 75% of their emergence or more the first year, and 21 had
90% or more. There were 6 species in which 100% of the germination occurred the first
year. These data are important to weed management for they suggest that a majority of
the seeds of many species may be prevented from becoming a part of the seedbank if
given conditions conducive to germination the year following their production.

Thus, there are clearly large differences among species in the survival of their
seeds under tillage. However, in all species, tillage hastens depletion of the seedbank.
The significance of this fact for weed management is considered in Chapter 15.

Effects of burial depth.   Much of the research on persistence of weed seeds in soil
has considered the effect of depth of burial. The most consistent result is that longevity

TABLE 4-6. Emergence of weeds from seed planted in a clay soil then alternately fallowed and cropped to cereals.

Weed	Accumulated Percent Emergence at the End of 5 Years[1]	First-Year Emergence as a Percentage of the Total[2]
Falseflax spp.	46.9	100.0
Corn cockle	77.0	100.0
Flatseed falseflax	48.7	100.0
Kochia spp.	26.0	100.0
Smallseed falseflax	17.7	100.0
Western salsify	30.9	100.0
Quackgrass	82.8	99.6
Russian thistle	87.6	99.5
Hare's ear mustard	88.2	99.4
Downy brome	33.4	98.6
Mexican dock	41.6	98.0
Green foxtail	75.5	97.9
Garden orach	66.8	97.6
Cow cockle	62.9	97.4
Indian mustard	35.8	97.4
Broadleaf plantain	28.7	97.2
Prickly lettuce var.	81.2	96.9
European sticktight	27.0	96.8
Canada thistle	70.0	96.4
Foxtail barley	76.0	96.1
Showy milkweed	47.8	94.5
Tumble mustard	8.8	91.8
Halberleaf orach	77.1	87.6
Prickly lettuce	86.0	86.8
Redroot pigweed	44.9	86.5
Stinking clover	82.2	82.2
Dog mustard	51.6	81.9
Tumble pigweed	65.0	78.7
Wild mustard	61.7	78.5
Wild oat	98.2	75.8
Prostrate pigweed	96.8	74.0
Beach cocklebur	79.6	73.1
Perennial sowthistle	10.4	72.0
Bladder campion	29.0	71.4
Tickseed spp.	27.0	66.4
Russian pigweed	56.4	66.4
Bindweed spp.	6.5	64.9
Corn spurry	11.1	64.9
Field pennycress	45.2	60.7
Nightflowering catchfly	43.7	60.6
Marshelder spp.	30.4	60.0
Common purslane	6.8	60.0
Wormseed mustard	31.7	55.0
Yellowflower pepperweed	15.8	51.9
Greenflower pepperweed	16.6	50.6
Povertyweed	9.1	47.8
Sheperdspurse	6.0	33.3
Dandelion	7.0	31.2
Evening primrose spp.	11.4	25.6
Lambsquarters	18.5	23.5
Greenweed spp.	11.7	22.6
Flixweed	10.3	9.7
Black medic	53.9	6.0

[1]Total number of seedlings emerged divided by the number of viable seeds planted.

[2]First-year emergence (not shown) divided by 5-year accumulated emergence (shown).

Source: Data from Chepil, 1946a.

is increased if seed is incorporated rather than left on the surface, although there are some exceptions (Williams, 1978; Gleichsner and Appleby, 1989). This effect can be seen for four species in Figure 4-10. The most striking effect was for seeds of all four species to lose much of their viability if left on the soil surface over winter. Additionally, survival was greater with seeds mixed in the top 15.2 cm than if mixed in only the top 6.4 cm (treatment 2 compared with treatment 6). A direct relationship between depth and viability at the end of 1 year was shown for seeds of three weeds (Figure 4-11). Part of the difference between 15.2 cm and shallower depths, of course, could be due to conditions in the shallower depths being more favorable for germination. The

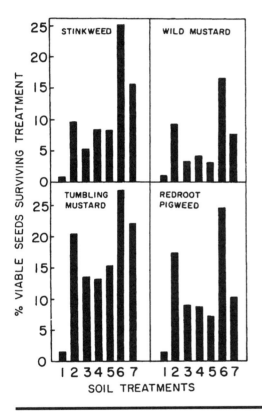

FIGURE 4-10. Effect of tillage on weed seed viability. This effect is shown by survival of seeds of four different weeds after 1 year of the following treatments: seeds scattered on the surface, no tillage (1); seeds mixed in the top 6.4 cm, no tillage (2); seeds mixed in the top 6.4 cm, cultivated 6.4 cm deep 4 times (3); same as 3 plus packed (4); same as 3 plus kept continuously moist (5); seeds mixed in the top 15.2 cm, no tillage (6); and seeds mixed in the top 15.2 cm, plowed June 1, cultivated 6.4 cm deep 3 times (7).

Source: After Chepil, 1946b. Reproduced with permission of Agricultural Institute of Canada.

15.2 cm depth is the approximate depth of the plow layer and 6.4 cm to 7.6 cm the depth of discing. Germination would serve to reduce the number of viable seeds remaining. Clearly, some incorporation of most weed seeds extends their life expectancy in soil. Also, the relationship is likely to approach linearity in the top 15 cm.

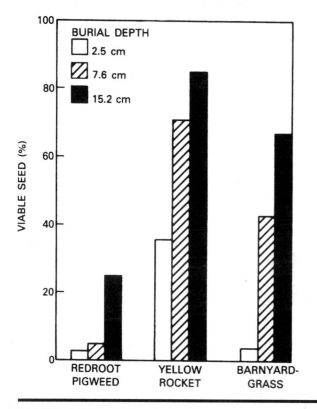

FIGURE 4-11. Retention of viability closely tied to depth of burial in some weed seeds.

Source: Data from Taylorson, 1970.

The effect on seedbank drawdown of burial depth due to type of tillage must not be confused with the effect of tillage on additions to the seedbank. As we will see in Chapter 14, additions to the seedbank frequently are greater with no tillage than with moldboard plowing.

Recovery and Importance of Current Weed Crop

In addition to knowing how long seeds may remain viable under different tillage or other treatment of the soil, it is important to know something about how quickly the seedbank rebuilds following drawdown and the relative importance of each year's addition to the seedbank in terms of competition for a crop. It was found for foxtail barley (*Hordeum jubatum*), that the major source for reinfestation is the current season's crop (Cords, 1960). In a more detailed study of blackgrass (*Alopecurus myosuroides*) in winter wheat, it was found that 80% to 90% of the weeds in the second year were from seeds produced in the first year where the wheat was direct-drilled—that is, where the soil was not plowed or cultivated to bring the previous year's weed seeds to the surface (Moss, 1980). Research in Colorado showed how quickly a redroot pigweed seedbank could rebuild following 3 years of effective control with atrazine in corn (Figure 4-12). The initial seedbank of 1,073 million ha^{-1} was reduced to 3 million ha^{-1} after 6 years of atrazine treatment. However, if atrazine was discontinued after 3 years, the

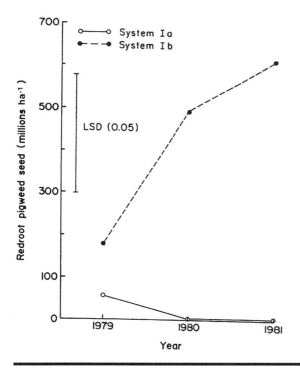

FIGURE 4-12. Number of redroot pigweed seeds in soil each spring for 3 years under two weed management systems. Under both systems, 2.2 kg ha^{-1} of atrazine was applied preemergence each of the first 3 years, beginning in 1975. This annual treatment was continued under system Ia but discontinued under Ib.

Source: Schweizer and Zimdahl, 1984. Reproduced with permission of the Weed Science Society of America.

seedbank increased to 648 million ha^{-1} in just 3 years. Similar work in Nebraska indicated that the seedbank was replenished in 1 year without weed control following 5 years of complete control (Burnside et al., 1986).

In considering the combined implications of weed seed production and persistence in the soil, it is clear that they pose sizable challenges for preventive weed management approaches. However, it appears that programs could be planned based upon current knowledge that would, in a reasonable time, lessen the problem posed by weeds. Some of these possibilities are examined in Chapter 15.

APPENDIX TO CHAPTER 4

Shape, Size, and Surface Characteristics of Various Weed Seeds

Source: Watson and Sampson, 1982. Reproduced with permission of the Department of Plant Science, Macdonald Campus of McGill University.

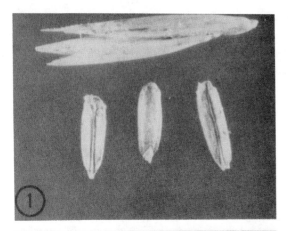

1. Quackgrass (*Elytrigia repens* [L.] Beauv.), 4.5×

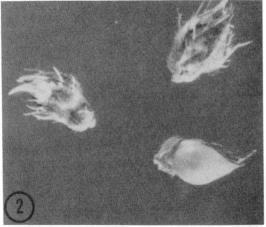

2. Barnyardgrass (*Echinochloa crus-galli* [L.] Beauv.), 7.5×

3. Yellow foxtail (*Setaria glauca* [L.] Beauv.), 7.5×

4. Wild buckwheat (*Polygonum convolvulus* L.), 6.7×

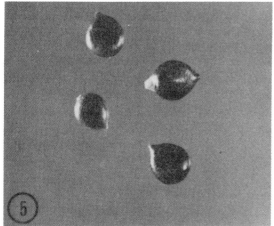

5. Ladysthumb (*Polygonum persicaria* L.), 7.5×

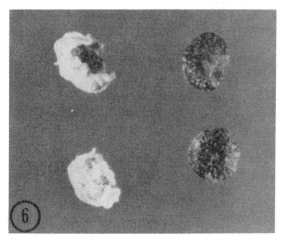

6. Lambsquarters (*Chenopodium album* L.), 11×

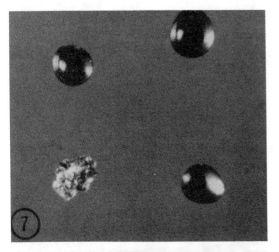

7. Redroot pigweed (*Amaranthus retroflexus* L.), 10×

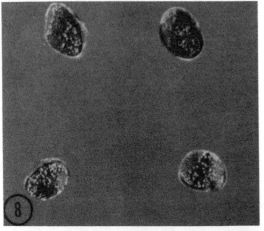

8. Corn spurry (*Spergula arvensis* L.), 11×

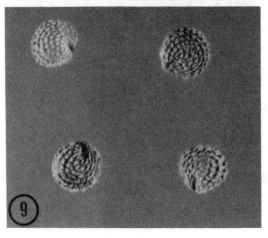

9. Common chickweed (*Stellaria media* [L.] Vill.), 11×

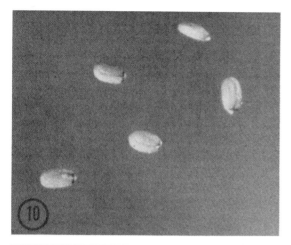

10. Shepherdspurse (*Capsella bursa-pastoris* [L.] Medic.), 10×

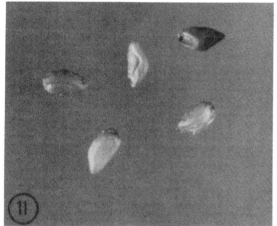

11. Wormseed mustard (*Erysimum cheiranthoides* L.), 9×

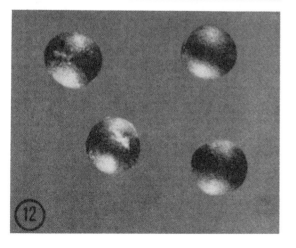

12. Crunchweed (*Sinapis arvensis* L.), 12×

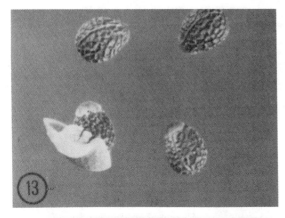

13. Leafy spurge (*Euphorbia esula* L.), 9×

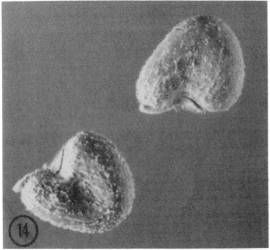

14. Velvetleaf (*Abutilon theophrasti* Medic.), 8×

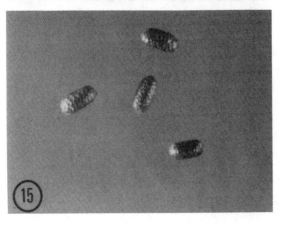

15. St. Johnswort (*Hypericum perforatum* L.), 9×

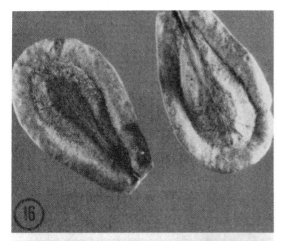

16. Common milkweed (*Asclepias syriaca* L.), 5.3×

17. Field bindweed (*Convolvulus arvensis* L.), 5×

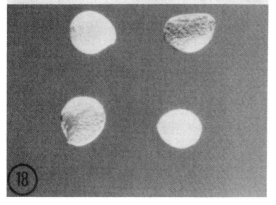

18. Dodder species (*Cuscuta* spp.), 10×

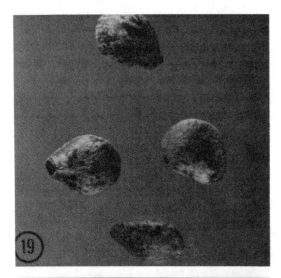

19. Hempnettle (*Galeopsis tetrahit* L.), 7×

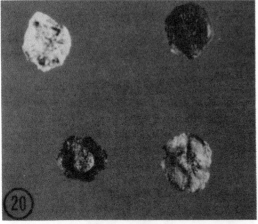

20. Yellow toadflax (*Linaria vulgaris* Mill.), 8×

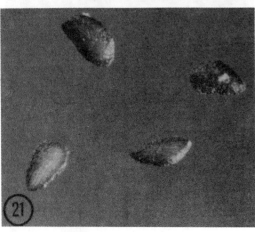

21. Broadleaf plantain (*Plantago major* L.), 9×

22. Spotted knapweed (*Centaurea maculosa* Lam.), 6.7×

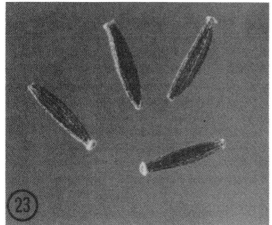

23. Common groundsel (*Senecio vulgaris* L.), 10×

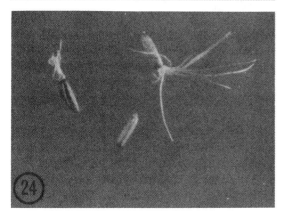

24. Canada goldenrod (*Solidago canadensis* L.), 9×

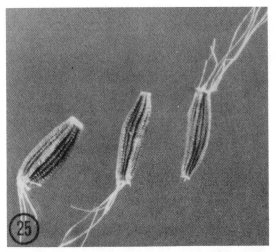

25. Perennial sowthistle (*Sonchus arvensis* L.), 8×

26. Dandelion (*Taraxacum officinale* Weber), 6.7×

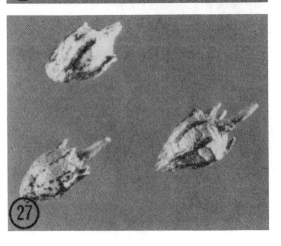

27. Common ragweed (*Ambrosia artemisiifolia* L.), 5.3×

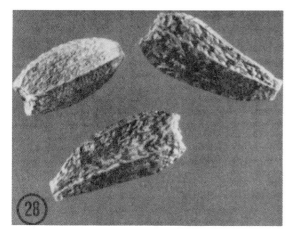

28. Common burdock (*Arctium minus* [Hill] Bernh.), 5×

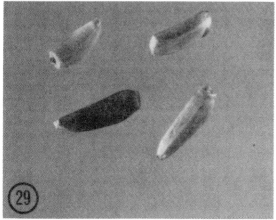

29. Canada thistle (*Cirsium arvense* [L.] Scop.), 6.7×

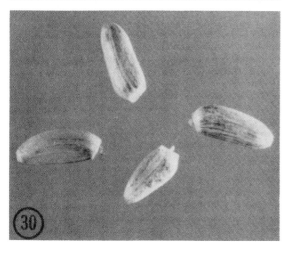

30. Bull thistle (*Cirsium vulgare* [Savi] Ten.), 4.5×

REFERENCES

Andersen, R.N., and E.A. Helgeson. 1958. Control of wild oats by prevention of normal seed development with sodium 2,2-dichloropropionate. Weeds 6:263-270.

Bleasdale, J.K.A. 1960. Studies on plant competition. In J.L. Harper, ed., The biology of weeds, pp. 133-142. Oxford: Blackwell Scientific.

Brenchley, W.E., and K. Warington. 1930. The weed seed population of arable soil, II. Influence of crop, soil, and methods of cultivation upon the relative abundance of viable seeds. J. Ecol. 21:103-127.

Burnside, O.C., R.S. Moomaw, F.C. Roeth, G.A. Wicks, and R.G. Wilson. 1986. Weed seed demise in soil in weed-free corn (*Zea mays*) production across Nebraska. Weed Sci. 34:248-251.

Chepil, W.S. 1946a. Germination of weed seeds, I. Longevity, periodicity of germination, and vitality of seeds in cultivated soil. Sci. Agr. 26:307-346.

————. 1946b. Germination of weed seeds, II. The influence of tillage treatments on germination. Sci. Agr. 26:347-357.

Cords, H.P. 1960. Factors affecting the competitive ability of foxtail barley (*Hordeum jubatum*). Weeds 8:636-644.

Egley, G.H., and J.M. Chandler. 1983. Longevity of weed seeds after 5.5 years in the Stoneville 50-year buried-seed study. Weed Sci. 31:264-270.

Ennis, W.B., Jr. 1977. Integration of weed control technologies. In J.D. Fryer and S. Matsunaka, eds., Integrated control of weeds, pp. 227-242. Tokyo: University of Tokyo Press.

Fisher, H.H., and L.C. Burrill. 1981. The increasing problem of itchgrass (*Rottboellia exaltata* L.f.). Abstract 1/8307. Proc. Annual Meeting of WSSA, Las Vegas, Nevada.

Forcella, F., R.G. Wilson, K.A. Renner, J. Dekker, R.G. Harvey, D.A. Alm, D.D. Buhler, and J. Cardina. 1992. Weed seedbanks of the U.S. corn belt: magnitude, variation, emergence, and application. Weed Sci. 40:636-644.

Gleichsner, J.A., and A.P. Appleby. 1989. Effect of depth and duration of seed burial on ripgut brome (*Bromus rigidus*). Weed Sci. 37:68-72.

Harper, J.L. 1961. Approaches to the study of plant competition. In F.L. Milthorpe, ed., Mechanisms in biological competition. New York: Academic Press.

———— 1977. Population biology of plants. New York: Academic Press.

Kelley, A.D., and V.F. Bruns. 1975. Dissemination of weed seeds by irrigation water. Weed Sci. 23:486-493.

Kivilaan, A., and R.S. Bandurski. 1981. The one-hundred-year period for Dr. Beal's seed viability experiment. Am J. Bot. 68: 1290-1292.

Knake, E.L. 1972. Effect of shade on giant foxtail. Weed Sci. 20:588-592.

Lueschen, W.E., and R.N. Andersen. 1980. Longevity of velvetleaf seeds in soil under agricultural practices. Weed Sci. 28:341-346.

Lueschen, W.E., R.N. Andersen, T.R. Hoverstad, and B.K. Kanne. 1993. Seventeen years of cropping systems and tillage affect velvetleaf (*Abutilon theophrasti*) seed longevity. Weed Sci. 41:82-86.

Moss, S.R. 1980. A study of populations of blackgrass (*Alopecurus myosuroides*) in winter wheat, as influenced by seed shed in the previous crop, cultivation system, and straw disposal method. Ann. Appl. Biol 94:121-126.

Palmblad, I.G. 1968. Competition studies on experimental populations of weeds with emphasis on the regulation of population size. Ecology 49:26-34.

Papay, A.I., and A. Thompson. 1979. Some aspects of the biology of *Cordeus nutans* in New Zealand pastures. In Proc. 7th Asian-Pacific WSS Conference, pp. 343-346. Sydney, Australia.

Roberts, H.A. 1968. The changing population of viable weed seeds in an aerable soil. Weed Res. 8:253-256.

———. 1970. Viable weed seeds in cultivated soils. In National Vegetable Research Station Annual Report 1969, pp. 25-38. Wellesbourne, Warwick, England.

Roberts, H.A., and P.A. Dawkins 1967. Effect of cultivations on number of viable seeds in soil. Weed Res. 7:290-301.

Sagar, G.R., and A.M. Mortimer. 1976. An approach to the study of the population dynamics of plants with special reference to weeds. Appl. Biol. 1:1-47.

Schweizer, E.E., and R. L. Zimdahl. 1984. Weed seed decline in irrigated soil after six years of continuous corn (*Zea mays*) and herbicides. Weed Sci. 32:76-83.

Standifer, L.C. 1979. Some effects of cropping systems on soil weed seed populations. Abstract 160. Proc 32nd Annual Meeting of SWSS.

Stevens, O.A. 1932. The number and weight of seeds produced by weeds. Am. J. Bot. 19:784-794.

Taylorson, R.B. 1970. Changes in dormancy and viability of weed seeds in soil. Weed Sci. 18:265-269.

Watson, A.K., and M.G. Sampson. 1982. Weeds: Biology and control. Quebec: Department of Plant Science, Macdonald Campus of McGill University.

Williams, E.A. 1978. Germination and longevity of seeds of *Agropyron repens* and *Agrostis gigantea* in soil in relation to different cultivation regimes. Weed Res. 18:129-138.

Wilson, B.J. 1978. The long-term decline of a population of *Avena fatua* with different cultivations associated with spring barley cropping. Weed Res. 18:25-31.

Wilson, R.G., Jr. 1980. Dissemination of weed seeds by surface irrigation in western Nebraska. Weed Sci. 28:87-92.

Young, J.A., and R.A. Evans. 1973. Mucilaginous seed coats. Weed Sci 21:52-54.

5

Reproduction from Vegetative Parts

CONCEPTS TO BE UNDERSTOOD

1. Confinement of perennating, or vegetative reproduction, parts near or physically attached to the parent plant maximizes the opportunities for the clone to utilize growth factors in a given area. Restricted dispersal is a major reason for the strong competitiveness of many perennial weeds.
2. The greater competitiveness from restricted dispersal is obtained at a sacrifice of protection against elimination by cultivation and the natural environment.
3. Until a plant originating from a perennating part separated from the parent plant begins to produce such parts on its own, removing the top growth is lethal. During this time, the perennial plant is no different than an annual in this respect.
4. The period after emergence when a perennial weed may be treated like an annual is commonly relatively short.
5. The production of perennating, or vegetative regeneration, parts is sensitive to competition for light.
6. The production of perennating parts is reduced by elevated nitrogen levels in the soil and by low temperatures.
7. The relative longevity of perennating parts in soil is less than that of most seeds.
8. Longevity of perennating parts varies directly with burial depth, due likely to protection against desiccation and low temperature.
9. The ability of perennating parts to support sprouting and growth is directly related to their level of food reserves.
10. The level of food reserves is rapidly reduced during sprouting and early growth. Thus, there is commonly a period of days immediately following the flush of sprouting when a perennial is most easily destroyed by tillage.

The ability of perennial weeds to regenerate from vegetative parts makes them highly competitive and difficult to control. Most annual weeds, with some notable exceptions, are destroyed if the plant is uprooted or severed at the soil surface. On the other hand, once a herbaceous perennial weed has developed the necessary tissue for vegetative regrowth, called the *perennating part,* it can regenerate, even if severed at or below the soil surface. In fact, severing the aboveground portion may increase the number of new shoots.

The dandelion serves as an extreme example of this phenomenon. As long as it is not disturbed, it does not vegetatively reproduce additional plant units. If, however, it is severed at or below ground level, numerous adventitious buds form at the cut surface from which several new shoots may arise. Even small segments of roots in the soil can reproduce new shoots in this way. Thus, it is important to understand the effects of the natural and human-imposed environment on the production and longevity of perennating parts.

TYPES OF PERENNATING PARTS

Vegetative reproduction of perennial weeds is achieved through the production of various types of perennating (vegetative) parts. Perennating parts can be produced in both

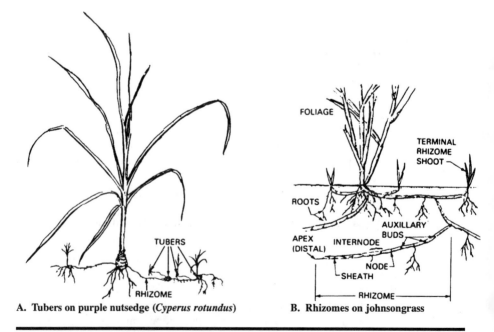

A. Tubers on purple nutsedge (*Cyperus rotundus*)   B. Rhizomes on johnsongrass

FIGURE 5-1. Stem tissue perennating parts.

Source: Part A from Agricultural Research Service, 1970. Reproduced courtesy of USDA; Part B from Beasley, 1970. Reproduced with permission of the Weed Science Society of America.

stem and root tissues of perennial weeds. The types produced in stem tissue include bulbs, corms, rhizomes, stolons, and tubers. Some of these reproductive parts and representative weeds possessing them are shown in Figures 5-1 and 5-2. A *bulb* is an underground bud consisting of a short stem axis with fleshy scales (leaves) enclosing a growing point. A *corm* is the swollen base of a stem axis and is distinguished from a bulb by its solid stem structure with distinct nodes and internodes. *Rhizomes* are specialized horizontal stems that grow belowground and produce adventitious roots when in contact with soil. *Stolon* is a general term for any of several specialized horizontal stems that take root at the nodes; it is best to restrict the term to aboveground stems to distinguish from rhizomes. A *tuber* is a specialized structure that results from the swelling of the subapical portion of an underground stem. The reproductive function in all of these stem structures is quite easily identified because each structure contains one or more buds.

Roots of some perennial plants have the capacity to produce regenerating buds adventitiously. The reproductive potential of such roots commonly is not readily apparent. Indeed, in many species, such as dandelion, the root assumes the reproductive

A. Clump of plants

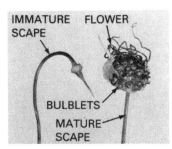

B. Aerial bulblets on a scape

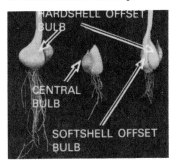

C. Different types of belowground bulbs

FIGURE 5-2. Stem tissue perennating parts on wild garlic.

Source: Reproduced courtesy of E.J. Peters, USDA, Columbia, Missouri.

function only following injury. In other species, such as leafy spurge shown in Figure 5-3, buds form on thickened roots, sometimes called *rootstocks,* in the soil and develop to send up new shoots or to extend laterally. Except for roots such as described for dandelion, all reproductive parts shown in Figures 5-1, 5-2, and 5-3 may be considered as being designed for regeneration. Thus, they are readily distinguished from segments of stems or leaves that for some species under proper conditions can form roots and develop into a new plant. Nonetheless, it is well known that many weeds, such as crab-

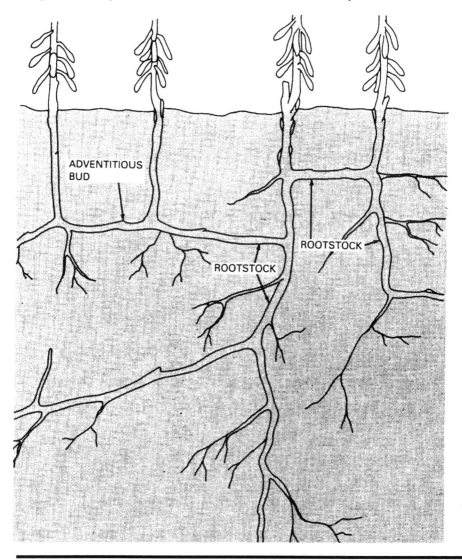

FIGURE 5-3. Root tissue perennating parts (rootstock) on a leafy spurge clone.

Source: Redrawn from Myers et al., 1964. Reproduced with permission of the Weed Science Society of America.

grass and purslane, have the ability to develop adventitious roots at nodes and, thus, become reestablished after the plant is severed at the soil surface. Therefore, vegetative specialization for regeneration must be recognized as a relative rather than an absolute characteristic.

Some aquatic plants have regenerating parts, termed *turions,* not found in terrestrial plants. Turions have very short internodes and specialized food-storing leaves or scales. Turions are detached when an abscission layer forms in *Hydrocharis morsusranae,* but in *H. verticillata,* they are released after the parent plant decays. They are covered by mucilage or a thick cuticle that protects them while dormant during winter.

Some perennial weeds utilize more than one type of perennating part for vegetative reproduction. Wild garlic, for example, reproduces from both aerial bulblets and underground bulbs. The bulbs may be central or offset, hardshell or softshell. Bermudagrass has both aboveground and belowground reproductive stems.

Potential Production Capacity of Perennial Weeds

A single perennial plant may produce a large number of regeneration units, as can be seen in Table 5-1. For each species, the amount of reproduction is for a single growing season or less. Except for those units moved away by tillage, burrowing animals, or other outside influences, all such units remain close to or physically attached to the parent. Thus, a dense stand capable of fully exploiting resources within the area occupied can soon develop if steps are not taken to prevent it.

Although the production capabilities shown in Table 5-1 are quite remarkable in view of the relatively large size of the units produced and the short time available to produce them, the numbers are still orders of magnitude less than the seed production capacity of some annual species (Table 4-1). Furthermore, lateral spread of even the most aggressive perennial seldom exceeds 3 m (10 ft) or so in a year. Seeds may be moved much greater distances by wind, surface water, birds, and other animals. Remember, the reference here is only to the distance of dispersal, without regard to the likelihood of establishment. Because the regeneration units of perennials remain close to the parent, whereas seeds of annuals may be more widely dispersed, it follows that the perennating parts have relatively more safe sites for reestablishment.

TABLE 5-1. **Vegetative reproduction capacity reported for some representative perennial weeds.**

Species	Extent of Production
Yellow nutsedge	One tuber resulted in 1918 new plants and 6864 tubers in 1 year.
Cattail (*Typha latifolia*)	One plant in 6 months developed a network of rhizomes 3.0 m (10 ft) in diameter.
Johnsongrass	One plant at 14 weeks had 25.9 m (85 ft) of rhizomes.
Quackgrass	One plant in a single season produced more than 200 new rhizome buds.
Canada thistle	One plant produced 111 m of roots 0.5 m in diameter or larger after 18 weeks of growth.

Source: Nutsedge data from Tumbleson and Kommedahl, 1961; cattail data from Yeo, 1964; johnsongrass data from Anderson, 1977; quackgrass data from Sagar and Mortimer, 1976; Canada thistle data from Nadeau and Vanden Born, 1989.

Perennating parts may be produced during much of the growing season unlike the relatively short flush of seed production for annual weeds. Furthermore, production of perennating parts may occur over a wide range of environmental conditions. For example, some quackgrass buds formed even when air temperatures were $-5°C$ (Lemieux et al., 1993). The extended time of production of perennating parts poses a very different need for prevention than does seed production of annual species.

Distribution in the Soil Profile

Perennating parts may be produced deep in the soil but tend to be concentrated in the top few centimeters. More than half of the roots of Canada thistle, on which adventitious buds may form, were in the top 40 cm, although some extended down 2 m (Figure 5-4). Additionally, roots in the top 20 cm produced nearly twice as many adventitious buds per meter of root than did those in the 20 to 40 cm depth. About 90% of viable buds in quackgrass were in the top 10-cm soil layer (Lemieux et al., 1993).

FACTORS AFFECTING PRODUCTION

An understanding of factors that influence production of perennating parts provides useful insights into ways to minimize or prevent losses from species having this capability. Age, light, plant density, nutrition, temperature, and growth regulators each have an effect, as do some interactions among these factors. The emphasis here is on the relationship with environmental factors and not on the biochemical or physiological processes themselves. However, we need to be aware of the fact that the production of vegetative reproductive parts involves the storage of carbohydrates. We would then expect factors that affect the production of carbohydrates—that is, factors that affect photosynthesis—to influence the production of these reproductive parts. In this respect, there is little difference between the seed and the vegetative offshoot because both depend, for a time, on food material supplied by the mother plant.

Age

The age at which a new perennial weed plant, either from seed or a vegetative part, begins to produce vegetative reproductive parts is very important. Until then, the plant is no different than an annual species in its sensitivity to top growth removal.

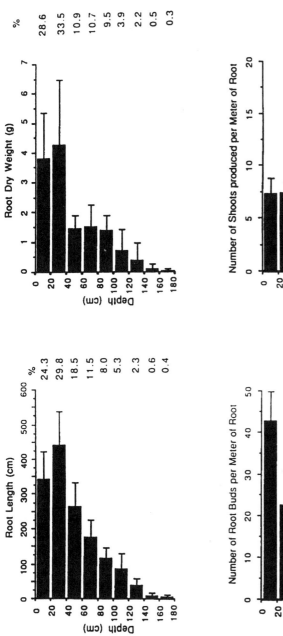

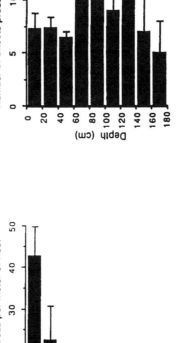

FIGURE 5-4. Root length, root dry weight, adventitious root buds m⁻¹ of root, and shoot production m⁻¹ of root of Canada thistle from a 10-year-old patch.

Source: Donald, 1994. From Nadeau and Vanden Born, 1989. Reproduced with permission of the Weed Science Society of America and of the Canadian Journal of Plant Science.

Sensitivity to early top growth removal is shown in Table 5-2 for johnsongrass. Plants from seed and rhizomes in the greenhouse were clipped weekly after the aboveground growth attained the height indicated. In only 3 weeks after emergence, plants from seed had produced sufficient storage in rhizomes, which first appeared 18 days after emergence, for them to survive weekly clippings thereafter. When only 2 weeks old, however, removing the top growth killed the plants. New rhizomes first appeared 21 days after emergence on plants from rhizome sections and removal of top growth 2 and 3 weeks after planting was lethal. If removal was delayed until 3 weeks, these plants survived as they did from seed.

TABLE 5-2.   Effect of interval between emergence and top growth removal on survival of johnsongrass seedlings from seed and from rhizomes.

	Dates of Clipping			
Plant Material	May 14 Ht. (in.)	May 21 Ht. (in.)	May 28 Ht. (in.)	July 9 Ht. (in.)
Plants from seed	6	Dead	—	—
		12	6	2
Plants from rhizome	3	1	Dead	—
		9	8	4

Note: Plants emerged from seed May 1 and from rhizomes May 2.
Source: Adapted from McWhorter, 1961.

Other studies with johnsongrass reported somewhat different ages when rhizome initiation commenced. Anderson et al. (1960) found that initiation began 4 to 5 weeks after emergence. Keeley and Thullen (1979) found that date of initiation was influenced by the date of planting and varied from 3 to 6 weeks. Observed variations in juvenile period may be due to presence of different ecotypes discussed in Chapter 3. Curly dock (*Rumex crispus*) plants were able to regrow from rootstock when clipped after about 40 days of growth (Monaco and Cumbo, 1972). Seedlings of common milkweed in the 1 to 1 3/4 leaf-pair stage were able to produce new shoots after being clipped (Jeffrey and Robison, 1971). In Canada thistle, seedlings 19 days old with 2 true leaves were able to resprout from mowing (Wilson, 1979). In yellow nutsedge, tuber initiation was found to begin 4 to 6 weeks after emergence (Keeley and Thullen, 1975) or 8 weeks after planting (Tumbleson and Kommedahl, 1961). The variable juvenile period with johnsongrass indicates that the age when vegetative reproduction is initiated in perennials is plastic. Nevertheless, this juvenile period is relatively short for most perennial weeds. Thus, there is a comparatively short time when new plants of most perennial species may be treated as annuals.

Density

To fully understand the effect of density on production of perennating parts, a distinction needs to be made between production on a per plant basis and on a population basis. In terms of behavior as a weed, the number of regenerating units produced per unit of surface (population) is the critical factor.

The effect of density on the individual plant basis appears to be quite clear cut. Figure 5-5 shows that for two species with very different types of vegetative reproduction, such reproduction is reduced as density increases. In fact, practically no rhizome production occurred in johnsongrass (Figure 5-5A) under the highest density of 8 plants per pot. If there was only 1 plant per pot, it produced about 10 g of rhizomes. The effect on purple nutsedge (*Cyperus rotundus*) tuber production (Figure 5-5B) was somewhat less drastic, but pronounced nonetheless.

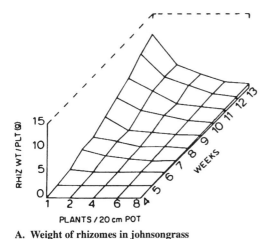

A. Weight of rhizomes in johnsongrass

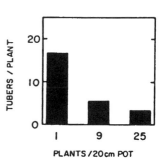

B. Number of tubers in purple nutsedge 9 weeks after planting

FIGURE 5-5. Effect of intraspecific density on vegetative reproduction per plant.

Source: Part A from Williams and Ingber, 1977; Part B from Williams et al., 1977. Reproduced with permission of the Weed Science Society of America.

When the same two species are examined on the population (area) basis, a very different effect is found (Figure 5-6). For johnsongrass (Figure 5-6A), the production of rhizomes decreased as density increased, just as it did on a per plant basis. Ogden (1974) observed decreased allocation of dry matter to vegetative reproduction as density increased in *Tussilago farfara,* which agrees with observations on johnsongrass. Just the opposite effect occurred for tubers in purple nutsedge (Figure 5-6B). Nearly twice as many tubers were produced in the pot containing 25 plants as in the pot with only 1 plant. Thus, in johnsongrass and *Tussilago farfara,* the density–vegetative reproduction relationship follows the general bell-shaped relationship typical of seed production. That is, vegetative reproduction increases as density increases up to a point, after which it decreases with further increases in density. In purple nutsedge, the relationship follows the straight-line density relationship typical of vegetative growth.

Not enough weed species have been studied to indicate the relative extensiveness of these two responses to density. However, there are probably many species in each

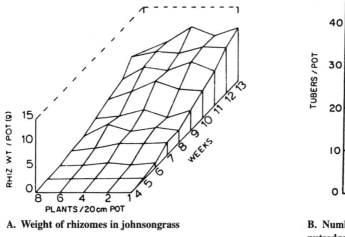

A. Weight of rhizomes in johnsongrass

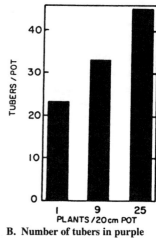

B. Number of tubers in purple nutsedge 9 weeks after planting

FIGURE 5-6. Effect of intraspecific density on vegetative reproduction on an area basis.
Source: Part A from Williams and Ingber, 1977; Part B from Williams et al., 1977. Reproduced with permission of the Weed Science Society of America.

category. Therefore, the implications for weed management need to be considered. Perennial weeds whose response to density is like that of purple nutsedge—that is, a straight-line relationship between density and numbers of regenerating units produced—can be expected to be somewhat less competitive toward crops than species like johnsongrass that shift toward vegetative growth at high densities. The reason is that the growing plant, not the vegetative reproductive part, is the immediate source of competition. In species such as johnsongrass, competition causes most of the resources to go into the production of roots and shoots to compete with desired plants. In species like purple nutsedge, some of the resources continue to go into tubers that in themselves do not compete with desired plants. Another way of looking at it is from the survival strategy perspective. Species like purple nutsedge express an r-strategy by giving priority to the production of regenerating units, but those like johnsongrass express a K-strategy by relying upon competitive ability.

From the weed's perspective, competition from desired plants can be expected to reduce production of perennating parts, regardless of whether production follows the relationship represented by nutsedge or by johnsongrass. The reason is that an increase in total density under field conditions is a result of the desired plants presence. That is, all of the weed's parts in position to resume growth in all likelihood will do so, whether or not the desired plant is present. When the desired plant is added, there is, of course, an increase in the combined number of plants occupying the area. Since the weed cannot increase in numbers, the production of vegetative reproduction units will be reduced like that observed on an individual plant basis in Figure 5-5. Such a response is readily understandable for species like johnsongrass in which production of perennating parts is decreased on both an individual plant and on an area basis by an increase in density.

Light

Quantity of light. Because carbohydrate production is dependent upon photosynthesis, we would expect light quantity to have an important influence on the production of vegetative reproduction parts. Indeed, this is the case, as shown in Figure 5-7. There is a direct and straight-line effect of shading on the production of tubers, in this case yellow nutsedge. With maximum shading, practically no tubers were produced. Because competition for light could be expected under increased density, the results with yellow nutsedge might at first seem to contradict the results for density effects on purple nutsedge presented in the previous section. Actually, they do not. In the case of shading on yellow nutsedge, it may be assumed that the population is constant. The results, therefore, essentially measure effects on a per plant basis. On a per plant basis, the results are similar to those shown for purple nutsedge in Figure 5-6B. A constant population is more representative of what would be faced under field conditions, although it must be remembered that numbers of plants can change in response to competition. The specific effects of competition are discussed in Chapter 7.

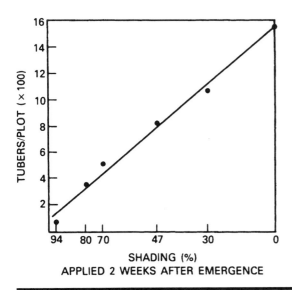

FIGURE 5-7. Relationship between shading and tuber production in yellow nutsedge.

Source: Keeley and Thullen, 1978. Reproduced with permission of the Weed Science Society of America.

Tuber production in purple nutsedge is also sensitive to shading (Patterson, 1982). Tuber numbers in this species were reduced 96% by 85% shade. Although precise data are lacking for other species, there is ample circumstantial evidence that light restricts production of vegetative reproductive parts in many species. The already mentioned effects of density on rhizome production in johnsongrass and *Tussilago farfara* implicate light as the factor involved. Light is also implicated in the sensitivity of leafy spurge to competition from a crested wheatgrass (*Agropyron cristatum*) and smooth bromegrass (*Bromus inermis*) sod (Morrow, 1979). Although moisture may have been the primary factor, leafy spurge in the sod did not spread during two growing seasons, whereas leafy spurge in plowed soil spread more than 3 m. Finally, the numerous reports of improved control of such perennials as Canada thistle (Derscheid et al., 1961; Hodgson, 1958), field bindweed (*Convolvulus arvense*) (Russ and Anderson, 1960; Derscheid et al., 1970), and leafy spurge (Derscheid et al., 1960), where so-called competitive crops are used in conjunction with tillage and herbicides, implicate light quantity since this factor is the most logical one to be influenced by the competing crops.

Shading may also decrease the size of the perennating part. In purple and yellow nutsedge, for example, 85% shade reduced the average size of tubers 60% and 55%, respectively (Patterson, 1982). These results have implications for management approaches that seek to prevent losses from such species since size influences the depth from which emergence occurs, as we shall see in the next chapter.

Quality of light. Evidence exists that light quality, in addition to quantity, is a factor in production of vegetative reproductive parts. White, red, and blue light inhibited rhizome formation in purple nutsedge, whereas far-red stimulated rhizome formation (Aleixo and Valio, 1976). As detailed in Chapter 7, a canopy of leaves tends to filter out the white, red, and blue and allow the far-red to transmit. Therefore, the reducing effect of shading on tuber production could be expected to be offset somewhat by the far-red passing through the canopy to understory nutsedge plants. Far-red enrichment, in part, may account for continued but limited production of tubers under increase in density, discussed in the previous section.

Day length. Day length has also been found to influence tuber production in yellow nutsedge (Jansen, 1971). Tuber formation was found to be inversely related to day length, with maximum production occurring at day lengths from 8 to 10 hours in yellow nutsedge. As the day length increased, tuber formation decreased.

Thus, there is extensive direct and indirect evidence that light has a marked effect on production of perennating parts in perennial weeds. The use of such information to develop effective weed management programs is considered in Chapter 15.

Nutrition

The general effect of added fertility on desired plants, especially of nitrogen (N), is to encourage vegetative growth rather than the storage of carbohydrate. Thus, we would expect production of perennating parts in weeds to be decreased by added fertility. Results are available from studies of the effects of added nitrogen but not other nutrients.

Figure 5-8 shows that nitrogen does indeed affect production of perennating parts. Reduction from added nitrogen was consistent for different forms of perennating parts represented by the species, even though the studies themselves differed greatly. The studies with yellow nutsedge (Figure 5-8A) were conducted in the greenhouse where nitrogen was controlled at the indicated fractions of Hoagland's nutrient solution. *Tussilago farfara* (Figure 5-8C) was studied outdoors, but in containers where high or low

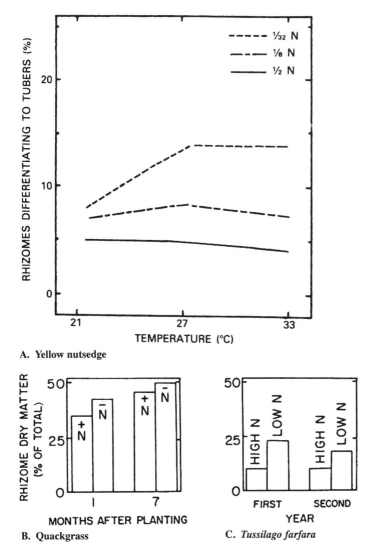

A. Yellow nutsedge

B. Quackgrass

C. *Tussilago farfara*

FIGURE 5-8. Effect of nitrogen (N) level on vegetative reproduction of three perennial weeds.

Source: Part A data from Garg et al., 1967; Part B data from Johnson and Dexter, 1939; Part C data from Ogden, 1974. Reproduced with permission of the Weed Science Society of America.

nitrogen was coincidental to high and low fertility obtained with different quantities of a planting compost medium. The quackgrass studies (Figure 5-8B) were conducted in the field where nitrogen was added to some plots and not to others.

Other results with quackgrass (Seyforth et al., 1978; McIntyre, 1971; Dexter, 1936) and with johnsongrass (Meyers and Caso, 1976) also show an inverse relationship between nitrogen supply and vegetative reproduction. The importance of this relationship for management of such weeds is based on the fact that rhizome production and shoot growth are alternative options open to the weed. This can be demonstrated in yellow nutsedge and johnsongrass. In nutsedge, the basal bulb area from Figure 5-1 is the area from which rhizomes are produced. These rhizomes may differentiate either into tubers or shoots. An adequate or excess supply of nitrogen can provide the amount needed for vegetative growth, thus supporting differentiation into shoots (Garg et al., 1967). A similar situation exists for johnsongrass, where the axillary buds can develop either as rhizomes or as tillers. Adequate nitrogen supported the development of the buds as tillers (Meyers and Caso, 1976)

Since the effect of nitrogen is so consistent, it should be possible to manipulate its supply in order to minimize production of perennating parts. Nitrogen manipulation is discussed in Chapter 15.

Interactions of Factors

Most of the research on the individual factors affecting vegetative reproduction shows the effects to be relative rather than absolute. That is, the degree of effect is modified by another factor or factors. Thus, we see in Figure 5-8A that fewer nutsedge rhizomes differentiated to tubers at 21°C than at higher temperatures under restricted nitrogen but not under the highest level of the nutrient. Fall temperatures and photoperiods favor production of adventitious root buds in Canada thistle (Donald, 1994).

Plant growth hormones interact with other factors to affect tuber production in nutsedge (Garg et al., 1967). Figure 5-9 indicates the complexity of these interactions. Recall from Figure 5-1A that rhizomes developing from basal bulbs may differentiate either to tubers or to shoots in yellow nutsedge. Each of the variables in Figure 5-9—photoperiod (hr), gibberellic acid (GA), nitrogen (N), and temperature (°C)—was influenced in its effect upon rhizome differentiation by its interaction with the other variables. The effect of nitrogen was greatest under the shortest photoperiod; at this same light level, gibberellin had an apparent inhibitory effect. Under the longer day length, gibberellin appeared to be stimulatory. At the shortest photoperiod, there was little effect of temperature on differentiation of rhizomes to shoots. At the longest photoperiod, differentiation was reduced with increases in temperature. Such interactive relationships increase the ways in which the environment may be manipulated to assist in weed management. Such manipulations are considered in Chapter 15. At this point, however, we need to recognize that manipulation of types of growth may offer contradicting results. For example, manipulation to encourage shoot growth, although effective in reducing an infestation the following year, also increases the number of shoots that compete with the current crop.

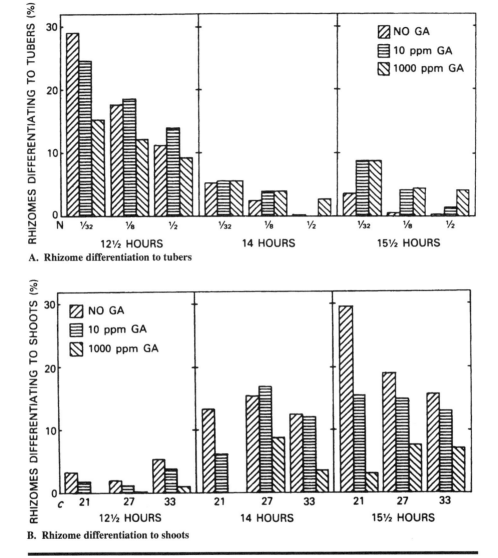

A. Rhizome differentiation to tubers

B. Rhizome differentiation to shoots

FIGURE 5-9. Interactions among photoperiod (hr), nitrogen (N), and gibberellin (GA) on yellow nutsedge development.

Source: Garg et al., 1967. Reproduced with permission of the Weed Science Society of America.

FACTORS AFFECTING LONGEVITY AND SURVIVAL

The length of life of vegetatively produced reproductive parts is clearly important to control and prevention of loss from perennial species. How much of one year's growth is continued into subsequent years varies greatly among herbaceous perennials. The range varies from plants that flower within a year or two after germination and then die, to those that may not flower for many years after germination, to those that flower over a period of several years, to those that have a single flush of flowering. The range also includes plants that renew their vegetative body every year and sluff the older tissues and plants in which the rhizomes persist for many years.

Most of the recent research on longevity and survival deals with rhizomes and tubers of grass or grasslike perennials, rather than with rootstocks common to broadleaf perennials. At least in the United States, this focus is partly explained by the earlier development of herbicides more effective in controlling broadleaf than grass weeds. This development, along with an increase in crop monoculture in some regions, has lessened problems posed by some broadleaf perennial species. The shift to minimum or reduced tillage, begun in the second half of this century and still on the increase, may well change this picture since there is evidence some broadleaf perennial weeds may increase under such tillage.

Longevity of reproductive organs for a particular species appears to differ from place to place. For example, quackgrass rhizomes may survive for many years in England (Palmer and Sagar, 1963), but Johnson and Buchholtz (1962) imply that the rhizomes live for only about 1 year in Wisconsin. Such differences in reports on longevity simply serve to point out the fact that longevity is a relative rather than a discrete matter and another example of plasticity on the part of the plants. Thus, it is not realistic to attempt to identify the precise term of longevity for vegetative reproductive parts. However, the relative longevity is important. In this sense, longevity of vegetatively reproduced parts for most perennial species that we deal with in weeds is relatively short and much shorter than for most seeds. The purpose here is to identify the overall relationships between the environment and relative persistence. An understanding of the relationship between such factors as temperature, desiccation, and depth of burial may well suggest ways of improving our ability to cope with such weeds.

Depth of Burial

Much evidence verifies a direct relationship between the depth to which reproductive parts are buried in the soil and their retention of viability. Figure 5-10 shows this relationship for three perennial species and for the two types of vegetative reproductive organs, tuber and rhizome. Quackgrass and johnsongrass (Figure 5-10A) were compared in a different study from nutsedge (Figure 5-10B), but burial of reproductive parts for all three was in November. Also, burial depths were comparable. Viability of quack-

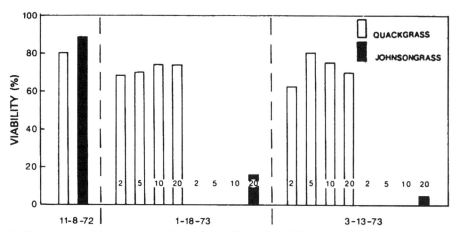

A. Quackgrass and johnsongrass rhizomes buried November 1972

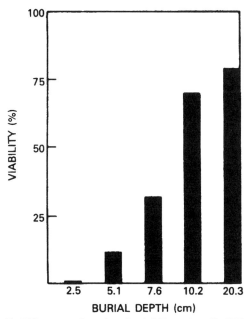

BURIAL DEPTH (cm)

B. Yellow nutsedge tubers buried November 13, 1970

FIGURE 5-10. Effects of depth of burial on viability of vegetative reproductive parts of three perennial weeds.

Source: Part A data from Stoller, 1977; Part B data from Stoller and Wax, 1973. Reproduced with permission of the Weed Science Society of America.

grass and johnsongrass rhizomes was measured the following January and March and that of nutsedge tubers the following May through September. Viability for quackgrass and johnsongrass was measured by the emergence from rhizome sections removed from the soil and germinated in the greenhouse. Viability for nutsedge was emergence in the field up to September 7 plus germination of the remaining tubers removed from the soil on that date and germinated in the greenhouse. For all three, burial below 2.0 to 2.5 cm increased viability, although quackgrass was only minimally affected. Quackgrass rhizomes retained much of their viability at all depths, although viability was lower on March 13 for rhizomes buried only 2 cm than if buried more deeply. Johnsongrass rhizomes lost all of their viability over winter except where buried 20 cm. Viability of nutsedge tubers was also increased by burying.

Duration of viability was also measured. Half-life for tubers buried 20.3 cm was calculated to be 5.8 months, and for 10.2 cm, 4.4 months. Even though these half-life studies indicate that most tubers germinate or lose their viability within 1 year from when they are produced, enough tubers will survive, especially with deep burial, to re-populate the area rather quickly.

Temperature

The effects of burial depth implicate temperature and moisture as causal factors. Indeed, examination of temperature effects on nutsedge tubers in Figure 5-11A shows tubers to be extremely sensitive to low temperatures. At a temperature of −4°C, all of the tubers survived, whereas at a temperature of −10°C, none survived. The soil temperature information in Figure 5-11B indicates that soil temperature at Champaign, Illinois, exceeded that critical minimum several times during the winter of 1972-1973 at the 2 cm depth. The lowest soil temperature recorded at 29 cm depth was about −8°C, whereas the lowest temperature at 2 cm was about −18°C. The relatively greater sensitivity of nutsedge tubers and johnsongrass rhizomes to low temperature than is true for most seeds should be noted.

Desiccation

Work with quackgrass rhizomes in the 1930s showed rhizomes to be sensitive to desiccation (Dexter, 1942). In the ensuing years, work on many other species has shown them also to be sensitive to drying. With johnsongrass, for example, drying to less than 40% of the original moisture content was lethal to all the rhizomes (Anderson et al., 1960). Reduction in moisture to 14% killed rhizomes of African feathergrass (*Pennisetum macrourum*) (Harradine, 1980). Desiccation of blue couch (*Digitaria scalarum*) for 4 days at 7°C to 27°C completely killed the rhizomes (Mshiu, 1978). Both yellow nutsedge tubers and bermudagrass buds rapidly lost viability in air-dry soil (Thomas, 1969).

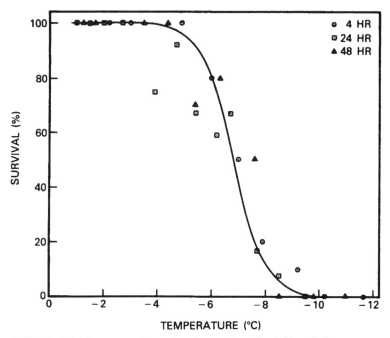

A. Survival of tubers exposed to various temperatures for 4, 24, and 48 hours in the laboratory

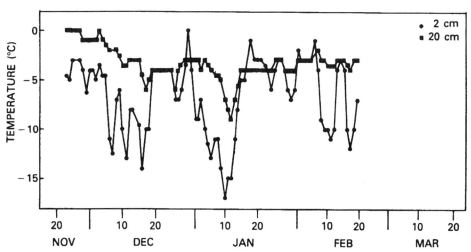

B. Soil temperatures recorded at 2 cm and 20 cm in pots at Urbana, Illinois, during winter 1972–73

FIGURE 5-11. Intolerance of yellow nutsedge tubers to cold.

Source: Stoller, 1977. Reproduced with permission of the Weed Science Society of America.

An interaction between desiccation and previous nitrogen fertilization was shown for quackgrass (Dexter, 1937). Rhizomes from plants grown under added nitrogen were damaged more from drying than were those from unfertilized plants. The widespread and pronounced sensitivity of some perennating parts to desiccation suggests opportunities for their control through management, which are considered in Chapter 15.

Food Reserves

Data obtained in conjunction with research on control of perennial weeds provide an insight into the relationship between the physiological age of the perennating part and its likelihood of survival. Prior to the widespread availability of effective selective herbicides, control of such weeds relied heavily upon the proper timing and frequency of tillage. Thus, data for many species exist showing that the extent to which food reserves (carbohydrates) are depleted in the perennating part greatly influences the part's ability to regenerate.

The relationship shown in Figure 5-12 for perennial sowthistle (*Sonchus arvensis*) is typical for species with rootstocks and rhizomes. As can be see in 5-12A, there is a small gradual decline in dry matter of the reproductive root piece (1a) during the winter and early spring, followed by a large rapid decline in mid-spring. The rapid decline coincides with initiation of growth of aerial shoots. The late May to early June period is when the reproductive root is least able to withstand burial, as shown both by rhizome weight in November of the same year (Figure 5-12B) and growth of aerial shoots during the following growing season (Figure 5-12C). The time during the spring when regrowth is initiated varies for different species. The sensitive period for vegetative parts varies accordingly. However, the rapid decline of stored foods associated with growth of new shoots, followed by the elaboration of new regeneration parts and buildup of food reserves in them, is the general pattern for carbohydrate utilization and buildup in perennating parts of all species. Further, that period when reserves are their lowest is when such parts are most easily destroyed.

Interactions of Factors

Several of the studies on temperature and moisture effects indicate an interaction between these two effects. Figure 5-13 demonstrates this relationship for yellow nutsedge. At 4°C, essentially no tubers retained their viability in air-dry soil. At 22°C, there was 20% to 40% survival, although viability was much lower in air-dry soil than where some moisture was present. Recall from Figure 5-11 that 4°C is well above the critical minimum for such tubers.

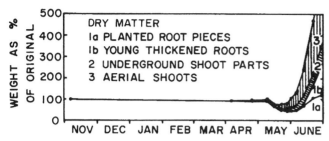

A. Changes in dry matter of undisturbed reproductive root pieces following planting on November 7, 1966

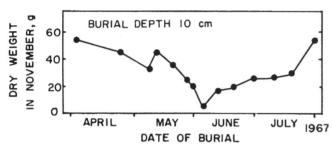

B. Effect of a single burial during 1967 on dry weight of reproductive roots in November 1967

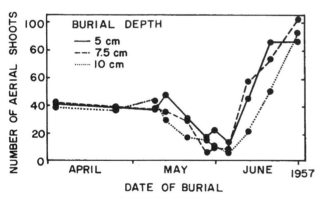

C. Effect on regrowth of time of burial of growing plants

FIGURE 5-12. Relationship among food reserves, burial, and regrowth in perennial sowthistle.

Source: Modified from Hakansson, 1969. Reproduced courtesy of S. Hakansson.

1 AIR-DRY SOIL
2 STEPWISE PROGRESSION TO 30% RELATIVE HUMIDITY
3 STEPWISE PROGRESSION TO 50% RELATIVE HUMIDITY
4 STEPWISE PROGRESSION TO 70% RELATIVE HUMIDITY
5 90% RELATIVE HUMIDITY THROUGHOUT

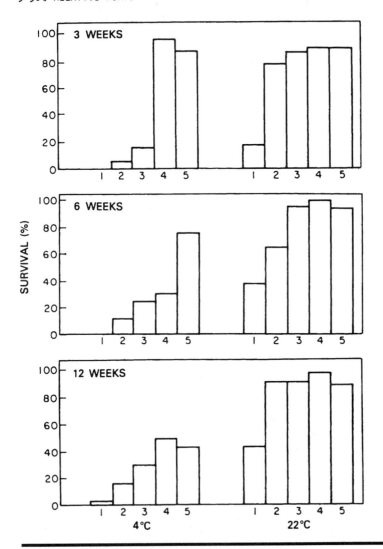

FIGURE 5-13. Percent survival of yellow nutsedge tubers under various desiccation levels, temperatures, and duration of desiccation.

Source: Thomas, 1969. Reproduced with permission of Blackwell Scientific Publications, Ltd.

Duration of exposure of rhizomes or tubers to either temperature extremes or desiccation also influences loss in viability. From Figure 5-13, it can be seen that desiccation of yellow nutsedge tubers for 3 weeks at 4°C had little effect under 70% relative humidity, whereas viability was markedly reduced under this regime if exposed for 6 or 12 weeks. In another species, surface exposure of johnsongrass rhizomes for 3 days at 45°C was lethal to all buds, whereas their viability was relatively unaffected if exposed for only 2 days at this temperature (McWhorter, 1972).

Size of the tuber or rhizomes may also affect longevity, although the evidence is indirect. Yellow nutsedge tubers averaging 120 mg in size produced 67% more shoots when buried at 20.3 cm than did tubers averaging 50 mg (Stoller and Wax, 1973). Similarly, emergence from johnsongrass rhizomes was greater from segments 152 mm long than from those 76 mm long when buried 22.9 cm.

Clearly, both the production and retention of viability of vegetative reproductive parts are influenced greatly by several factors. Some factors may be usable in developing management approaches to minimize problems from such weeds.

SEED VS. VEGETATIVE REPRODUCTION

Plants that depend solely or nearly so on vegetative reproduction do so in spite of obvious drawbacks to this form of reproduction compared with reproduction by seed. The most obvious drawbacks are greater susceptibility to elimination by cultivation, less effective dispersal, and greater genetic uniformity, with attendant lessened plasticity in response to climatic variations or to disease and insect pests. A species solely dependent upon vegetative reproduction will be eliminated from a given area once all vegetative parts are destroyed. Such eradication can be accomplished for even the most persistent species in only 2 or 3 years of intensive tillage (Phillips, 1961; Derscheid et al., 1961). As discussed in the previous chapter, many years tillage may be required to eliminate the seedbank in the soil. Although the importance of seed to survival and spread varies for perennial weeds, as shown in Table 5-3, the fact remains that most such weeds produce at least some seed. Therefore, it is helpful to understand something of the relationship between these two strategies for reproduction.

TABLE 5-3.   Relative importance of seed production in survival and spread of selected perennial weeds.

Species	Vegetative Reproductive Part	Seed Production
Austrian fieldcress	Rhizomes	Unimportant
Canada thistle	Creeping rootstock	Fairly important
Dandelion	Adventitious buds on taproot	Very important
Field bindweed	Creeping rootstock	Very important
Hoary cress	Creeping rootstock	Important
Johnsongrass	Rhizomes	Very important
Kikuyu grass	Stolons and rhizomes	Rarely produced
Purple nutsedge	Tubers	Mostly infertile
Quackgrass	Rhizomes	Fairly important
Wild garlic	Bulbs and bulblets	Unimportant

At the outset, we need to recognize a very basic difference between sexual and vegetative reproduction. In the strictest sense, vegetative reproduction may more appropriately be termed *regrowth* since it results from the development of existing meristems in root or stem tissue. Such regrowth leads to a new plant quite like that from which it developed. In contrast, sexual reproduction begins with a single new cell and leads to the development of a new individual unique in some respects from its parent(s). Thus, sexual reproduction involves recombination of the genetic material, whereas vegetative regrowth involves merely repeating the existing genetic material. Nevertheless, vegetative regeneration is clearly an important means of increasing the number of plants with which we must deal. In this sense, it acts as a reproductive process and has been treated as such. The basic difference from true reproduction should be kept in mind, however.

Allocation of Resources

An important question is the relative priority of each process on the plant's resources. That is, does the production of seed, once the flowering process has begun, have priority on photosynthate over the production of vegetative reproduction units? The answer has important implications for preventive approaches to weeds. For example, if a species gave priority to seed production over production of perennating parts, we might look for ways of promoting seed production as a way of limiting vegetative reproductive growth and thus competition.

The limited studies have yielded variable answers. In johnsongrass, as shown in Figure 5-14, more of the dry weight was allocated to rhizomes than to panicles (seed production). However, under the stress of increased density, rhizome production suffered relatively more. At 13 weeks, as the density increased from 1 to 8 plants per pot (32 to 256 plants m^{-2}, respectively), the percent of dry weight in panicles dropped only from 4% to 3%, whereas the dry weight in rhizomes dropped from 27% to 15%. In purple nutsedge (Williams et al., 1977), the opposite effect was observed. As the density increased from 32 to 800 plants m^{-2} for plants 9 weeks old, the percentage dry weight in tubers increased from 4% to 15%, respectively, compared with a decrease from 15% to 8% in inflorescences.

Of course, dry matter by itself may not adequately measure the value to the species of resources allocated to each type of reproduction. Because of the large difference in size, a gram of dry matter committed to seed production may yield many more potential new plants than a gram committed to rhizomes. This situation is illustrated in Table 5-4 by the numbers of rhizome nodes and seeds produced by johnsongrass. Except for the very earliest planting, the number of reproduction units per plant was from 30 to 150 times greater for seed than for rhizome nodes.

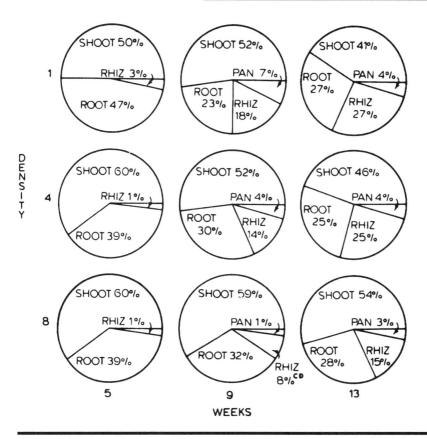

FIGURE 5-14. Influence of plant density and plant age on partitioning of dry weight in johnsongrass plants.

Source: Williams and Ingber, 1977. Reproduced with permission of the Weed Science Society of America.

TABLE 5-4. Numbers of seed and of rhizome nodes per plant in johnsongrass 12 weeks after planting.

Date Planted	Number of Rhizome Nodes per Plant	Number of Seeds per Plant
March 1	7	0
April 1	71	3,909
May 1	140	20,870
June 1	145	19,767
July 1	156	11,416
August 1	101	13,282
September 1	8	250
October 1	—	—

Source: Data from Keeley and Thullen, 1979.

Competing Processes

Much of the data discussed suggest that seed and vegetative reproduction are alternative and competing processes. However, interpreting data within a species is often complicated by the fact that the processes occur at somewhat different times in the plant's development and, thus, are subjected to different environments. Further, it is difficult to fully evaluate the success of each process in producing new plants because of dormancy and environmental influences on germination. It is the number of new individuals generated by each type that is the ultimate measure of its efficiency in utilization of resources.

A comparison between two species of *Agropyron* (now identified as *Elytrigia*), one that relies mainly on seed and the other on rhizomes, is especially revealing (Harper, 1977). The seed of the two species are closely similar in weight and appearance. When grown as an isolated plant, quackgrass (*Agropyron repens*) was observed to produce 30 seeds and 215 rhizome buds, for a total of 245 reproductive units. *Agropyron caninum* produced only seeds, with a total of 258. In other words, the total number of regenerating units was quite similar. Thus, it appears that clonal reproduction and seed production are alternative processes.

Relationship of Reproduction to Competitiveness

The production of seed and of any vegetative reproductive part is a resource utilization process. Resources allocated to the production of seed clearly detract from the competitiveness of the plant producing them. Resources allocated to vegetative reproduction may or may not contribute to the competitiveness of the plant producing them.

In general terms, resources allocated to produce vegetative regeneration units that grow and become established the year in which they are produced do contribute to competitiveness. For example, the main reason why rhizomatous-producing weeds, such as johnsongrass and quackgrass, are so competitive is that their regeneration units frequently grow the year they are produced. Resources allocated to production of units that grow only in subsequent years represent a sacrifice of competitiveness. Wild garlic bulbs produced one year usually do not germinate until the following year, which is one reason this weed is not a strong competitor with such crops as winter wheat, in which it is found. Rather, it is a problem weed because its bulblets may lower wheat quality. Possibly, the extent of contribution to competitiveness that the resource utilization process makes can be exploited in management to reduce the survival and competitiveness of species possessing such capabilities.

REFERENCES

Agricultural Research Service, USDA. 1970. Selected weeds of the United States, Agricultural Handbook no. 366.

Aleixo, M.D., and I.F. Valio. II. 1976. Effect of light, temperature, and endogenous growth regulators on the growth of *Cyperus rotundus* tubers. Zeitschrift fur Pflanzenphysiologie 80:336-337.

Anderson, L.E., A.P. Appleby, and J.W. Weseloh. 1960. Characteristics of johnsongrass rhizomes. Weeds 8:402-406.

Anderson, W.P. 1977. Weed Science: Principles. St. Paul: West Publishing.

Beasley, C.A. 1970. Development of axillary buds from johnsongrass rhizomes. Weed Sci. 18:218-222.

Derscheid, L.A., K.E. Wallace, and R.L. Nash. 1960. Leafy spurge control with cultivation, cropping, and chemicals. Weeds 8:115-127.

Derscheid, L.A., R.L. Nash, and G.A. Wicks. 1961. Thistle control with cultivation, cropping, and chemicals. Weeds 9:90-102.

Derscheid, L.A., J.F. Stritzke, and W.G. Wright. 1970. Field bindweed control with cultivation, cropping, and chemicals. Weed Sci. 18:590-596.

Dexter, S.T. 1936. Response of quackgrass to defoliation and fertilization. Plant Physiol. 11:843-851.

———. 1937. The drought resistance of quackgrass under various degrees of fertilization with nitrogen. Agron. J. 29:568-576.

———. 1942. Seasonal variations in drought resistance of exposed rhizomes of quackgrass. Agron. J. 34:1125-1136.

Donald, W.W. 1994. The biology of Canada thistle (*Cirsium arvense*). Rev. Weed Sci. 6:77-101.

Garg, D.K., L.E. Bendixen, and S.R. Anderson. 1967. Rhizome differentiation in yellow nutsedge. Weeds 15:124-128.

Hakansson, S. 1969. Experiments with *Sonchus arvensis* L., I. Development and growth, and the response to burial and defoliation in different developmental stages. Lantbrukshogskolans Annaler 35:989-1030.

Harper, J.L. 1977. Population biology of plants. New York: Academic Press.

Harradine, A.R. 1980. The biology of African feathergrass (*Pennisetum macrourum*) in Tasmania, II. Rhizome biology. Weed Res. 20:171-175.

Hodgson, J.M. 1958. Canada thistle control with cultivation, cropping, and chemical sprays. Weeds 6:1-11.

Jansen, L.L. 1971. Morphology and photo response in yellow nutsedge. Weed Sci. 19:210-219.

Jeffrey, L.S., and L.R. Robison. 1971. Growth characteristics of common milkweed. Weed Sci. 19:193-196.

Johnson, A.A., and S.T. Dexter. 1939. The response of quackgrass to variations in height of cutting and rates of application of N. Agron. J. 31:67-76.

Johnson, B.G., and K.P. Buchholtz. 1962. The natural dormancy of vegetative buds on the rhizomes of quackgrass. Weeds 10:53-57.

Keeley, P.E., and R.J. Thullen. 1975. Influence of yellow nutsedge competition on furrow-irrigated cotton. Weed Sci. 23:171-175.

———. 1978. Light requirements of yellow nutsedge (*Cyperus esculentus*) and light interception by crops. Weed Sci. 26:10-16.

———. 1979. Influence of planting date on the growth of johnsongrass from seed. Weed Sci. 27:554-558.

Lemieux, C, D.C. Cloutier, and G.D. Leroux. 1993. Distribution and survival of quackgrass (*Elytrigia repens*) rhizome buds. Weed Sci. 41:600-606.

McIntyre, G.E. 1971. Apical dominance in the rhizome of *Agropyron repens* in isolated rhizomes. Can. J. Bot. 49:99-109.

McWhorter, C.G. 1961. Morphology and development of johnsongrass plants from seeds and rhizomes. Weeds 9:558-562.

————. 1972. Factors affecting johnsongrass rhizome production and germination. Weed Sci. 20:41-45.

Meyers, E.J., and D.H. Caso. 1976. The effect of nitrogen supply on the growth of *Sorghum halepense*. Malezas 5:3-12.

Monaco, T.J., and E.L. Cumbo. 1972. Growth and development of curly dock and broadleaf dock. Weed Sci. 20:64-67.

Morrow, L.A. 1979. Studies on the reproductive biology of leafy spurge (*Euphorbia esula*). Weed Sci. 27:106-109.

Mshiu, E.P. 1978. Studies on *Digitaria scalarum*, Technical crop production, master's thesis. Reading, England: University of Reading.

Myers, G.A., C.A. Beasley, and L.A. Derscheid. 1964. Anatomical studies of *Euphorbia esula* L. Weed Sci. 12:291-295.

Nadeau, L.B., and W.H. Vanden Born. 1989. The root system of Canada thistle. Can. J. Plant Sci. 69:1199-1206.

Ogden, J. 1974. The reproductive strategy of higher plants, II. The reproductive strategy of *Tussilago farfara* L. J. Ecol. 62:291-324.

Palmer, J.H., and G.R. Sagar. 1963. *Agropyron repens* L. Beauv. J. Ecol. 51:783-794.

Patterson, D.T. 1982. Shading response of purple and yellow nutsedge (*Cyperus rotundus* and *C. esculentus*). Weed Sci. 30:25-30.

Phillips, W.M. 1961. Control of field bindweed by cultural and chemical methods, USDA technical bulletin no. 1249. Washington, D.C.

Russ, O.G., and L.E. Anderson. 1960. Field bindweed control by combinations of cropping, cultivation, and 2,4-D. Weeds 8:397-401.

Sagar, G.R., and A.M. Mortimer. 1976. An approach to the study of the population dynamics of plants with special reference to weeds. Appl. Biol. 1:1-47.

Seyforth, W., W. Kreil, and O. Knobe. 1978. Investigations into the reserve carbohydrate balance in *Agropyron repens* at various levels of N fertilization of grassland. Soils Fert. 42:5722.

Stoller, E.W. 1977. Differential cold tolerance of quackgrass and johnsongrass rhizomes. Weed Sci. 25:348-351.

Stoller, E.W., and L.M. Wax. 1973. Yellow nutsedge shoot emergence and tuber longevity. Weed Sci. 21:75-81.

Thomas, P.E.L. 1969. Effects of desiccation and temperature on survival of *C. Esculentus* and *Cynodon dactylon* rhizomes. Weed Res. 9:1-8.

Tumbleson, M.E., and T. Kommedahl. 1961. Reproductive potential of *Cyperus esculentis* by tubers. Weeds 9:646-653.

Williams, R.D., and B.F. Ingber. 1977. The effect of intraspecific competition on the growth and development of johnsongrass under greenhouse conditions. Weed Sci. 25:293-297.

Williams, R.D., P.C. Quimby, Jr., and K.E. Frick. 1977. Intraspecific competition of purple nutsedge (*Cyperus rotundus*) under greenhouse conditions. Weed Sci. 25:477-481.

Wilson, R.G., Jr. 1979. Germination and seedling development of Canada thistle (*Cirsium arvense*). Weed Sci. 27:146-151.

Yeo, R.R. 1964. Life history of common cattail. Weeds 12:284-288.

6

Resumption of Growth

CONCEPTS TO BE UNDERSTOOD

1. Germination of seeds involves a precise sequence of events; this sequence may either be delayed in its initiation or stopped along the way by what is termed dormancy.
2. Dormancy, although not itself a factor in competition, is important in the persistence and survival of weeds.
3. Many seeds, and possibly vegetative regenerating parts, have an apparent season-anticipating characteristic that assures resumption of growth when chances are optimal for survival and completion of their life cycle.
4. Although there are quite wide differences among species, the general pattern is that emergence from seed is inversely related to depth of burial; the top 2.5-cm soil layer contributes the most new seedlings.
5. Many weed seeds require light for germination; the far-red, which passes through leaves, in fact often inhibits germination.
6. Many weed seeds that do not have innate dormancy acquire dormancy when subjected to shading by burial in the soil and thus are protected against rapid loss of viability.
7. Temperature is a modifying rather than a triggering factor in germination. Thus, the curve relating temperature and germination percentage is S-shaped.
8. Germination of seeds of weeds common to temperate regions is improved by exposure to alternating temperatures and by accumulated exposure to temperatures at or below freezing.
9. Emergence for perennating parts is also inversely related to depth of burial, but in general terms, such emergence is from greater depths than for seeds.
10. Growth-regulating substances are most likely important modifiers of resumption of growth of both seeds and perennating parts.
11. High nitrogen levels in soil are conducive to regrowth from perennating parts produced by weeds grown under such conditions.

12. An inverse relationship exists between the length of the rhizome and the number of buds that sprout.
13. It is common for one bud to have an inhibiting effect on regrowth of other buds; the apical bud commonly possesses this inhibiting effect.

Production of seed or vegetative parts, covered in Chapters 4 and 5, and their resumption of growth are the two phases in a weed's life history of primary concern to weed management. When growth resumes in relationship to the desired plant's life cycle determines the severity of competition or whether there is competition at all. The numbers are influenced by the number of propagules added to the soil bank in previous years. As discussed in Chapters 4 and 5, both the numbers of propagules produced and their longevity are influenced by a number of factors. We are now ready to consider factors that influence resumption of growth.

Germination can be defined as the resumption of growth of a seed or a vegetative part. Germination of seeds involves a precise sequence of events: (1) imbibition of water, (2) marked increase in respiration, (3) mobilization of food reserves, and (4) digestion of reserved foods. For many crop seeds, this sequence begins promptly after planting and proceeds in an orderly fashion to the emergence of the young seedling. The time required is modified only by soil moisture and temperature conditions. The situation with many weed seeds and vegetative parts is quite different. The process either does not start promptly upon their introduction into the soil or it is stopped along the way. That is to say, the seed or vegetative part has a period of metabolic quiescence—usually termed *dormancy*—after it is produced. When in this state, the seed or the vegetative part does not resume growth, even though all environmental conditions seem to be favorable.

DORMANCY

Dormancy, so common in weeds, needs to be briefly examined, especially relative to environmental influences, before proceeding to a discussion of germination. Similarities are found in the relationships between the environment and onset of dormancy in seed and vegetative parts. Therefore, the following discussion of such principles for seeds is assumed to be applicable to vegetative parts also. The perspective of dormancy here is that of a pause in the normal sequence of events leading to germination and emergence. By itself, it is not a factor in competition. However, it is an important factor in persistence and survival of species. In this regard, dormancy plays a very significant function indeed in the life of the weed by protecting the embryo or meristem during dissemination and during periods unfavorable for germination and successful establishment of a new plant. It is outside the purposes of this text to examine the biochemical processes involved or the embryonic development per se. Refer to Chapters 3,4, and 5 in *Dormancy and Development Arrest* (Clutter, 1978) for a detailed discussion of biochemical reactions in the regulation of dormancy and to any good text on plant anatomy for a discussion of embryology.

Our interest here is in dormancy as it affects longevity and resumption of growth. As discussed in Chapter 4, seeds may survive many years in soil. This survival is the direct result of dormancy. Dormancy has many definitions. A useful one for weed–crop ecology is that state in which growth is not resumed even though the environment supports germination and seedling growth of other apparently identical, but nondormant, tissues of the same species or plant. Among other things, this definition recognizes that dormancy is a relative matter and can best be evaluated by comparison with very closely related tissues.

The following development timetable of a seed places dormancy in perspective with the other processes:

Stages	Characteristics
Water Loss	
Cleavage and histo-differentiation	Cell division and differentiation of all major tissues, but little growth
Growth	Rapid expansion and division of cells
Maturation	Cessation of cell division and expansion; synthesis and storage of food reserves
Dormancy	Developmental arrest
Water Uptake	
Germination	Resumption of cell division and expansion

The first three stages, with a few exceptions, occur while the developing seed is physically attached to the parent. The final stage, germination, normally occurs after the seed is separated from the parent. The dormant condition may be acquired while the seed is still physically attached to the parent or after it has been released. Thus, the dormant condition may be a result of genetic messages and biochemical reactions between the maternal tissue and the embryo or meristem. Or, the condition may be brought on by the environment to which the reproductive part is exposed.

Types of Dormancy

There are three broad types of dormancy: (1) innate, (2) induced, and (3) enforced. *Innate dormancy,* sometimes referred to as primary dormancy, is that present in the seed or vegetative part when released from the parent. *Induced dormancy,* sometimes referred to as secondary dormancy, is a result of conditions to which the seed or vegetative part is exposed after release from the parent. Once such dormancy is induced, germination or regrowth usually does not commence immediately when the condition is removed. *Enforced dormancy* is that imposed by conditions unfavorable for resumed growth, most commonly a shortage of water or unsuitable temperature. Stated simply: "Some seeds are born dormant, some acquire dormancy, and some have dormancy thrust upon them" (Harper, 1977).

Innate dormancy. Innate dormancy may be imposed and maintained by several mechanisms. Among them are: immaturity of the embryo, seed coats impermeable to water, seed coats that inhibit gaseous exchange, mechanical resistance of the seed coat to embryo growth, and growth substance imbalance within the embryo. All of these mechanisms are genetically controlled, with the degree of expression, of course, influenced by growing conditions. The most consistent modifier appears to be maturity. Thus, with many weeds, dormancy increases as seeds mature. In one study with wild oat, for example, germination of mature seed was only 8% compared to 50% germination for immature seed (Richardson, 1979). Since maturity is subject to control with management, there may be opportunities to use this fact to reduce the longevity of weed seed in the seedbank. Treatment of the parent plant as a possible approach to reducing the seedbank is considered in Chapter 13.

Seed *polymorphism,* the production of different kinds of seeds by the same plant, is a characteristic of many weeds that further complicates efforts to deal with innate dormancy. Plants of many weed species produce both dormant and nondormant seed. Dormancy may differ among seeds that look alike. For example, dark seeds from the same plant of common purslane varied from 0% to 100% in germination (Egley, 1974). Or, dormant and nondormant seeds may be quite different in appearance. In cocklebur, for example, two seeds are produced in each capsule. Commonly, but not always, one seed occupies the upper position and the other the lower position in the capsule. The upper one is usually dormant and the lower one nondormant.

Induced dormancy. Seeds of some species, following a period of exposure to unfavorable conditions, do not germinate readily when provided proper conditions. Most commonly, a light requirement is induced. For example, buckhorn plantain, corn spurry, and field poppy seeds that did not require light to germinate before burial would not germinate without light after burial for 50 weeks (Wesson and Wareing, 1969b). A variety of treatments in the laboratory can induce secondary dormancy. A general summary of such treatments is that secondary dormancy can be induced in imbibed seeds by exposure to high temperatures coupled with restriction in the supply of oxygen (Villiers, 1972). Induced dormancy in itself has implications for resumption of growth and for weed management. The more important aspect, however, may be its interaction with innate dormancy. As we know, the vast majority of weeds produce seeds with at least some innate dormancy. Burial may induce a secondary dormancy that greatly extends the longevity of the seed in soil. Because agriculture, even of perennial plants and with reduced tillage, involves tillage that provides burial, the implications for resumption of growth and weed management are substantial.

Enforced dormancy. Many seeds fail to germinate simply because there is insufficient moisture for the imbibition needed to initiate the process. In temperate regions, low temperature prevents germination. Under these circumstances, dormancy is enforced. That is, seeds remain dormant only because conditions necessary to support growth are absent. As soon as proper moisture and temperature are provided, germination occurs.

Propagule–Environment Interactions in Dormancy

The seed, or propagule, commonly carries information about dormancy: All three major components—embryo, reserves, and seed coat—may have a role. The environment interacts with this inherent information and influences the seed to alter those messages or influences its biochemical processes to change directions or initiate new directions. Level 1 of Figure 6-1 schematically summarizes the relationship between the seed and the environment in dormancy and shows how complex it may be.

All factors in the physical environment may have an effect. Photoperiod (light) would appear to be a primary factor in innate dormancy. Temperature also has an effect, apparently largely as a modifier of the photoperiod effect. Similarly, mineral deficiency and lack of moisture may induce the initial stages of dormancy in buds. For induced dormancy of seed, carbon dioxide and moisture appear to have the greatest impact. Moisture is involved in the development of so-called hard-seed, widespread in members of the Leguminosae. In this case, induced dormancy involves the hilum act-

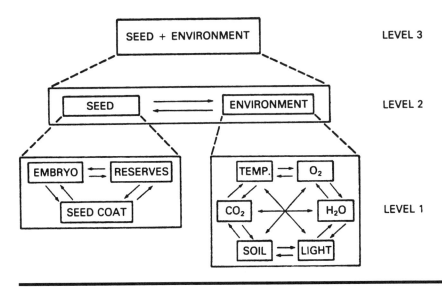

FIGURE 6-1. Levels of organization of a seed–environment system. Each component of the seed and of the environment, plus interactions among them (Level 1), determines seed dormancy.

Source: Simpson, 1978. Reproduced courtesy of Academic Press, Inc.

ing as a hydrostatic control valve. The *hilum* is a scar on the surface of the seed left after its detachment from the stalk. Under dry conditions, the hilum opens and allows water to escape. Under humid conditions, it closes, thus preventing water uptake. In effect then, the embryo dries progressively to a level equal to the driest environment it has experienced. As long as the integrity of the seed coat is present and the hilum is functioning, such seeds do not imbibe water and thus remain in the "hard" condition.

Carbon dioxide. Carbon dioxide (CO_2) deserves special comment because it may be the most controllable of the environmental factors, in the sense that plant residues, which are a potential source of CO_2 enrichment, can be incorporated at the discretion of the farmer. It has been shown (Kidd, 1914; Kidd and West, 1917; Hart and Berrie, 1966) that elevated levels of CO_2 induce dormancy in seeds of some species. Further, the induction of dormancy by CO_2 was related to both temperature and oxygen levels. At 3°C, 2% CO_2 caused about a 90% reduction in germination, whereas 36% CO_2 was required for the same germination reduction at 20°C. Studies of CO_2 in the soil profile (deJong and Schappert, 1972) and in soil with large amounts of green plant materials added (Kidd, 1914) showed levels in the range that might cause dormancy (2% to 20%). It can be speculated that incorporation of weed seeds with large quantities of weed and crop residues may result in CO_2 levels, at the vicinity of the seed, sufficiently high to induce dormancy.

Onset of dormancy related to plant hormones. Naturally occurring plant hormones appear to be intimately involved with entry into and release from dormancy in both buds and seeds. Abscisic acid would appear to be the dormancy-inducing hormone involved. Production of abscisic acid is apparently triggered by a stimulus from mature leaves under short days. This type of induced dormancy may be viewed as a seasonal dormancy in that it occurs in a fairly predictable fashion following the induction of the stimulus by the right photoperiod. In seeds, production of abscisic acid seems to be related to water loss that occurs as the seed passes from histo-differentiation to growth to maturation. Here, too, photoperiod appears to be involved (Gutterman, 1978). Thus, our concern for the influence of the environment likely comes down to the effect upon the synthesis, transport, and activity of the regulating plant hormones. The precise mechanism involved is yet to be determined.

To summarize, seed structure, environment, and metabolism are all involved in dormancy and probably in an interactive fashion, as shown in Figure 6-1. The fact that the environment affects onset of dormancy suggests that there may be ways of manipulating the environment to either bring on or prevent dormancy. Where CO_2-induced dormancy is involved, proper timing of the incorporation of the weed seed and the associated plant residues may be a useful approach. Where innate dormancy is involved, such manipulation must likely focus on ways of blocking formation of abscisic acid and possible other growth-inhibiting substances or on increasing the levels of growth-promoting substances. Dormancy, indeed, is fortuitous to the weed involved since it spreads resumption of growth over a period of time; in some cases, even over a period of several years. Thus, one unfavorable growing season does not eliminate the species.

Developing a clear understanding of the mechanisms involved is particularly important to weed management because it may be easier to avoid the onset of dormancy than to release the propagule from dormancy once it is established.

ECOLOGICAL RELATIONSHIPS IN RESUMPTION OF GROWTH

The same factors that bring about dormancy are involved in germination, or the resumption of growth. In other words, the seed–environment relationships given in Figure 6-1 for dormancy are also involved in germination. Although there are many similarities in the effects of environment on their resumption of growth, obvious differences also exist between seeds and vegetative regenerating parts. The most notable difference may be the inhibiting effect of one bud on another in many vegetative reproduction parts that of course is not a factor with seeds. Also, there are differences in crop associations in the difficulty of control and in the basic approaches to preventing losses from them. Thus, resumption of growth is considered separately for seeds and vegetative parts.

RESUMPTION OF GROWTH: SEED

Season Anticipation

Seeds of many weeds germinate at a time when conditions can be expected to be favorable for establishment of the new plant and completion of its life cycle. That is, the seed seems to have the ability to predict the right season of the year in which to germinate. This season-anticipating characteristic is demonstrated by the fact that seeds of many annual weeds in temperate regions germinate only in the spring and summer when there is sufficient time to mature and produce seed. Germination occurs only at these specific times, even though there most certainly are times in the late summer and early fall when the seed is exposed to the same moisture, temperature, and light conditions as in the spring and early summer. Of course, this situation is fortunate for weeds. If their seeds germinated in late summer and early fall, they could be killed by frost before producing seed, and the species would soon disappear.

As we have already learned, in many species, dormancy prevents germination for a time after the seed is produced. Our concern now is with those factors that trigger an end to dormancy and with the fact that this change occurs when conditions are favorable for establishment. Further, an end to dormancy does not occur during a brief period of favorable conditions but rather occurs over time. If this were not so, a brief period of favorable conditions, such as a warm period in midwinter, could produce a flush of germination with the seedlings being destroyed by the return of low temperatures. It is this season-anticipating ability that provides the seed a maximum chance for successful germination, seedling establishment, maturation, and seed production. Recall the discussion of safe sites from Chapter 4 where it was pointed out the environment is heterogenous from the perspective of the weed seed. Because the environment is so heterogeneous, we expect seed to respond to many different signals from the environment. Relative to this matter, Koller (1972) makes the comment: "Species which occupy the same habitat rarely have common denominator environmental control of their germination ... since different species are able to share the same habitat only by occupying different microenvironmental niches in it." This statement adds to our understanding of why we commonly have a mixture of weed species. It also helps to explain why all species of weeds do not germinate at the same time and why individuals of a single species do not. The seed's perception of the environment and the different triggering factors among species are important to our examination of the individual factors affecting germination. The discussion of individual factors, by necessity, emphasizes general patterns.

Periodicity. Not only do seeds of many weeds germinate only when conditions for survival are favorable, but many also have a flush of germination at a given time in the growing season, termed *periodicity*. The germination flush for eight weeds is shown in Figure 6-2. Two generalizations regarding periodicity are supported by Figure 6-2. One is that species may have more than one flush of germination, as is the case of wild oat. The second is that the discreteness of periodicity varies with species. For example, periodicity for black nightshade is restricted to the late spring and summer periods, whereas annual bluegrass germinates throughout the year, even though it has a peak in spring and fall. Periodicity, particularly where it is discrete, offers possibilities for preventive approaches.

Effect of growth-regulating substances. Research has conclusively established that seeds contain growth-regulating substances that influence germination. Although the chemistry involved is outside the scope of this text, there is evidence to suggest that such substances influencing germination involve three broad kinds: (1) those that are promoting in their effect (gibberellins or gibberellin-like substances); (2) those that are inhibiting in their effect (quite possibly abscisic acid); and (3) those that are antagonistic to the endogenous inhibitors (possibly cytokinins). All have been found in weed seeds. Further, it has been found that strains of wild oat vary in the levels of cytokinins and gibberellins contained in their seeds (Taylor and Simpson, 1980). Thus, mounting evidence suggests that germination is dependent upon the balance between promoting and inhibiting growth substances.

Quantity produced of each growth regulator is subject to influence by environmental factors to be discussed in the next section. It should be noted here that the interaction between growth regulators and environmental factors can be expected to modify the period and extent of weed seed germination. Further, there is evidence that

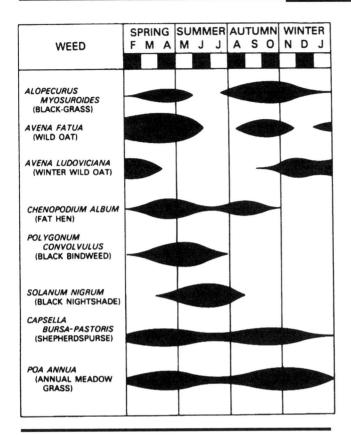

WEED	SPRING			SUMMER			AUTUMN			WINTER		
	F	M	A	M	J	J	A	S	O	N	D	J
ALOPECURUS MYOSUROIDES (BLACK-GRASS)												
AVENA FATUA (WILD OAT)												
AVENA LUDOVICIANA (WINTER WILD OAT)												
CHENOPODIUM ALBUM (FAT HEN)												
POLYGONUM CONVOLVULUS (BLACK BINDWEED)												
SOLANUM NIGRUM (BLACK NIGHTSHADE)												
CAPSELLA BURSA-PASTORIS (SHEPHERDSPURSE)												
POA ANNUA (ANNUAL MEADOW GRASS)												

FIGURE 6-2. Period of maximum germination of seeds of eight weeds.

Source: Hill, 1977. Reproduced with permission of Blackwell Scientific Publications, Ltd.

leaching (washing) can remove these substances and thus either inhibit or promote germination. For example, as shown in Table 6-1, washing lambsquarters seeds increased their germination, whereas leaching of fumitory (*Fumaria officinalis*) achenes reduced germination, as shown in Figure 6-3. It is significant for weed management that the inhibiting and promoting effects are influenced by such factors as leaching, which may be controllable under some circumstances.

Environmental Effects on Resumption of Growth

Perception and characterization of the environment by the seed obviously depend upon relatively few factors—mainly temperature, light, and water—but the many possible

TABLE 6-1.   Germination of common lambsquarters following different times of washing in running tap water.

Duration of Washing (hours)	Germination(%)
0	27.8
8	39.0
70	49.0

Source: Chu et al., 1978. Reproduced with permission of the Weed Science Society of America.

combinations and sequences among these factors provide many potential signals. Our interest here is in signals from the environment rather than in the processes involved. The present discussion, therefore, focuses primarily on effects of environmental factors on germination with only as much discussion of the processes involved as may be needed to make the effect clear. The environmental effects to be covered include depth of burial in soil, canopy/light, temperature, soil type and condition, soil chemicals, and the several factors interacting with one another.

Depth of burial.   Of all the environmental effects on germination, that of depth of burial is most consistent. First, emergence for many weed seeds is inversely related to depth of seed in the soil from about 1 cm down. For some, emergence is best for seeds on top of the soil, but most emerge better when incorporated in the very shallow surface layer. The relationship shown in Figure 6-4 is representative of the effect of bur-

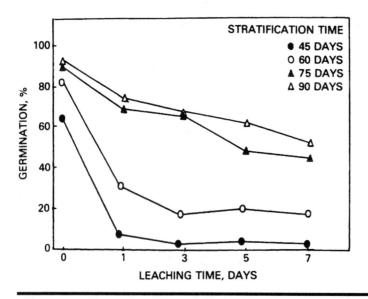

FIGURE 6-3. Effect of leaching achenes of fumitory (*Fumaria officinalis* L.).

Source: Jeffrey and Nalewaja, 1970. Reproduced with permission of the Weed Science Society of America.

ial depth for many weed seeds. Temperature, of course, influenced emergence, but it did not alter the effect of burial depth. The results depicted allow three broad conclusions about depth: (1) best emergence occurs from seeds buried shallowly; (2) species differ in their response to burial depth; and (3) some emergence is possible beyond the ideal depth. Although the depth at which the break occurs varies among species and for environmental conditions, the sharp drop-off below this depth is typical of most seeds.

A second aspect to depth of burial is also quite consistent with a wide variety of species, soil conditions, and other environmental effects: Emergence is delayed in a direct relationship with depth of burial. Possible explanations for the effect of burial depth are discussed after considering other individual factors.

At this point, the question remains of whether it is emergence or germination that is affected by burial depth. While the data are insufficient for an unequivocal answer,

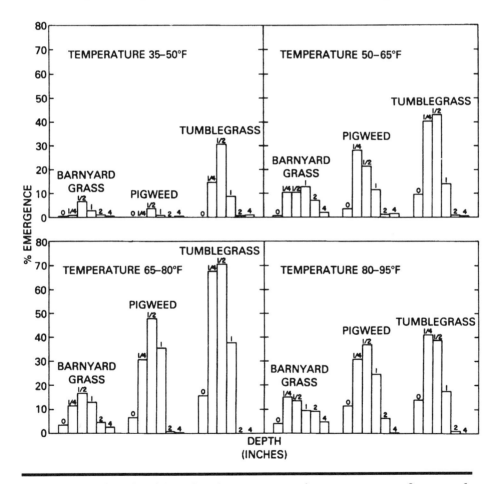

FIGURE 6-4. Effect of burial depth and temperature regimes on emergence of some weeds from seed.

Source: Wiese and Davis, 1967. Reproduced with permission of the Weed Science Society of America.

the preponderance of evidence indicates that depth has little effect on germination if the conditions for germination have been met. As we see later, these conditions, which influence metabolic processes, include adequate moisture, right temperature, and for some, light. Thus, burial depth is symptomatic of these factors rather than itself the factor. Its effects, though, are real and may be usable in dealing with weeds.

Canopy/light. We have long known that germination of seeds of many weed species is affected by light. Figure 6-5 shows how pronounced the light requirement may be. Maximum germination after 30 days in a germinator occurred only when the pusley (*Richardia scabra*) seeds received 12 hours of light each day. Ten hours of light produced more than a threefold increase in germination compared with 8 hours. Practically no germination occurred without at least 4 hours of light.

Table 6-2 lists some weeds that are known to require light for germination. The list includes a broad cross section of plant families.

For many years, this effect was assumed to be simply that of the presence or absence of light. Only within the last half of this century have data accumulated to show that there are three qualitative aspects of light that can affect germination: (1) intensity, (2) spectral composition, and (3) duration. Since all three may vary considerably under natural conditions, it is not surprising that results from studies of light are somewhat variable. The interaction of light with other factors to be discussed later further complicates the picture. Our purpose is to identify and understand the effects of light for those environmental factors over which we may exercise some measure of control. Thus, although intensity and duration can affect germination, it is the effect of spectral composition that is of particular importance. This is so because spectral composition is

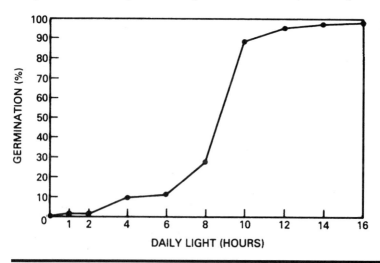

FIGURE 6-5. Effect of light on germination of Florida pusley (*Richardia scabra*).
Source: Data from Biswas et al., 1975.

TABLE 6-2.   Partial list of weed seeds that require light for best germination.

Birdseye speedwell	Corn spurry	Lambsquarters	Sorrel
Black knapweed	Crunchweed	Manyseeded	St. Johnswort
Brazil callalily	Curly dock	goosefoot	Thymeleaf
Bristly foxtail	Fall panicum	Mouse-ear	sandwort
Broadleaf dock	Field pennycress	chickweed spp.	Triple-awned
Buckhorn	Field pepperweed	Nightflowering	grass spp.
plaintain	Fingergrass spp.	catchfly	Tumble mustard
Carline thistle	Florida pusley	Oxeye daisy	Virginia
Common	Giant foxtail	Pennsylvania	pepperweed
chickweed	Hairy beggartick	smartweed	Wild carrot
Common	Hawkbit	Perennial	Wild marigold
cinquefoil	Healall	sowthistle	Wild marjoram
Common	Hedge mustard	Prostrate	Wild mustard
purslane	Hemp	knotweed	Wild parsnip
Common ragweed	Hoary plaintain	Redroot pigweed	Wormseed
Common yarrow	Johnsongrass	Rock cress	mustard
Corn marigold	Kochia	Smallflower	Yellow bedstraw
Corn poppy	Ladysthumb	galinsoga	Yellow rocket

largely determined by the extent of leaf filtering by the canopy that in turn, can be manipulated. Although intensity is also influenced by the canopy, spectral composition may have a relatively greater effect, as we see in the next section.

Light filtered by leaves.   Table 6-3 shows the relative importance of leaf shade. The intensity of light, measured by a light meter, reaching dishes containing the seed was the same under banana leaf and neutral shade. Germination of 16 of the 18 species was inhibited—6 completely—by leaf shade. Neutral shade inhibited germination of 10 species, but in all but one species (*Aristida adscensionis*), the inhibition was less than under leaf shade. Note that there are more instances where the effect is a matter of degree than where it is complete.

In other work, it was found that leaf canopy inhibition of germination was greater among wild than among cultivated herbaceous species and greater in those species common to open habitats than in those of cultivated areas (Gorski et al., 1977). Does this suggest that those species common to cultivated areas in Table 6-3 may lose such sensitivity in time?

Dormancy acquired by shading.   An important question is whether or not seeds whose germination is inhibited by light filtered through a canopy of leaves acquire a dormancy that inhibits germination after removal of the canopy, at least for a time. This question is important both from the standpoint of survival of the species and for approaches for preventing losses from them. Apparently, there are conflicting answers. Silvertown (1980) found no evidence that seeds of 17 species in which dormancy was induced by leaf shade had acquired a light requirement for germination. On the other hand, Fenner (1980a) found that hairy beggarticks (*Bidens pilosa*) seed inhibited from germinating by leaf shade required light for germination. Further, this light requirement was acquired in only 1 hour of shade.

TABLE 6-3. Effect of light and of neutral or leaf shade on seed germination of selected species.

Species*	Light	Dark	Banana-Leaf Shade	Neutral-Paper Shade
		Percent Germination (each of two dishes)		
Achyranthes aspear	97	91	91	95
Ageratum conyzoides	99	0	0	56
Amaranthus caudatus	100	93	6	98
Aristida adscensionis	40	5	20	17
Bidens pilosa	97	81	0	99
Chloris pycnothrix	91	3	52	78
Conyza bonariensis	44	38	25	47
Cynoglossum lanceolatum	97	0	0	16
Ethulia "sp. A"	50	1	0	27
Galinsoga parviflora	93	3	0	80
Launaea cornuta	78	15	20	70
Richardia brazilienis	29	14	0	92
Schkuhria pinnata	62	15	15	33
Setaria verticillata	49	23	29	61
Sonchus oleraceus	86	51	6	86
Tagetes minuta	94	57	2	74
Vernonia hindei	44	33	1	10
Vernonia lasiopus	42	16	3	26

*Because most are not common in this country, only the scientific name is given.
Source: Data from Fenner, 1980b.

Under the analogous situation of burial in soil, many seeds apparently acquire a dormant condition. Light-absent dormancy was suspected as being responsible for the flush of weed seedlings observed following cultivation of a field that had been in pasture for 6 years and before that was in cultivated crops (Wesson and Wareing, 1969a). The area had 12.5 seedlings m^{-2} before disturbance and 300 seedlings m^{-2} after. Tests were made to determine whether light was the factor responsible. Holes 75 cm by 75 cm wide by 5 cm, 15 cm, and 30 cm deep were dug in the field. The holes were dug at night and the openings either covered or left open. Figure 6-6 shows the results. As can be seen, no germination occurred where light was excluded. Where light was provided, numerous seeds germinated. There were more seedlings at the 5 cm depth than at the 15 cm or 30 cm depth, which Wesson and Wareing attributed to the larger initial population of seeds at the shallow depth.

Sufficient information has now accumulated to indicate that species differ in their light sensitivity following burial. Although the majority likely need light to germinate after burial, ivy leaf morning-glory (*Ipomoea hederacea*) and cocklebur did not (Stoller and Wax, 1973). The effect of burial on those seeds sensitive to light helps explain why such species are able to survive even though their seeds may not possess innate dormancy.

Furthermore, there is evidence that the burial effect may extend to species not considered to be light sensitive. As mentioned earlier, Wesson and Wareing, in additional work (1969b), found that following 50 weeks of burial, three species not originally sensitive to light needed light to germinate. In addition, common chickweed, which was inhibited by light before burial, needed light after being buried for 50

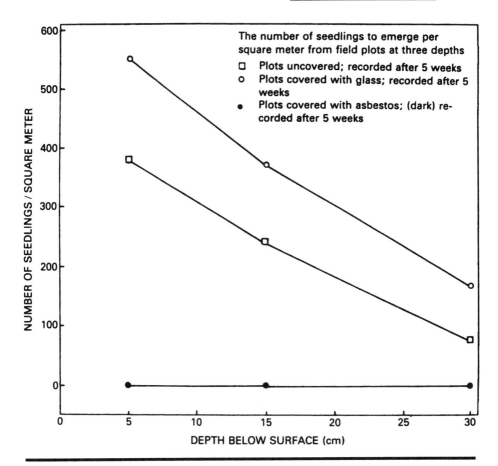

FIGURE 6-6. Light requirement of buried weed seeds. Numbers of seedlings m⁻² were counted after 5 weeks at the bottom of holes 5 cm, 15 cm, or 30 cm deep whose openings were treated as follows: (1) uncovered, (2) covered with glass, (3) covered with asbestos (dark), and (4) asbestos cover replaced with a glass cover after 5 weeks and counted 3 weeks later.

Source: Wesson and Wareing, 1969a.

weeks. These results suggest that the mechanism for acquiring dormancy is different with burial than with exposure to light filtered through a leaf canopy.

Germination affected by phytochrome. It is now known that the responsiveness of seed germination to light is tied to a pigment in the seed called *phytochrome.* Phytochrome exists in the seed in two forms. One form promotes germination and the other does not. The quantities of each present at a given time are determined by light, more precisely by the red to far-red ratio of the light.

Light in the red (650 nm) portion of the spectrum pushes the phytochrome toward the active form.[1] That in the far-red (730 nm) pushes it toward the inactive form. This photoreaction is of the form:

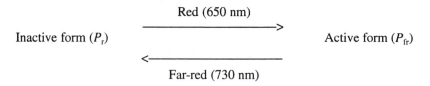

Red (650 nm)

Inactive form (P_r) ————————————————> Active form (P_{fr})

<————————————————

Far-red (730 nm)

Note: Diagram from Taylorson and Borthwick, 1969. Reproduced with permission of Weed Science Society of America.

In the dark, the inactive form predominates. Unfiltered light, which contains a preponderance of red (650 nm range), would shift it to the active form. Since chlorophyll absorption is strong for light in the 650 nm range and weak for that beyond 700 nm, a leaf canopy shifts the red/far-red relationship in the transmitted light toward the far-red. This shift in turn causes an increase in inactive phytochrome, thus inhibiting germination. The red/far-red relationships may vary greatly under different microsites, as shown in Table 6-4. There is more than a threefold range of ratios for the different types of shade.

TABLE 6-4.   Red/far-red relationships in different microsites.

Type of Leaf Canopy	Ratio of Red/Far-Red
Shaded bare ground	0.85
Moderate shade, edge of tussock	0.80
Deep shade, middle of tussock	0.67
Bromegrass (*Bromus erectus*) (2 layers)	0.20
Linden (*Tilia europaea*) leaves (2 layers)	0.25
Tobacco (*Nicotiana Tabaccum*) leaves (2 layers)	0.28

Source: Data from Silvertown, 1980.

A statement by Toole et al. (1955) in their review of light effects provides a fitting summarization: "The germination process, which depends on respiration, is controlled at different points by the several factors. The photoreaction, while possibly present in all seeds, is not obligatory for germination for all seeds. It controls the levels of two compounds which are also under control by other reactions subject to influence by temperature." This statement serves to emphasize the complexity of the light relationship in germination. Nonetheless, it is clear that light is needed by enough species and in sufficient quantities that there may be opportunities in the use of crop cover and in the timing and depth of tillage to minimize emergence with the crop and over time the amount of viable seeds in the soil. Chapter 15 explores these opportunities in greater depth.

———————

[1]A nanometer (nm) is a unit for measuring wavelength. One nanometer is 1/1000 of a micron.

Temperature.   It is common knowledge that temperature affects germination. Over the years, much work has been done with seeds of desired plants, primarily in terms of identifying the best soil temperature for germination and seedling establishment. In view of the effect of temperature on physiological processes, we would expect temperature to have a modifying effect on germination rather than a triggering effect. That is to say, we might expect germination to occur with most species over a fairly wide temperature range. This, indeed, is the case.

Data for Florida pusley in Figure 6-7 are quite representative of the temperature effect on germination. As can be seen, germination approaching 50% or better occurred over a temperature range from 20°C to 35°C, with some germination occurring even at 15° and 40°C. Of course, species differ in the range of temperatures for best germination, but the general relationship is bell shaped, as for Florida pusley. Thus, wild oat germinates over a range from 10°C to 30°C, with the maximum at about 20°C (Friesen and Shebeski, 1961); tall morning-glory over a range from 15°C to 35°C, with the maximum at 25°C to 30°C (Cole and Coats, 1973); and common mullein over a range from 15°C to 40°C, with the maximum at 30°C (Semenza et al., 1978).

Temperature also affects the rate of germination. The responses of redroot pigweed, tumblegrass (*Schedonnardus paniculatus*), and barnyardgrass, shown in Table 6-

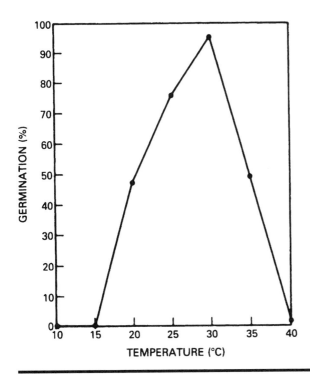

FIGURE 6-7. Effect of temperature on germination of weed seed.

Source: Data from Biswas et al., 1975.

5, are representative of the pronounced effect on rate. Emergence was 6 to 8 times as rapid under the highest as under the lowest regimes. The message for us in weed–crop ecology is that there may be ways of managing our cultivation practices so as to avoid providing problem weeds with the temperature range needed for seed germination. These ways are explored in Chapter 15.

TABLE 6-5. Days required for one-half of total weed emergence from seeds planted 0.5 in. deep.

Weed	Temperature Ranges (°F)			
	35-50	50-65	65-80	80-95
Redroot pigweed	24	21	6	3
Tumblegrass	21	21	7	3
Barnyardgrass	31	25	8	5

Source: Wiese and Davis, 1967. Reproduced with permission of the Weed Science Society of America.

Alternating temperatures.   Laboratory and field studies show that many weed seeds germinate better under alternating rather than under constant temperatures. Germination at 20°C to 30°C in Table 6-6 indicates how striking this effect may be. Of the 85 weed species involved in these same studies, about 80% showed better germination under alternating rather than under constant temperatures.

TABLE 6-6. Effect of alternating temperatures on the germination of seeds of wormseed mustard.

Treatment	Percent Germination at 5 Days
20°C constant, water	0
30°C constant, water	0
20°C-30°C, water	67

Source: Adapted from Steinbauer and Grigsby, 1957.

The effect of alternating temperatures may explain the observed effect of depth of burial on germination (Koller, 1972). As we saw earlier, many weed seeds emerge best from shallow depths. Because of the damping effect on temperature by the soil itself, temperature alterations are greater closer to the surface than farther down in the soil. Thus, better germination under alternating temperatures, he speculates, may be a matter of the seed's perception of the environment, in this case indicating that the seed is close enough to the surface for the seedling to emerge and become established. Irrespective of the reason, the fact that many weed seeds do germinate better under alternating temperatures and from shallow depths offers possibilities for managing our production practices to control germination of such weeds.

Need for low temperatures.   Another temperature effect common to many weeds is

for their seeds to require temperatures at or below freezing before germination can occur. Further, an accumulation of several days at such temperatures is often necessary to
satisfy the low temperature requirement of the seed. Results with soapwort (*Saponaria
officinalis*), shown in Table 6-7, are representative of this phenomenon that is characteristic for many temperate zone weeds.

**TABLE 6-7. Exposure of seeds of soapwort (*Saponaria offici-
nalis*) to low temperature for best germination.**

Temperature	Percent Germination
15°C constant	0
20°C constant	0
30°C constant	0
20°C 16 hours, 30°C 8 hours	12
5°C 1 week, then 20°C-30°C	97

Note: Based on 4 × 100 seeds on two blotters in petri dish with water and
darkness. Final counts at 21 days.
Source: Data from Steinbauer and Grigsby, 1957.

The low temperature requirement has a clear-cut value for survival of such
species in temperate climates. It precludes germination of seeds in the late summer or
early fall when the seedling might soon be killed by freezing temperatures. Thus, it is
a season-anticipating characteristic on the part of the seed. The physiological/bio-
chemical processes that operate to explain the effect are still unclear. At this point, we
can ask if there are ways to keep the seed from receiving the necessary accumulation
of near-freezing days.

Water. As discussed earlier, imbibition of water is the first step in the germination
process. It is needed continuously thereafter. Thus, water obviously has a critical effect on germination. The important question for us in terms of weed management is:
What can be done about the environmental factors that determine the dynamics of
water relations from the perspective of germinating weed seeds? Included among the
environmental factors are physical properties of the soil, soil compaction, soluble material, the rate of supply and loss of water from the soil, and the seed contact. In this
regard, we need to keep in mind that weeds evolved under uncertain soil moisture
conditions. Thus, we expect their seeds to be somewhat more tolerant of moisture extremes than those of desired plants, whose development and production have occurred because of our efforts to assure adequate moisture for germination and establishment.

Soil. The soil may have an effect on germination. This effect appears to be mainly
upon emergence rather than upon germination itself. Reduction in emergence is the
result of: (1) obstruction to penetration by the coleoptile/hypocotyl under compacted
soil conditions or (2) inability of the plant shoot to penetrate a surface crust when it
forms (Thill et al., 1979; Wiese and Davis, 1967). Indirectly, of course, the soil can
influence a number of factors that in turn affect germination and emergence. For ex-

ample, both water-holding capacity and temperature are markedly affected by soil properties. In addition, as pointed out in the section on depth of burial, the soil acts as a damper on temperature extremes. For the most part, soil may be viewed as having a modifying effect on germination and seedling emergence.

Nitrate. For some time, a weak nitrate solution used as a moistening agent has been known to improve germination of many crop seeds under laboratory conditions. It was also found that about half of 85 different weed species tested germinated better in dilute nitrate solution than in water alone (Steinbauer and Grigsby, 1957). It follows that nitrate might have some effect on germination under field conditions. Limited work suggests that it does. Germination of fresh yellow foxtail seed in soil was slightly stimulated by nitrate (Schimpf and Palmblad, 1980). Lambsquarters seed harvested from plots that received up to 336 kg ha^{-1} of nitrogen in the form of ammonium nitrate germinated 34% compared with 3% from unfertilized plots (Fawcett and Slife, 1978). In this regard, they found a close correlation between rates of nitrogen applied and the concentration of nitrate in the lambsquarters seed. They did not see any effect of nitrate in the soil on number of weeds that emerged in the field.

Interactions of factors. It is difficult to visualize field conditions under which only one of the factors discussed above would be involved as an influence on germination. That is to say, it is not possible to hold them at some constant level. Temperature changes with time of day and season, but so do light quality, intensity, and duration. Moisture level influences soil temperature, and so forth. Thus, interactions are the rule rather than the exception under field conditions, and the effects, therefore, are of interest to weed–crop ecology. The interrelationships between and among temperature, light, and nitrate (KNO$_3$) shown in Table 6-8 not only show that interactions do occur,

TABLE 6-8. Interactive effects of alternating temperature, nitrate, and light on germination of seeds of wormseed mustard.

Treatment[a]	Percent Germination at 5 Days
20°C constant, water[b]	0
20°C constant, 0.2% KNO$_3$[b]	0
30°C constant, water[b]	6
30°C constant, 0.2% KNO$_3$[b]	7
20°-30° C, water[b]	67
20°-30°C, 0.2% KNO$_3$[b]	98
20°-30°C, water, light[c]	94
20°-30°C, water, dark[d]	53
20°-30°C, 0.2% KNO$_3$, light[c]	95
20°-30°C, 0.2% KNO$_3$, dark[d]	86

[a]Based on 4 × 100 seeds on two blotters in petri dishes.
[b]Darkness except during transfers and counts.
[c]100 foot candle illumination from white fluorescent bulb.
[d]Complete darkness by wrapping petri dishes in aluminum foil, final counts only.
Source: Steinbauer and Grigsby, 1957.

but also that the effects are of a modifying rather than discrete nature. For example, alternating temperatures improved germination markedly over constant temperatures. Nitrate further improved germination, and light added an additional increment. In fact, studies involving only one factor, especially those conducted in the laboratory, may have relatively limited application to field conditions (Vincent and Roberts, 1977). This conclusion is based on their finding several first- and second-order interactions among light, alternating temperatures, nitrate, and chilling. The greatest effect occurred when all three factors were involved.

RESUMPTION OF GROWTH: PERENNATING PARTS

Periodicity

As we saw with seeds, perennating parts also commonly exhibit periodicity. In yellow nutsedge, for example, it was found that emergence usually occurs from early May to the end of June in Illinois (Stoller and Wax, 1973). In quackgrass, regrowth usually occurs in early spring and again in late summer and early fall (Johnson and Buchholtz, 1962). This observed periodicity is no doubt related to the interactions of temperature, light, moisture, growth regulators, and any inherent dormancy. Regardless of the explanation, the existence of periodicity needs to be taken into account in any approach to control or manage such weeds. It is important to know also that periodicity is a relative matter. That is, no sharp line demarks the initiation and termination of regrowth. Rather, there is a time when relatively more growth occurs than at other times.

Effects of Correlative Inhibition

Apical dominance effect.   The inhibiting effect of one bud on another, termed *correlative inhibition*, is widespread among vegetative regenerating parts of weeds. About 95% of the axillary buds on quackgrass rhizomes did not germinate unless the rhizome was fragmented (Johnson and Buchholtz, 1962). In this instance, there is an inhibiting effect of the apical meristem, called the *apical dominance* effect, on buds at the nodes along the rhizomes. Purple nutsedge (Smith and Fick, 1937), ironweed (*Vernonia baldwini*) (Davis and McCarty, 1966), bermudagrass (Moreira and Rosa, 1976), and johnsongrass (Hull, 1970; Beasley, 1970) are among other species in which apical dominance has been demonstrated.

The inhibiting effect of the apical meristem, although not usually that of complete inhibition of other buds, is nevertheless very pronounced. The magnitude of this effect in johnsongrass, which is fairly representative of the magnitude in bermudagrass and quackgrass, is shown in Table 6-9. Presence of the apical meristem allows only 5% to 7% germination of the axillary buds. Removing the apical meristem allowed a three- to fivefold increase in germination of the axillary buds.

TABLE 6-9. Influence of the apical meristem on the sequence of bud germination of three-node johnsongrass rhizome pieces.

	Bud Position		
	Apical	Middle	Basal
	% germination[a]		
Apical meristem present (74 pieces)	87.8	5.4	6.8
Apical meristem absent (67 pieces)	55.2	28.4	16.4

[a]The first bud to produce a shoot is regarded as the one germinating.
Source: Hull, 1970. Reproduced with permission of the Weed Science Society of America.

This phenomenon has been extensively studied in quackgrass. Results of this work have greatly expanded our understanding of the environmental effect, leading to these conclusions: (1) separating the rhizome from the parent increases the apical dominance effect (McIntyre, 1969); (2) dividing the rhizome into single-node sections reduces, but does not completely eliminate, the dominance effect (McIntyre, 1972; Hull, 1970; Moreira and Rosa, 1976); and (3) apical dominance is reduced by high nitrogen fertilization of the parent plant (Robinson, 1976; McIntyre, 1965, 1971, 1972; Leakey et al., 1978). Furthermore, nitrate in the germinating medium either reduces or offsets the apical dominance effect.

Shoot dominance effect. Extensive work with quackgrass also shows a correlative inhibition effect among shoots. This effect is for one or a few shoots to inhibit or completely prevent growth of other buds on a rhizome fragment. In Figure 6-8, the shoot originating from the number 3 node has restricted growth of buds at the 1, 2, and 4 nodes and almost completely prevented growth at the 5, 6, and 7 nodes.

Light, temperature, nitrogen, and growth regulators have been examined for their possible effects on shoot dominance. Light appears to have two distinct effects (Leakey, 1978b). One effect inhibits all buds for a short time (4 weeks) on some fragments. Far-red light appears to increase this incidence. The second effect releases shoot dominance in some fragments. Temperature effects are different for the two extremes (Leakey et al., 1978). Permanent dominance apparently is not established when temperatures are maintained at about 3°C. At 33°C, very little shoot dominance occurs. In the range from 13°C to 26°C, shoot dominance does occur. Nitrate in the soil may influence the dominance effect. This effect has been demonstrated in some studies in the laboratory (Leakey et al., 1978), although there are other studies in which it has not (Chancellor, 1974). It appears that the nitrogen content of the rhizome has a greater influence than nitrogen content exogenous to the rhizome.

Correlative inhibition theories. A detailed consideration of the mechanisms involved in correlative inhibition is not given here. Some understanding of such mechanisms is important, however, for possible use of the observed effects to prevent losses

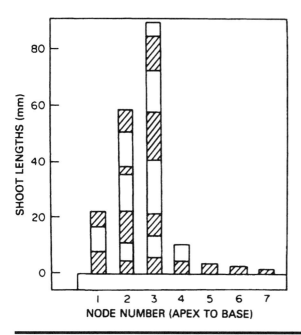

FIGURE 6-8. Shoot dominance effect in quackgrass among shoots on a rhizome fragment. Horizontal lines on bars represent lengths after 3, 5, 7, 10, 12, 14, 17, and 20 days, respectively.

Source: Chancellor, 1974. Reproduced with permission of Blackwell Scientific Publications, Ltd.

from weeds. We need to recognize at the outset that both apical and shoot dominance serve to protect some buds against destruction from an environmental catastrophe such as cultivation. Two theories exist to explain the observed result.

The nutritive control theory holds that the observed results are an expression of competition, especially for nitrogen (McIntyre, 1965, 1969). Indeed, it has been shown that there is a gradient of nitrogen in rhizomes in the direction of the apical meristem. Further, it is known that nitrogen enhances utilization of carbohydrate reserves. Thus, growth resumption by the apical bud or nearby buds under this theory is explained by nitrogen-enhanced mobilization of food materials in this area of the rhizome. Once growth of this bud or buds becomes established, there is a flow of food materials to the developing shoot or shoots.

The growth regulator theory holds that inhibition is the result of growth regulators that affect translocation of nutrients and inhibition of axillary buds. Many studies show that growth regulators have such a modifying effect in yellow nutsedge (Tumbleson and Kommedahl, 1961; Garg et al., 1967; Aleixo and Valio, 1976), in ironweed (Davis and McCarty, 1966), in quackgrass (Leakey, 1978a, b), and in johnsongrass (Beasley, 1970). On the basis of available evidence, both theories are likely to be involved, although the exact mechanism is yet to be determined. Not understanding the mechanism further compounds our problems of dealing with such weed species. Nevertheless, the overwhelming evidence showing many effects of factors that can be manipulated is encouraging relative to the potential these factors offer for preventing losses from such species over time.

Correlative inhibition has important implications for management of perennial weeds. New plants arise when axillary, apical, or adventitious buds grow into independent, rooted, upright shoots. Where correlative inhibition is present, the regenerating unit must be disconnected from related units for regeneration to occur. Such disconnection is a natural consequence of tillage. Tillage may thus increase the number of plants on a given area.

Age Spectrum

We have thus far considered resumption of growth without regard to the age spectrum of perennating parts. However, a moment's pause to reflect that growth is an on-going process in perennial weeds tells us that rhizomes or rootstocks of different ages are present at any given time. It is this entire population of rhizomes, rootstocks, tubers, and the like on a given land area at a particular point in time that must be dealt with in minimizing competition for our desired plants.

Once new growth begins in the spring, the relationship between new and old perennating parts in many species is likely to be like that shown for quackgrass in Figure 6-9. That is, sprouting on old parts steadily depletes their viability. This effect is offset by the development of new parts. The result is a shift in the age spectrum as the season progresses. The total in an established stand remains fairly constant except for a period in early June, when the total goes up because new rhizomes are being produced more rapidly than old ones are being depleted. The important point for weed management is that the population of reproductive parts contains parts of different physiological ages. As a result, they are expected to vary in their response to control attempts.

Growth Regulators

The understanding of growth regulator effects on resumption of growth of perennating parts is quite limited. That work which has been done suggests that endogenous growth regulators affect germination.

Leaching of 70-day-old purple nutsedge tubers markedly decreased growth of the buds compared with those tubers that were not leached (Aleixo and Valio, 1976). This effect lessened with age of the tubers until at 150 days of age, there was no difference in growth between leached and nonleached tubers. The authors suggest that this result was due to the leaching out of growth-promoting gibberellins; the tissues become less leachable with age. Under chromatographic analysis, they identified three types of growth-regulating substances, including gibberellin-like substances, cytokinin-like

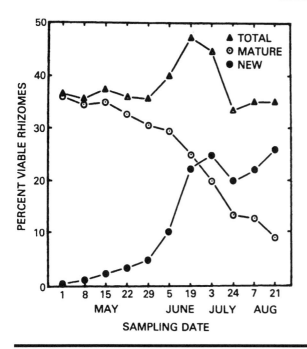

FIGURE 6-9. Shifts in the age spectrum of quackgrass rhizomes during the growing season.

Source: Johnson and Buchholtz, 1962. Reproduced with permission of the Weed Science Society of America.

substances, and an indolic substance. They also found that drying the tuber for up to 96 hours destroyed the gibberellin activity, did not affect the indole activity, and increased the cytokinin activity.

Two studies on yellow nutsedge yielded conflicting results. Tumbleson and Kommedahl (1961) found that leaching the tubers under tap water increased germination about 10 times, thus implicating water-soluble growth inhibitors. Tames and Vietz (1970), working with the same species, obtained the opposite results. Clearly, additional work is needed before any generalized conclusions can be drawn as to the precise effects of growth-regulating substances on resumption of growth. However, abscisic acid appears to be the primary hormone that causes bud dormancy and apparently is day-length sensitive (Salisbury and Ross, 1978). As more is learned, this area, too, may offer opportunities for manipulation in managing perennial weeds.

Environmental Effects on Resumption of Growth

As in the preceding section on seeds, factors affecting germination and emergence from perennating parts are first examined individually and then collectively to provide as much insight as possible about the factors involved. It must be emphasized, however, that under field conditions, no one factor is apt to be acting independently of at least one or more other factors. As with seeds, the physiological and biochemical processes involved in explaining the observed results are not presented except when they are needed to clarify the effect of a factor.

Depth of burial. An inverse relationship exists between depth of burial and emergence for most perennating parts, as can be seen for two species in Figures 6-10 and 6-11. For both types of perennating parts—tubers and rhizomes—emergence dropped off rather sharply with increases in depth. Planting depth also influences the rate of emergence. For example, in Figure 6-10, we see that 33 nutsedge shoots had emerged by the end of 1 week from tubers planted 2.5 cm deep, whereas it took 4 weeks for 32 shoots to emerge from 22.9 cm. Further, in quackgrass (Figure 6-11), the size of the rhizome section directly influenced the number of shoots, with emergence from the 15.2 cm (6 in.) depth for 30.5 cm (12 in.) segments being somewhat greater than emergence from 2.5 cm (1 in.) for 2.5 cm (1 in.) segments.

The above study of burial depth with yellow nutsedge was done in the greenhouse, where winter survival was not a factor. In the field, it has been shown for yellow nutsedge (Figure 6-12) that winterkill at shallower depths may cause emergence to

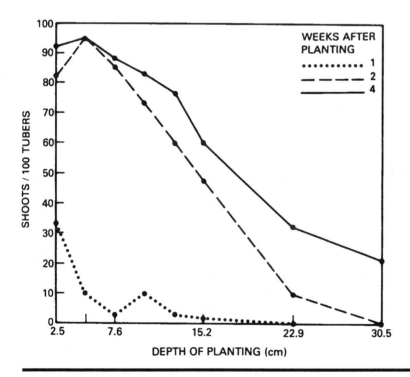

FIGURE 6-10. Effect of burial depth on emergence from yellow nutsedge tubers.

Source: Data from Tumbleson and Kommedahl, 1961.

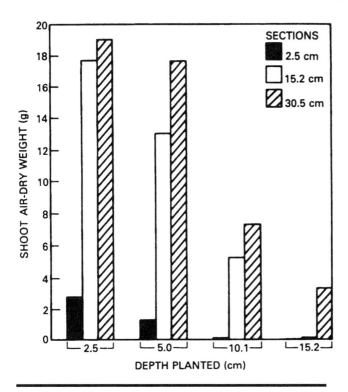

FIGURE 6-11. Effect of size of rhizome section on regrowth of quackgrass from different depths of burial in soil.

Source: Data from Vengris, 1962.

increase with depth of burial down to at least 10.2 cm. In this study, less than 4% of the tubers down to 7.6 cm survived the low soil temperatures experienced in the winter. The quackgrass rhizomes in Figure 6-11 were held over winter under conditions intended to simulate exposure in the field. As can be seen, they are quite tolerant of low temperatures.

In relative terms, it appears that emergence from perennating parts occurs from greater depths than from seed. Indeed, some perennials successfully emerge from perennating parts several meters deep. Offsetting this advantage of perennating parts is the relatively greater resistance to loss of viability at shallow depth on the part of seeds. The net result for preventive control is that depth of burial offers fewer opportunities for effectively dealing with perennating parts than it does with seed. This conclusion will be expanded upon in Chapter 15.

Temperature.   The limited work to measure temperature effects indicates that regrowth can occur over a fairly wide range, with a relatively narrow optimum range. The temperature effect is shown for two different species in Figure 6-13. The optimum ranges for the two species no doubt determine their distribution. Quackgrass (Figure 6-

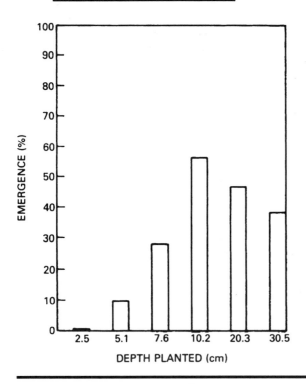

FIGURE 6-12. Effect of burial depth on overwinter survival of yellow nutsedge tubers.

Source: Data from Stoller and Wax, 1973.

13B), which is a common weed in the northern temperate regions, has a lower temperature optimum than ironweed (Figure 6-13A), which is more common somewhat south of quackgrass in northern latitudes.

It appears that buds, as do seeds, have a season-anticipating mechanism and that this mechanism is determined by photoperiod exposure of the parent plant. With respect to temperature, it would be helpful to know to what degree alternating temperatures affect regrowth since this variable can be influenced by time of tillage and depth of burial in the soil. Work with oxalis (*Oxalis cernua*), a perennial with scaly bulbs, showed regrowth to be better under temperature variation (Jordan and Day, 1967). Also, the fact that regrowth from perennating parts is better from shallow depths in the soil—in places where winterkill does not occur—is at least circumstantial evidence that alternating temperatures may be important since temperature variations are greater closer to the surface than deeper down in the profile.

Light. Light may be a factor in the resumption of growth, especially in the pattern of such growth. Light inhibited resumption of growth of axillary buds in rhizome segments of quackgrass but if buds sprouted in darkness, light prevented onset of correla-

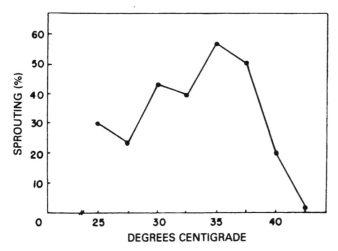

A. Ironweed sprouting

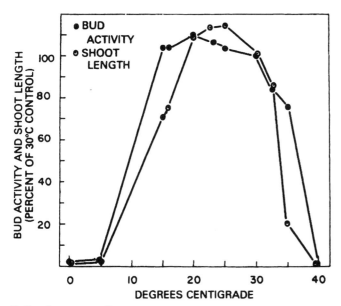

B. Quackgrass sprouting

FIGURE 6-13. Fairly wide temperature range for regrowth of rhizome buds, with relatively narrow optimum range.

Source: Part A from Davis and McCarty, 1966; Part B from Meyer and Buchholtz, 1963. Reproduced with permission of the Weed Society of America.

tive inhibition (Leakey et al., 1978). Thus, several shoots may be produced by segments of rhizomes if they emerge before dominance is established. Adventitious buds on roots of Canada thistle made no further growth unless exposed to light (Peterson, 1975). Wavelength influenced response of buds, thereby implicating the phytochrome system in regulation (Torrey, 1958). Table 6-10 provides evidence for this effect in purple nutsedge. Buds on tubers exposed to white, blue, and red (visible) light failed to initiate rhizomes, although roots and leaves were produced. Far-red light promoted rhizome initiation, as well as initiation of roots and leaves. This also suggests that the phytochrome system is involved.

TABLE 6-10.   Effect of light of different wavelengths on the growth pattern of tuber buds of purple nutsedge.

	Pattern of Growth		
Treatments	Root	Leaf	Rhizome
Darkness	+	−	+
White light	+	+	−
Blue light	+	+	−
Red light	+	+	−
Far-red light	+	+	+

Note: Results after one week of incubation.
+ = present.
− = not present.
Source: Data from Aleixo and Valio, 1976.

Phytochrome is present in vegetative reproductive parts of horticultural and ornamental plants and it concentrates in such areas of growth as buds and cambial regions (Koukkari and Hillman, 1966). Further, the total amount of phytochrome can be influenced by exposure to red light. Measurable levels of phytochrome in johnsongrass rhizomes, and the amount, depended upon location on the rhizome (Duke and Williams, 1977). In general, the phytochrome level decreased basipetally from the apex to the eighth node.

It can be speculated that some root buds may not resume growth unless exposed to light by cultivation. More needs to be learned about the effect of light on vegetative regrowth to know if it may offer an opportunity for preventive approaches through tillage and other production practices.

Nitrogen.   In early work, nitrogen fertilizer was found to markedly increase sprouting of buds of quackgrass rhizomes growing in such soil (Dexter, 1937). Further studies showed that soil nitrogen level was associated with concentration of nitrogen in the rhizomes. In addition, rhizomes with a high concentration of nitrogen sprouted more than did those with a low concentration (Dexter, 1942). More recently, nitrogen has also been found to encourage sprouting of lateral buds of johnsongrass. Sprouting was 74% for high nitrogen treatment versus 33% at the lowest level of nitrogen (Myers and Caso, 1976). It is not clear whether the effect observed is that of nitrogen or of no nitrogen. That is, is there a type of dormancy associated with no nitrogen, or is sprouting being promoted by available nitrogen under the high nitrogen levels? Irrespective of the mechanism involved, it seems clear that the nitrogen level in the soil in which the perennating part is produced influences resumption of growth from that part. This information should be usable in developing preventive approaches for such weeds.

Size.   Work on nutsedge, quackgrass, and johnsongrass shows that the size of a vegetative reproductive part influences germination and emergence. For both johnsongrass and quackgrass, the percent of buds that sprouted (germinated) decreased as the length of rhizome increased. This relationship for johnsongrass is shown in Figure 6-14.

A similar pattern was found for quackgrass (Vengris, 1962), although the decline with length was steeper. With quackgrass, 80% of the buds on 2.5 cm (1 in.) rhizome sections produced shoots. This percentage dropped to 31% for 5.2 cm (2 in.) rhizomes. The effect of rhizome length on germination of buds is significant for management and control since it is something that can obviously be affected by the type and amount of tillage. The explanation for this observed effect is likely to lie in the fact that as the length of rhizomes is reduced, the dominance relationship is reduced.

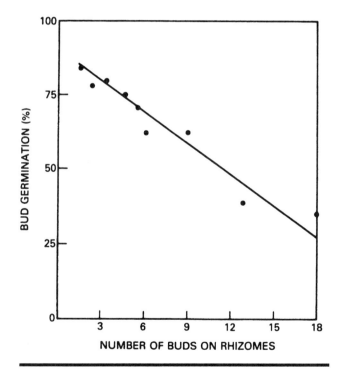

FIGURE 6-14. Inverse relationship of johnsongrass bud germination to the number of buds on the rhizome.

Source: McWhorter, 1972. Reproduced with permission of the Weed Society of America.

Work with nutsedge tubers, shown in Table 6-11, shows that size of tuber does not affect germination as such but does affect emergence. The larger tubers are more successful in producing shoots that emerge. Chapter 5 pointed out that tubers produced by plants in shade are smaller than those on unshaded plants (Patterson, 1982). The combination of shading and deep burial could reduce an infestation by preventing emergence from the small tubers in the population.

TABLE 6-11.   Germination versus emergence of yellow nutsedge tubers as affected by size of tuber.

Tuber Fresh Weight (mg/tuber)	Germination[a] (%)	Tuber Dry Weight (mg/tuber)	Emergence[b] (%)
294	92	120	44.6
201	91	50	29.7
128	83		
61	86		

[a]Adapted from Stoller et al., 1972.
[b]Adapted from Stoller and Wax, 1973.

REFERENCES

Aamisepp, A. 1966. Herbicide effects on plants from seeds from treated plants. Vaxtodling 22:1-147.

Aleixo, M.D., and I.F. Valio, II. 1976. Effect of light, temperature, and endogenous growth regulators on the growth of *Cyperus rotundus* tubers. Zeitschrift fur Pflangenphipiologic 80:336-347.

Beasley, C.A. 1970. Development of johnsongrass rhizomes. Weed Sci. 18:218-222.

Biswas, P.K., P.D. Bell, J.L. Crayton, and K.B. Paul. 1975. Germination behavior of Florida pusley seeds. I. Effects of storage, light, temperature, and planting depths on germination. Weed Sci. 23:400-403.

Chancellor, R.J. 1974. The development of dominance amongst shoots arising from fragments of *Agropyron repens* rhizomes. Weed Res. 14:29-38.

Chu, C.C., R.D. Sweet, and J.L. Ozbun. 1978. Some germination characteristics in common lambsquarters (*Chenopodium album*). Weed Sci. 26:255-258.

Clutter, M.E., ed. 1978. Dormancy and development arrest: Experimental analysis in plants and animals. New York: Academic Press.

Cole, A.W., and G.E. Coats. 1973. Tall morning glory germination response to herbicides and temperature. Weed Sci. 21:443-446.

Davis, F.S., and M.K. McCarty. 1966. Effect of several factors on the expression of dormancy in Western ironweed. Weeds 14:62-69.

deJong, E., and H.J.V. Schappert. 1972. Calculation of soil respiration and activity from CO_2 profiles in the soil. Soil Sci. 113:328-333.

Dexter, S.T. 1937. The drought resistance of quackgrass under various degrees of fertilization with nitrogen. Agron. J. 29:568-576.

——— 1942. Seasonal variations in drought resistance of exposed rhizomes of quackgrass. Agron. J. 29:568-576.

Duke, S.O., and R.D. Williams. 1977. Phytochrome distribution in johnsongrass rhizomes. Weed Sci. 25:229-232.

Egley, G.J. 1974. Dormancy variations in common purslane seeds. Weed Sci. 22:535-540.

Fawcett, R.S., and F.W. Slife. 1978. Effects of field applications of nitrate on weed seed germination and dormancy. Weed Sci. 26:594-596.

Fenner, M. 1980a. The inhibition of germination of *Bidens pilosa* seeds by leaf canopy shade in some natural vegetation types. New Phyto. 84:95-101.

———. 1980b. Germination tests on 32 East African weed species. Weed Res. 20:135-138.

Friesen, G., and L.H. Shebeski. 1961. The influence of temperature on the germination of wild oat seed. Weeds 9:634-638.

Garg, D.K., L.E. Bendixen, and S.R. Anderson. 1967. Rhizome differentiation in yellow nutsedge. Weeds 15:124-128.

Gorski, T., K. Gorska, and J. Nowicki. 1977. Germination of seeds of various species under leaf canopy. Flora, Morphologie, Geobotanik, Oekophysiologie 166:249-259.

Gutterman, Y. 1978. Germinability of seeds as a function of the maternal environment. Acta Horticulturae 83:41-55.

Harper, J.L. 1977. Population biology of plants,. New York: Academic Press.

Hart, J.W., and A.M.M. Berrie. 1966. The germination of *Avena fatua* under different gaseous environments. Physiol. Plant 19:1020-1025.

Hill, T.A. 1977. The biology of weeds. London: Edward Arnold.

Hull, R.J. 1970. Germination control of johnsongrass rhizome buds. Weed Sci. 18:118-121.

Jeffrey, L.S., and J.D. Nalewaja. 1970. Studies of achene dormancy in fumitory. Weed Sci. 18:345-348.

Johnson, B.G., and K.P. Buchholtz. 1962. The natural dormancy of vegetative buds on the rhizomes of quackgrass. Weeds 10:53-57.

Jordan, L.S., and B.E. Day. 1967. Effect of temperature on growth of *Oxalis cernua* Thumb. Weeds 15:285.

Kidd, R. 1914. The controlling influence of carbon dioxide in the maturation, dormancy, and germination of seeds, Part II. Proc. R. Soc. London, Ser. 13.87:609-625.

Kidd, R., and C. West. 1917. The controlling influence of carbon dioxide. Ann. Bot. 31:457-487.

Koller, D. 1972. Environmental control of seed germination. In T.T. Kozlowski, ed., Seed biology, vol. II, pp. 1-101. New York: Academic Press.

Koukkari, W.L., and W.S. Hillman. 1966. Phytochrome levels assayed by in vivo spectrophotometry in modified underground stems and storage roots. Physiol. Plant 19:973-1078.

Leakey, R.R.B. 1978a. Regeneration from rhizome fragments of *Agropyron repens* (L.) Beauv., IV. Effects of light on bud dormancy and development of dominance amongst shoots on multinode fragments. Ann. Bot. 42:205-212.

———. 1978b. Regeneration from rhizome fragments of *Agropyron repens,* I. The seasonality of shoot growth and rhizome reserves in single-node fragments. Ann. Appl Biol 87:423-431.

Leakey, R.R.B., R.T. Chancellor, and D. Vince-Prue. 1978. Regeneration from rhizome fragments of *Agropyron repens*, III. Effects of N and temperature on the development of dominance amongst shoots on multinode fragments. Ann. Bot. 42:197-294.

McIntyre, G.I. 1965. Some effects of the nitrogen supply on the growth and development of *Agropyron repens*. Weed Res. 5:1-12.

———. 1969. Apical dominance in the rhizome of *Agropyron repens*: Evidence of competition for carbohydrate as a factor in the mechanism of inhibition. Can. J. Bot. 47:1189-1197.

————. 1971. Apical dominance in the rhizome of *Agropyron repens* in isolated rhizomes. Can. J. Bot. 49:99-109.

————. 1972. Studies on bud development in the rhizomes of *Agropyron repens,* II. The effect of nitrogen supply. Can J. Bot. 50:393-401.

McWhorter, C.G. 1972. Factors affecting johnsongrass rhizome production and germination. Weed Sci. 20:41-45.

Meyer, R.E., and K.P. Buchholtz. 1963. Effect of temperature, carbon dioxide, and oxygen levels on quackgrass rhizome buds. Weeds 11:1-7.

Moreira, I., and M.L. Rosa. 1976. The effect of nodal position on the sprouting of buds on *Cynodon dactylon.* In II Simposio Nacional de Herbalogia, Oeircis, vol. 1, pp. 37-43. Lisbon, Portugal.

Myers, E.J., and O.H. Caso. 1976. The effect of nitrogen supply on the growth of *Sorghum halepense.* Malezas 5:3-12.

Patterson, D.T. 1982. Shading response of purple and yellow nutsedge (*Cyperus rotundus* and *C. esculentus*). Weed Sci. 30:25-30.

Peterson, R.L. 1975. The initiation and development of root buds. In The development and function of roots. J.G. Torrey and D.T. Clarkson, eds., p. 125. Academic Press: New York.

Richardson, S.G. 1979. Factors influencing the development of primary dormancy in wild oat seeds. Can. J. Plant Sci. 59:777-784.

Robinson, E.L. 1976. Yield and height of cotton as affected by weed density and nitrogen level. Weed Sci. 24:40-42.

Salisbury, F.B., and C.W. Ross. 1978. Plant physiology, 2nd ed. Belmont, Calif.: Wadsworth.

Schimpf, D.J., and I.G. Palmblad. 1980. Germination response of weed seeds to soil nitrate and ammonium with and without simulated overwintering. Weed Sci. 28:190-193.

Semenza, R.J., J.A. Young, and R.A. Evans. 1978. Influence of light and temperature on the germination and seedbed ecology of common mullein (*Verbascum thapsus*). Weed Sci. 26:577-581.

Silvertown, J. 1980. Leaf-canopy-induced seed dormancy in a grassland flora. New Phyto. 85:109-118.

Simpson, G.M. 1978. Metabolic rate of dormancy in seeds—A case history of the wild oat. In M.E. Clutter, ed., Dormancy and developmental arrest: Experimental analysis in plants and animals. New York: Academic Press.

Smith, E.V., and G.L. Fick. 1937. Nutgrass eradication studies: 1. Relation of the life history of nutgrass (*Cyperus rotundus* L.), to possible methods of control. J. Am. Soc. Agron. 29:1007-1013.

Steinbauer, G.P., and B. Grigsby. 1957. Interaction of temperature, light, and moistening agent in the germination of weed seeds. Weeds 5:175-182.

Stoller, E.W. 1974. Dormancy changes and fate of some annual weed seeds in the soil. Weed Sci. 22:151-155.

Stoller, E.W., and L.M. Wax. 1973. Yellow nutsedge shoot emergence and tuber longevity. Weed Sci. 21:76-81.

Stoller, E.W., D.P. Neva, and V.M. Bhan. 1972. Yellow nutsedge tuber germination and seedling development. Weed Sci. 20:93-97.

Tames, R.S., and E. Vietz. 1970. Estudios sobre la brotadura de tubereulos de *Cyperus sculentus* Tem. Var. Aureus Richt, I. Accion de factores fisicos y quimicos. Anales Edafologia Agrobiology 29:775-781.

Taylor, J.S., and G.M. Simpson. 1980. Endogenous hormones in after-ripening wild oat seed. Can. J. Bot. 58:1016-1024.

Taylorson, R.B., and H.A. Borthwick. 1969. Light filtration by foliar canopies: Significance for light-controlled weed seed germination. Weed Sci. 17:48-51.

Thill, D.C., R.D. Schirman, and A.P. Appleby. 1979. Influence of soil moisture, temperature, and compaction on the germination and emergence of downy brome (*Bromus tectorum*). Weed Sci. 27:625-630.

Toole, E.H., V.K. Toole, H.A. Borthwick, and S.B. Hendricks. 1955. Interaction of temperatures and light in germination of seeds. Plant Physiol. 30:473-479.

Torrey, J.G. 1958. Endogenous bud and root formation by isolated roots of *Convolvulus* grown in vitro. Plant Physiol. 33:228.

Tumbleson, M.E., and T. Kommedahl. 1961. Reproductive potential of *Cyperus esculentus* by tubers. Weeds 9:646-653.

Vengris, J. 1962. The effect of rhizome length and depth of planting on the mechanical and chemical control of quackgrass. Weeds 10:71-74.

Villiers, T.A. 1972. Seed dormancy. In T.T. Kozlowski, ed., Seed biology, vol. II, pp. 219-281. New York: Academic Press.

Vincent, E.M., and E.H. Roberts. 1977. The interaction of light, nitrate, and alternating temperatures in promoting germination of dormant seeds of common weed species. Seed Sci. and Tech. 5:659-670.

Wesson, G., and P.F. Wareing. 1969a. The role of light in the germination of naturally occurring populations of buried weed seeds. J. Exp. Bot. 20:402-413.

———. 1969b. The induction of light sensitivity in weed seeds by burial. J. Exp. Bot. 20:414-425.

Wiese, A.F., and R.G. Davis. 1967. Weed emergence from two soils at various moistures, temperatures, and depths. Weeds 15:118-121.

7

Nature of
Weed Competition

CONCEPTS TO BE UNDERSTOOD

1. The combined dry matter production on a given area tends to be constant irrespective of composition; therefore, it is impractical to try and provide enough of a competed-for growth factor to meet the needs of both the desired plant and the weeds.
2. The leaf is the site of competition for light.
3. Light can neither be transferred nor stored within the plant; therefore, plant height is of paramount importance in determining competition for light.
4. Most weeds, especially annuals, are very intolerant of shade.
5. Competition for CO_2 is not likely to occur under field conditions.
6. The root is the site of competition for water and mineral nutrients.
7. Competition for a soil factor cannot occur until the root depletion zones of neighboring plants overlap.
8. Relative competitiveness of weeds and of desired plants for soil factors is largely determined by the soil volume occupied by the roots of each.
9. Plants differ greatly in the moisture extraction profiles of their roots.
10. Nitrogen moves mainly as the result of mass flow with soil water to the plant's roots; movement may be quite rapid, with the resulting depletion zone being roughly comparable to that for water.
11. Potassium, phosphorus, and some other minerals move mainly as the result of diffusion, which is a slow process.
12. Competition is seldom restricted to a single growth factor because of the interrelationships between competition and plant growth form and rate.
13. Weeds commonly take up added nutrients (fertilizer) more rapidly and in larger quantities than do desired plants.

169

14. A relatively scarce supply of a growth factor encourages earlier onset of competition for that factor.

COMPETITION PARAMETERS

"Competition occurs when each of two or more organisms seeks the measure it wants of any particular factor or things and when the immediate supply of the factor or things is below the combined demand of the organisms." This definition of competition by Donald (1963) represents a condensation of earlier definitions and still fits weed–desired plant competition relationships. In particular, it recognizes that both the desired plant and the weed are involved—not just the weed—in a competitive relationship. It should be noted that the demands need not necessarily be at the same time. A dense stand of weeds present before the desired plant is seeded could utilize soil water and nutrients resulting in an inadequate supply for the desired plant when seeded.

By restricting the definition to competition for some factor that is in limited supply, the definition distinguishes competition from interference; thus, allelopathy, which is discussed in detail in Chapter 8, is not considered to be a factor in competition since it adds "something" to the environment. This definition also helps to provide a proper perspective relative to space. Except in unusual circumstances, such as might exist with root crops, competition is not for space but rather for the things that space contains— that is, for one or more of the *five* growth factors: nutrients, water, light, carbon dioxide, and oxygen.

Inspection of much of the literature on weed research indicates that the concept of competition is often misused relative to weed–desired plant relationships. Competition is frequently used to identify the period when weeds are present with the desired plant, even though there may be no evidence that "each ... seeks the measure of ... things below the combined demand." Such usage of the term competition can be misleading relative to the ecological interrelationships at work in agroecosystems. In effect, it tends to focus only on the weed's effects on the desired plant, although as we have already learned, there are important effects of desired plants upon the weed. As we shall see later, we can draw upon desired plant's effects on the weed in building management systems to minimize effects from weeds. Thus, the term *competition* as used in this book is restricted to interactions in which some factor is in insufficient supply to meet the needs of both the weed and the desired plant. This usage implies removal of something from the environment, thus excluding allelopathy.

COMMUNITY YIELD

At the outset, it is important to consider applicability of the ecological principles of limiting factors to competition. Chapter 3 explained that there is a limit to how many individuals can occupy a given area. Does this have implications for the total yield of the crop plus weeds that can be produced on a given area? The answer is important to weed management. If the combined (community) yield exceeds the yield of the weed-free crop, adding the competed-for factor is an alternative to removing the weed. Of course, this may not be practical. If the combined yield does not exceed the yield of the weed-free crop, then the weed must be removed to obtain maximum crop yield.

Combinations of crops have been used occasionally with field and horticultural crops for many years. The practice is quite common with forages. Results of extensive studies of community yield of such mixtures provide principles to help answer the question of weed–crop community yield. Trenbath (1976) summarized and interpreted results of many of these studies. Information on yield related to a single growth factor is discussed first to provide a basis for better understanding the discussion of density and community yield relationships that follows. At this point, we need only recognize that competition frequently occurs for light, water, and nutrients in mixtures of species.

Uptake of Growth Factors by Mixtures of Species

By definition, *weeds* are interfering associates with desired plants. Thus, the proportionate uptake of available resources, or growth factors, determines the extent of weed competition. In fact, as we see later, a reduction in crop yield by weeds is a result of the weeds obtaining a disproportionate share of the available resources. Here, a review of the theory involved in competitive uptake by a mixture of species helps to clarify why the community yield usually does not exceed unity.

Figure 7-1 shows the relationship between biomass and uptake of the competed-for factor Q. In this figure, it is assumed that A uses Q more efficiently than does B. When grown together, therefore, per plant uptake by A (resulting in Y_A yield) increases over the uptake when grown alone (resulting in Y_{AA} yield) because there are fewer A plants competing. The per plant uptake by B (resulting in Y_B yield) decreases over the uptake when grown alone (resulting in Y_{BB} yield) because B is competing with the more efficient user. The difference in uptake is F. Thus, the per plant yields in the intercrop are

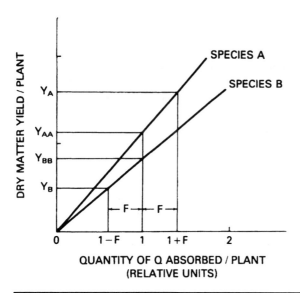

FIGURE 7-1. Model of the biomass yields of the two components in a 1:1 intercrop competing for a growth factor Q in limited supply.

Source: Trenbath, 1976. Reproduced from Multiple cropping, ASA Special Publication no. 27, 1976, with permission of the American Society of Agronomy, Crop Science Society of America, and Soil Science Society of America.

$$Y_A = Y_{AA}(\text{i.e., } 1 + F)$$

$$Y_B = Y_{BB}\ (\text{i.e., } 1 - F)$$

The relative yield total (RYT) in this 1:1 intercrop can be calculated by an appropriate formula (Trenbath, 1976) that reduces to

$$RYT = \tfrac{1}{2}(1 + F) + (1 - F) = 1$$

Thus, in this model, the RYT is unity. That is, the combined per plant yield of the two grown together does not exceed that of the two grown separately.

Density Relationships

Our interest in a weed–crop situation is in the yield for a given land area. Trenbath has shown that the relative yield total is equivalent to a land equivalent ratio of 1.0 where the total density is constant. Land equivalent ratio (LER) is a mathematical way of expressing productivity in terms of a land unit. Thus, an LER greater than 1 indicates that an intercrop produces more on a given land area than either crop alone.

When LER is measured without total density held constant, the value can be expected to vary, shown by examination of Figure 7-2. If the D1 density is well below the optimum density for each of two species, a 1:1 mixture of the two gives a total yield of $2 \times Y_2$ since the yields of each will be Y_2, assuming equal competitive ability. Thus, the area that produced Y_1 in the sole crop produces much more in the intercrop. Even if the species differ in competitive ability, the total yield will be greater than unity. However, under D3 density, which is assumed to be close to the optimum for each species, the total yield of the intercrop ($2 \times Y_4$) from the graph can be seen to be very close to the yield of the sole crop (Y_3). Thus, the area yield of the intercrop is quite comparable to that of the sole crop.

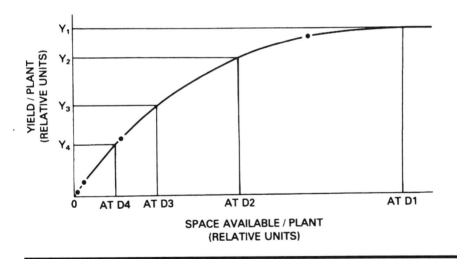

FIGURE 7-2. Typical response of per plant biomass to density.

Source: Trenbath, 1976. Reproduced from Multiple cropping, ASA Special Publication no. 27, 1976, with permission of the American Society of Agronomy, Crop Science Society of America, and Soil Science Society of America.

From this discussion of model and mathematical treatment of density–yield relationships, it can be concluded that if the density is optimum, combined yield will not exceed unity where there is competition for a growth factor in deficient supply. Data from a large number of actual crop mixtures, shown in Figure 7-3, support this conclusion. The preponderance of the LERs is close to 1.0. In other words, the usual situation is that species in a mixture compete for the resources available.

Annidation.   The complementary use of resources is called *annidation.* There are many instances where mixtures outyield the individual species grown alone. The combination of grasses and legumes, widely used in forage production, is one example. The Inca Indians in South America apparently intercropped their corn and beans, a method still being used in that area. In Missouri, young walnut plantations are sometimes intercropped with forages. In these and other instances, increased total production occurs

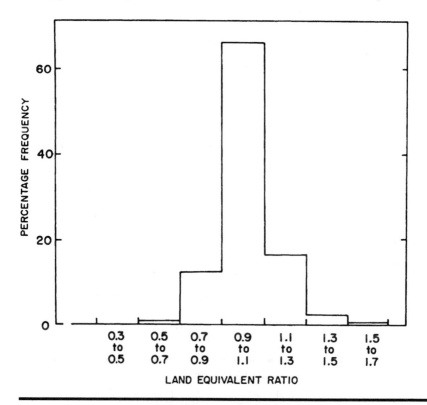

FIGURE 7-3. Land equivalent ratios calculated for either two species of legumes or two species of nonlegumes from experiments where intercrops and sole crops were grown at the same density.

Source: Trenbath, 1976. Reproduced from Multiple cropping, ASA Special Publication no. 27, 1976, with permission of the American Society of Agronomy, Crop Science Society of America, and Soil Science Society of America.

in the absence or avoidance of competition. That is, the use of resources is complementary not competitive.

The general principles of annidation were discussed in Chapter 3. Annidation may occur for space, nutrients, and time (Trenbath, 1976). Annidation in space may involve both the leaf canopy and root structures. Thus, species with different light requirements and light tolerances growing together are not competing but, rather, are making more effective use of the total light available than either species does alone.

Harper (1977) describes an experiment with *Avena* species that dramatically illustrates the annidation in space with respect to roots. In a study in which *Avena strigosa, Avena fatua, Avena sativa,* and *Avena ludoviciana* were grown in pure stands and in mixtures, RYT values much larger than 1.0 were found when *A. fatua* and *A. strigosa* were grown together on deep soil. On shallow soil, the RYT values approached 1.0. Later studies of rooting depth showed that in the deep soils, the roots of *A. strigosa* developed mainly in the upper layers of the soil profile, whereas *A. fatua* contributed most to roots deeper in the profile.

There are many examples of annidation in time in weed–crop relationships. One example, found in the central United States, is infestation of wild garlic in the early fall following corn and soybeans, with at least partial completion of the weeds' life cycle prior to planting the crop the next spring.

An example of annidation in time, which is of increasing interest in the southern part of the Corn Belt and in the Cotton Belt, is the interplanting of soybeans in wheat prior to wheat harvest. Competition for resources is avoided since the soybean plants do not reach a size to be competitive with the wheat while the wheat still needs such resources.

Annidation with respect to nutrients is exemplified by grass–legume mixtures. Competition for nitrogen is avoided because the legume has the ability to fix its own. The presence of the legume black medic (*Medicago lupulina*) in pastures is an example of such weed–crop annidation.

Effect of changing weed numbers on combined density. The separate and combined yields of weeds and crops have been measured in many weed studies. For the most part, these measurements were taken in studies designed to evaluate only the effects of the weeds on the crop. The evaluation is commonly done either by: (1) adding increasing increments of weeds or (2) thinning a dense weed stand to desired levels of decreasing weed density per unit area. The study from which data for Figure 2-1 in Chapter 2 were obtained was of the first type. Similarly, herbicide effectiveness is commonly measured in studies where weights or numbers of weeds and crops are recorded for different herbicides and rates.

In all such studies, the combined density changes for the various treatments. The changes may be very large. For example, recall from Figure 2-1 that the combined density of cotton and of tall morning-glory at 32 plants per 15 meter row was about double that of cotton by itself. Figure 7-3 and its discussion showed that total yield may be greatly influenced by such large changes in combined density. Therefore, caution must be followed in interpreting community yields of weed–crop research. Such studies tell us the effect of weeds on crop yield but do not explain the effect. Thus, they are of little use in identifying the nature of competition since density is confounded with competition for growth factors.

Soybeans. Measurements of community yield in weed research have produced a preponderance of results approaching unity. The results with soybeans deserve special comment. Frequently, combined yields of weeds plus soybeans exceed unity (Moulani et al., 1964; Knake and Slife, 1969), although sometimes they do not (Knake and Slife, 1962). Even though the relationship is not consistent, combined yields exceeding unity have been observed frequently enough with soybeans and with some other weed–crop situations to warrant an attempted explanation.

In view of what is known about community yield of two crops growing together, it seems quite likely that the occasional greater community yield of soybeans plus weeds, as well as similar occurrences for other weed–crop mixtures, is due to annidation for space. That is, the crop stand may be below its optimum density. As discussed in the preceding section, if the stand is below the optimum density, the total yield is expected to exceed unity. This situation is not improbable with soybeans since they were planted in wide rows in most of the studies cited. Also, if a legume is present, competition may not occur for nitrogen because the legume can fix its own. Thus, occasional greater combined yield of crops plus weeds than of the crop alone may not necessarily detract from the conclusion that a maximum, or plateau, in combined yield of crops plus weeds can be expected for a given area.

As we turn to a consideration of competition between weeds and desired plants, we should keep firmly in mind that annidation may also be involved in such situations. Further, although weed density is important in determining crop yield loss, if density is not held constant, care must be exercised in using results to understand competition.

Competition Parameter

If the entire crop growing season is taken into account, there are likely to be few instances under field conditions where competition is for only one factor, even though individual factors may be separately involved at certain times and under specific circumstances. The reason is that the relationship between the competing plants, and with the environment, is a dynamic one—not static. As discussed in Chapter 3, weeds are active participants in the ecological game. Thus, competition for one factor can be expected to alter the growth form of competing plants, thereby changing their ability to sample the environment for other growth factors. Further, these effects are complicated by density relationships that change with time and by the impact of temperature.

Although not itself a growth factor, temperature may have a pronounced effect on competition through its effect on plant growth rate and plant growth form. Soil temperatures may also affect the availability of nutrients, especially nitrogen. Of particular significance for weed–crop competition is the more rapid recovery of some weeds from cold temperatures. For example, it was found that velvetleaf and spurred anoda recovered in growth more completely than did cotton after all three were exposed for 3 days to cold temperatures ($17°C$ day/$13°C$ night) (Patterson and Flint, 1979). We can speculate that other weeds may also recover more rapidly from chilling than other crops besides cotton. This speculation is based on weeds' evolvement under stress conditions and consequent production of general-purpose genotypes, discussed in Chapter 3. Because occasional periods of below-normal temperatures are not uncommon during the early growing season for annual crops, it can be seen that this modifying effect of temperature on competition may be fairly widespread.

Even though there are only a few resources, or growth factors, that plants compete for, many interacting forces serve to make competition a complex phenomenon. The dynamics and complexities of weed–crop competition are illustrated in Figure 7-4. Characteristics of both weeds and crops are involved with weather, soil treatment, and other pests acting as modifiers. Even though competition is indeed complex, examination of influences on uptake of individual growth factors yields concepts and principles that can be applied in weed management. To do this, growth factors need to be examined both independently and in terms of their interactions. First, we consider competition that occurs belowground and, second, that which occurs aboveground. Only the concepts and principles involved in competition are considered.

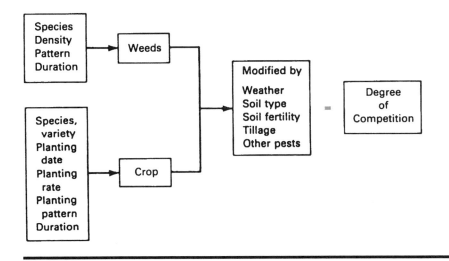

FIGURE 7-4. Interacting forces in weed–crop competition.
Source: After Bleasdale, 1960.

BELOWGROUND COMPETITION

Roots take up nutrients, water, and oxygen. The rate of uptake may be influenced by other factors, such as temperature, inherent soil properties, and root growth form, but competition can develop only for the three growth factors. Further competition between plants occurs only when there is an actual overlap of their depletion zones for one or more of these soil-supplied growth factors. Each of these three factors obtained from the soil is mobile—although to different degrees—both within the soil and within the plant itself. Indeed, in a sense, a plant can create a supply of one of the soil-obtained growth factors by creating a concentration gradient along which nutrients move toward the root. This concept is described diagrammatically in Figure 7-5. As can be seen in the figure, evaporation is the driving force for establishing a water gradient that, in turn, serves as the carrier for readily soluble nutrients and oxygen for which gradients are then established.

In considering the nature of competition for the soil-supplied factors, we need to remember that the surface area of roots may be many times that of tops. Thus, the soil can soon become crowded, even though competition cannot occur until there is an actual overlap of the depletion zones.

It is difficult to directly demonstrate competition between plants for individual soil growth factors, although such competition obviously occurs. This is because competition for one factor commonly alters growth form leading to competition for another factor. Plant characteristics that can affect competitiveness for such growth factors and

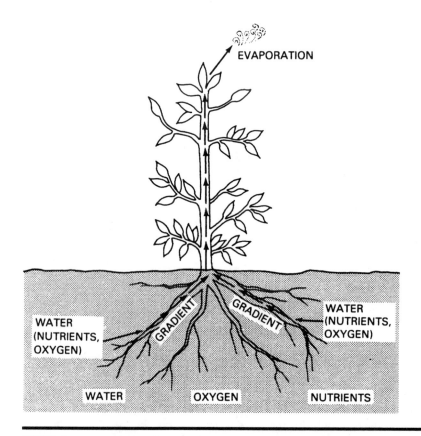

FIGURE 7-5. Gradient for movement of soil-supplied growth factors created by water evaporation.

of the general pattern of their use will be examined to provide perspective of the processes and possibilities involved. Characteristics that could impart competitiveness for soil factors to a plant include: (1) early and fast root penetration of the soil, (2) high root density, (3) high root:shoot ratio, (4) high root length:root weight, (5) high proportion of root system actively growing, (6) long root hairs, and (7) high uptake potential for the nutrient (Trenbath, 1976). The uptake of a nutrient beyond its efficient utilization may give a competitive advantage to the plant possessing that capability. It may also explain the competitiveness of some weeds since research has shown that some weeds may accumulate a nutrient well beyond their apparent need. For example, pigweed had a total P content 7 times that of snap bean (Chambers and Hold, 1965). Such uptake on the part of a weed may be especially significant in view of the fact that successful competition for one factor may well lead to successful competition for others.

Competition for Water

As depicted in Figure 7-5, water moves from the soil to the root, then into the plant, where 1% to 3% is used in the process of photosynthesis. The remaining 97% to 99% of the water entering the plant on average is lost through evaporation. Evaporation, thus, is a principle driving force in the establishment of the water gradient. Both time and distance are important aspects in competition for water. Moisture may begin to move toward a root from several centimeters away in a matter of days. Trenbath (1976) refers to movement from 12 cm away in 6 days under laboratory conditions with a calculated depletion zone extending outward to 25 cm.

Root volume. The degree of competition for water between a desired plant and a weed is determined primarily by the relative root volume occupied by each. Moisture extraction profiles provide a good indicator of root volume. Species differ in both the depth and breadth of moisture extraction. Profiles of individual plants of several different species are shown in Figure 7-6. As can be seen, there is more than a twofold difference between kochia 1.9 m^2 (10 ft^2 in cross section) and cocklebur 4.1 m^2 (44 ft^2 cross section), suggesting that the density of weed infestation necessary to cause a given degree of competition for moisture varies from weed to weed.

Figure 7-6 provides a perception of both the lateral and vertical distribution as factors in determining the moisture extraction capacities. Depending upon the types of root systems of the desired plant and the weed and the supply or distribution of water in the soil profile, either extensive lateral or extensive vertical distribution can impart a competitive advantage for one species over the other. For example, a kochia plant in the crop row is probably more competitive for water than a plant in between crop rows because its lateral root distribution is relatively narrow. By contrast, cocklebur, because of its extensive lateral root distribution, might be competitive even if growing only in between rows. Russian thistle (*Salsola kali* L. var. *Tenuifolia tausch*) might be relatively more competitive for moisture in deep soils than other weeds because of its deeper vertical root distribution.

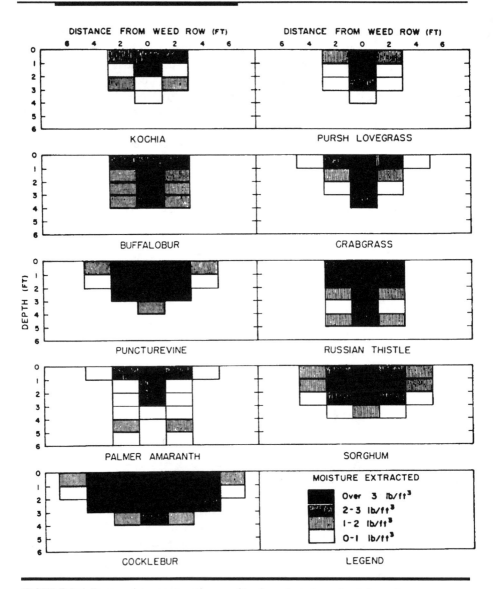

FIGURE 7-6. Root moisture extraction profiles for selected weeds and sorghum.

Source: Davis et al., 1967. Reprinted from Agron. J. vol. 59, 1967, p. 556, with permission of the American Society of Agronomy.

The above emphasis on evaporation and root volume does not mean that there are no other important aspects. Differences among species in their inherent efficiency of water use are well known. Also, as already mentioned, C_4 plants are relatively more efficient in water utilization than C_3 plants. Studies involving weeds in crops have shown that some weeds may be able to produce more dry matter per unit of water than other weeds and than some crops. Nevertheless, because of the 97% to 99% pass-through of water, differences in water use efficiency tend to be overshadowed by the root volume–evaporation effect.

When competition occurs.   More than one-half of the maximum amount of water needed by soybeans in Missouri is needed by 7 weeks after emergence (Figure 7-7). By that time soybean yields may be reduced about 50% by weed presence.

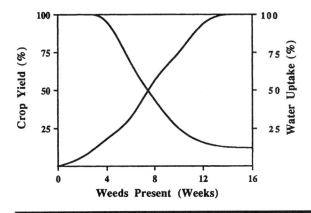

FIGURE 7-7. Soybean response to weed presence (*left axis*) and relative water uptake by the crop (*right axis*).

Source: Modified from Aldrich, 1987.

The combination of water usage, continual evaporative loss, and potential for a soil depletion zone extending beyond the roots suggest that competition for water may frequently occur. There is a time immediately after emergence when the limited extent of the root systems makes competition improbable. However, in view of the rather intimate space relationships often encountered between weeds and crops in the field, overlap of depletion zones for water, if in limited supply, could occur relatively early in the crop's life cycle.

Relative competitive ability of weeds and crops.   Species differ in response to competition for water. This situation is shown for 10 species in Figure 7-8. The species were grown together in the greenhouse. The growth (dry matter produced) of each under the three moisture regimes indicates that the species responded differently. The soil for the wet level was maintained at about field capacity; soil for the medium level was brought to about one-half field capacity when plants began to wilt; and for the dry level, plants were allowed to wilt severely, then watered to slightly above wilting. In general, species that produced the most growth under wet conditions—corn, barnyardgrass, and cocklebur—were hurt most by competition under dry conditions. Those

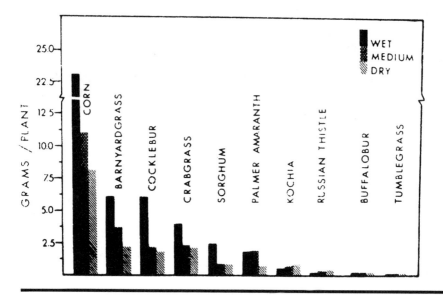

FIGURE 7-8. Competitive ability for water as measured by dry matter production of 10 species grown together under three moisture levels.

Source: Wiese and Vandiver, 1970. Reproduced with permission of the Weed Science Society of America.

species that produced relatively little under wet conditions—kochia, Russian thistle, buffalobur (*Solanum rostratum*), and tumblegrass—were not hurt by dry conditions. In fact, the first two produced more under dry conditions. Palmer amaranth (*Amaranthus palmeri*) was intermediate.

Competition for Nutrients

Two phenomena are involved in competition for nutrients in soil: (1) mass flow and (2) diffusion. *Mass flow* is the movement of water through the soil to a root, from the root up to the aboveground portion of the plant, and finally into the atmosphere by evaporation. Movement of nutrients with the water can basically be viewed as a passive process for those nutrients readily released by soil particles and soluble in water. *Diffusion* is the process whereby nutrients move from the soil particles, or ionize if in a chemical compound, into the soil water and then disperse throughout the soil water. Diffusion into the soil water is determined by the tightness with which the nutrient is held on the soil particle or by ionization and solubility if in a chemical compound. Nutrients strongly adsorbed on soil particles move mainly by diffusion, which is a relatively slow process.

Nitrogen.   Applied nitrogen, and that present in soil, may be in many different forms. Some forms are readily water soluble and others are not. Sooner or later, the organic and insoluble forms undergo nitrification to soluble inorganic forms. Water-soluble nitrogen is freely mobile in soil. Its depletion zone is the same as that for water, providing the nitrogen is utilized as it arrives at the root. Thus, as with water, relative competitiveness of weeds and of desired plants for nitrogen is largely determined by the soil volume occupied by the roots of each. Because it moves freely with soil water, nitrogen is frequently a competed-for nutrient. Typical nitrogen deficiency symptoms, such as firing of lower leaves in corn, are common for crops with heavy weed infestations.

Even though nitrogen competition can be mainly determined by relative root volumes and spatial distribution of the weed and the crop, the differences among species in their rate of utilization may also be a factor. Weeds usually take up fertilizer more rapidly than crops (Alkamper, 1976). The relative ability of corn and redroot pigweed to compete for nitrogen shows a striking advantage for the pigweed. Corn plants growing with pigweed contained only 58% as much nitrogen as weed-free corn plants (Vengris et al., 1955).

Crop varieties also vary in their relative competitiveness for nitrogen, as shown in Figure 7-9. In competition with yellow foxtail, the early corn hybrids as a group were relatively more competitive for nitrogen than the late hybrids. Under 157 kg N ha^{-1}

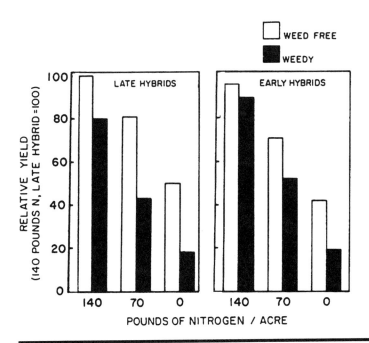

FIGURE 7-9. Relative competitiveness for nitrogen of late- and early-maturing corn hybrids.

Source: Data from Staniforth, 1961.

(140 lb per acre), weeds reduced yields of late-maturing varieties 20%, but early-maturing varieties only 6%. A possible explanation is given in the discussion of interactions later in the chapter. Here, we need only be aware that such differences may be usable in designing competitive production systems.

As we have seen, competition for nitrogen can either be the result of relatively passive aspects associated with crop–weed density or the result of relatively active aspects associated with nitrogen utilization.

Phosphorus, potassium, and other cations. Phosphorus (P), potassium (K), and other cations are relatively immobile nutrients in soil, but for different reasons. Because of their positive charge, K and other cations (positively charged ions) are held on the negatively charged clay particles. P is commonly applied in the phosphate form, which is negatively charged; however, it readily forms insoluble salts with calcium, iron, and other cations in the soil. Thus, movement of such salts is dependent upon diffusion. Because diffusion is a relatively slow process, the depletion zones are small and develop slowly. Trenbath (1976), in his review, cites 0.7 cm as the extent of a depletion zone from a root for P after 1 week. Mycorrhizae could extend this depletion zone somewhat (Chiariello et al., 1982). *Mycorrhizae* are fungal growths that attach to roots of many plant species and are a factor in nutrient uptake, among other things. Chiariello et al. showed that the common connection between neighboring plants provided by mycorrhizae could be a route for P movement. In effect, mycorrhizal connections might extend the root's depletion zones. There is also evidence that the plant root itself actively affects uptake of P and K as the result of the cation exchange capacity of the root; thus, this capacity becomes a factor in competition for these nutrients.

When competition occurs. A corn crop uses relatively little nitrogen during the first 5 to 6 weeks after emergence even though yield losses due to weeds may exceed 25% (Figure 7-10). In view of the usage pattern, competition for nitrogen is apt to occur

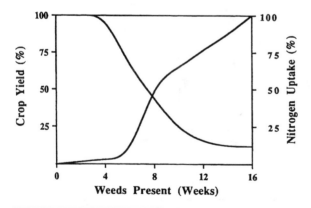

FIGURE 7-10. Corn response to weed presence (*left axis*) and relative nitrogen uptake by the crop (*right axis*).

Source: Modified from Aldrich, 1987.

somewhat later than for water. Competition for nitrogen may account for a significant part of the crop yield reduction from weeds but likely accounts for little of the earliest observed yield reductions in nonleguminous crops.

Usage of P and K during the first 5 to 6 weeks after corn and soybean emergence is less than 25% of the total (Figures 7-11 and 7-12) but still somewhat greater than what we saw for nitrogen. Since movement of P and K is slow and for comparatively short distances, competition early in the crops' life cycle is less likely than for water and nitrogen. Competition for such nutrients is most apt to occur after the crop and weeds are well established, when chances for extensive root development and over-lapping are at a maximum.

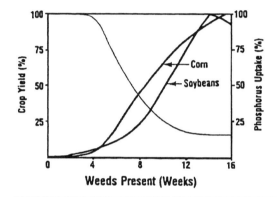

FIGURE 7-11. Crop response to weed presence (*left axis*) and relative phosphorous uptake (*right axis*).
Source: Modified from Aldrich, 1987.

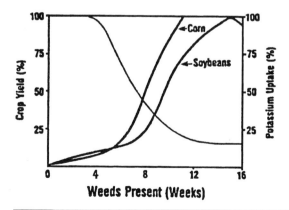

FIGURE 7-12. Crop response to weed presence (*left axis*) and relative potassium uptake (*right axis*).
Source: Modified from Aldrich, 1987.

Competition for Oxygen

There is no evidence of competition for oxygen even though it is theoretically possible. That is, there are conditions under which insufficient oxygen limits plant growth. For example, oxygen is a limiting factor for the growth of plants in very wet soils. However, there is no research to show that competition occurs between plants under such conditions. Under most conditions, oxygen in the soil and available to roots is adequate for respiration of roots.

ABOVEGROUND COMPETITION

Plants obtain light and carbon dioxide through their aboveground parts, primarily the leaves. From the discussion of limiting factors in Chapter 3, it is inappropriate to view one growth factor as being any more important to growth than another. However, from a competition standpoint, light is much more frequently involved than is carbon dioxide, if in fact the latter occurs at all under field conditions. This is not to suggest that carbon dioxide may not restrict growth under certain conditions, only that competition for it is not likely to occur.

Competition for Light

The photosynthesis process in plants is driven by light. In this process, light energy is transformed into chemical energy in the green leaf. Thus, it is the leaf, not the plant as such, that is the site of potential competition for light. This point is especially significant relative to competition since light cannot be transferred or stored within the plant. If light is kept from one leaf, that leaf cannot get light from another that is in the light. The end result of this fact is demonstrated by the death of lower leaves and branches on a tall tree in a dense forest. Anytime one leaf is shaded by another, the shaded leaf suffers competition for light.

Anything that affects absorption of light by the leaf can affect competition for light. In the field, both variations in light itself and variations in the plant may affect absorption. Light may vary in intensity (morning versus midday), duration (early spring versus midsummer), quality (clear versus cloudy), direction, and angle of incidence, depending upon the particular circumstances at the time, as well as the time of year. Although these aspects of light themselves may theoretically enter into competition, they are relatively insignificant compared with the effects of plant characteristics.

Plant characteristics affect competition in both the horizontal and the vertical dimensions. The horizontal dimension is influenced mainly by leaf characteristics and the vertical by plant height. Leaf and height characteristics together determine the relative competitiveness of a species for light. Those characteristics that enable a plant to be competitive for light include the following: (1) rapid expansion of a tall canopy, (2) leaves horizontal under overcast conditions and plagiotropic under sunny conditions, (3) large leaves to minimize penumbra effects, (4) leaves with the C_4 photosynthesis pathway and low transmissivity, (5) leaves forming a mosaic leaf arrangement, (6) a climbing habit, (7) a high allocation of dry matter to building a tall stem, and (8) rapid stem extension in response to shading.

Effect of leaf area. The leaf area of a plant clearly affects its potential for light absorption and thus its competitiveness for light. In view of its importance, a way of relating leaf area to land area is needed for evaluating competition for light. The leaf area index (LAI) provides such a measure. As discussed earlier, LAI is the ratio of surface area of leaves to a given area of ground. The larger the index number, the more leaf area for a given land area. Index numbers as high as 8 are common for many plants. LAI identifies the interception potential for light and also indicates the amount of light available to successively lower levels within a canopy. Figure 7-13 graphically depicts

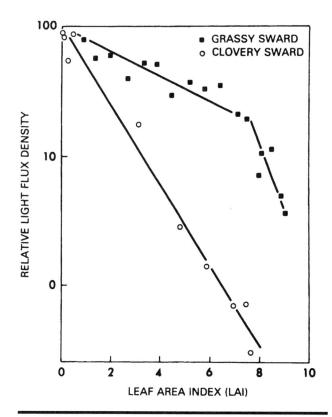

FIGURE 7-13. Effect of leaf inclination on light interception.

Source: Trenbath, 1976. Reproduced from Multiple cropping, ASA Special Publication no. 27, 1976, with permission of the American Society of Agronomy, Crop Science Society of America, and Soil Science Society of America.

these effects in a mixed grass–clover sward. As can be seen, light intensity drops steadily with the increase in accumulated LAI associated with progressively lower levels in the canopy. Reductions in light intensity within a canopy as a result of interception are usually exponential.

Effect of leaf angle and arrangement. The plant can also influence light interception through its determination of the angle of inclination of the leaves toward the sun, as well as by its pattern of leaf arrangement. As shown in Figure 7-13, leaf inclination can have a pronounced effect on light interception. The relatively horizontal leaves of clover intercept much more light than the upright grass leaves at any LAI level. Thus, weeds that have leaves more or less horizontal to the ground, such as velvetleaf, are relatively more competitive for light than weeds that have leaves more or less upright, such as giant foxtail. Similarly, weeds with opposite leaves, such as common milkweed, may be less competitive for light than those with alternate leaves, such as kochia, that form a mosaic whorl. Although LAI identifies the potential for light interception, it fails to take into account other plant characteristics and aspects of the light itself that may affect competition.

Effects within the canopy. Leaves absorb those wavelengths (400–700 nm) most effectively utilized in photosynthesis. Thus, light that reaches progressively lower levels in the canopy is not only of lower intensity but also of inferior quality (Figure 7-14). In the figure, velvetleaf allowed to overtop soybeans by 25% reduced photosynthetic photon flux density (PPFD) at the top of the soybeans by 26% and at the newest trifoliate leaf by about 50% when measured 3 weeks after emergence. In simple terms, PPFD is a measure of photosynthetically effective light. The significance of this fact emphasizes further the effect of competition for light on lower levels within the community canopy and for the lower-story species. The significance for competition relationships is that a very slight height advantage of the desired plant over the weeds can result in a strong competitive edge for the desired plant or vice versa.

The wavelength of leaf-filtered light affects axillary branch production in soybean (Table 7-1). Recall from Chapter 6 that chlorophyll in green leaves tends to capture red and allow far-red to pass through. The high proportion of far-red in transmitted light, comparable to shading by weeds, reduced number of branches 46%. In the field, velvetleaf 25% taller than soybeans reduced number of branches per soybean plant 80%, number of pods per soybean plant 44%, and soybean yield per hectare 68% (Begonia et al., 1991). It now appears that far-red-enriched radiation enhanced apical dominance thereby suppressing axillary bud growth (Begonia and Aldrich, 1990). Further, inhibition of lateral bud growth in velvetleaf-shaded soybeans apparently is mediated by the sequential action of both indole-3-acetic acid and abscisic acid. The direct effect of weed shading during the first weeks of the growing season on production of branches helps explain why soybean yields are reduced by weeds present during this early period.

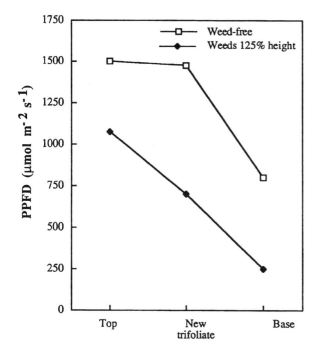

Soybean canopy position

FIGURE 7-14. Photosynthetic photon flux density (PPFD; 400–700 nm) at different levels in a soybean canopy as affected by height of competing velvetleaf.

Source: Modified from Begonia et al., 1991.

TABLE 7-1. Reduction of soybean branching from exposure of axillary buds to far-red radiation.

Treatment	Branches per Plant
Control	4.8
Red (terminal bud)	4.8
Far-red (terminal bud)	4.0
Red (axillary buds)	4.5
Far-red (axillary buds)	2.6

Note: Band-pass acetate filters were used to allow selective transmission of red or far-red radiation to reach the target organs. Filters were in place from 2 to 6 weeks after soybean emergence.

Source: Begonia et al., 1988.

The potential impact of a height difference was strikingly demonstrated in a study in which small and large seeds of subterranean clover were planted together (Black, 1958). Plants from the small seeds, although not genetically different from the large-seed plants, were sufficiently less vigorous, and therefore shorter, at the outset so that they obtained only 2% of the incident light after 82 days.

Effects on the whole plant.   Although the leaf is the site of competition, it is the effect of competition for light on the whole plant that is of concern to us in the weed–crop context. That is, the combined reduction in production of photosynthate by all leaves on a plant tells us how much loss that plant has suffered from competition for light. On a whole-plant basis, the leaves in the upper story are of primary importance in photosynthesis. It follows that this level in the canopy is where the extent of competition between a crop and weeds is largely determined.

The effect of canopy level is clearly evident in Figure 7-15. As can be seen, even at midday, those leaves in the bottom one-third of the canopy are net users rather than net producers of energy. Those in the top one-third are the main source of net production.

Effect within a weed–crop community.   Light effects on an individual plant basis translate into sizable effects in a weed–crop community in the field. For example, soy-

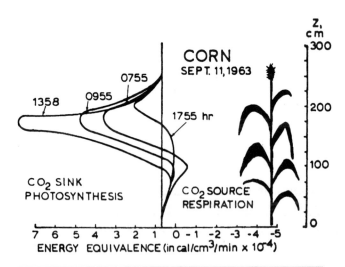

FIGURE 7-15. Effect of level in a corn canopy on photosynthesis and respiration. Positive values indicate net photosynthesis, whereas negative values indicate net respiration. At no time during the day are leaves in the lower one-third of the canopy net producers of photosynthate.

Source: Lemon and Wright, 1969. Reproduced from Agron. J., vol. 61, 1969, pp. 405-411, with permission of the American Society of Agronomy.

beans in 50 cm rows had only 28% as much dry weed weight 16 weeks after planting as soybeans in 100 cm rows (Felton, 1976). Since the soybeans were irrigated throughout the season, it is assumed competition for light accounts for the pronounced difference in weed yield. Referring back to Figure 2-7, we see that a 3-week head start by the crop markedly reduces dry matter production of giant foxtail. In soybeans, practically no growth occurred with this head start. Fertilizer was provided and sufficient rainfall occurred to minimize the likelihood that either of these factors could have accounted for the reduction in foxtail. Actual measurement showed that only 2.5% to 3.0% of the incident light reached the ground under the soybean canopy once it was established. It has been shown at least 50% reduction in light is needed for crops to effectively suppress weeds (Sweet, 1976). More than 60% reduction was readily obtained by sweet corn, potatoes, and tomatoes. Crop variety, or cultivar, has also been shown to affect competitive ability for light (Sweet 1976, 1979; Smith, 1974; Burnside, 1972; McWhorter and Hartwig, 1972; Staniforth, 1961). Clearly, field results indicate that characteristics imparting competitive ability for light can be built into varieties and production programs to enhance a crops' competitive ability for light. Possibilities for using such characteristics in weed management are examined in greater depth in Chapter 15.

Competition for Carbon Dioxide

Carbon dioxide (CO_2) and water are the basic raw materials involved in the photosynthetic capture of light and its transfer into chemical energy. CO_2 is obtained from the atmosphere. Studies have shown that the concentration is reduced within vegetation. Further, it has been shown that CO_2 supplementation increases the rate of photosynthesis under enclosed conditions. Thus, competition for CO_2 seems possible. However, as Trenbath (1976) points out, competition is not likely to occur in the field because turbulence within the canopy is so great that it causes rapid mixing between the interior and exterior atmospheres. Further, a reduced level of CO_2 is hard to visualize without an attendant reduced-light intensity, which, since it is nontransferable within the plant, might be expected to be the limiting factor.

CO_2 fixation: C_3 vs. C_4 pathway. Although competition for CO_2 may not be an important direct factor in competition between crops and weeds, the C_3 versus C_4 pathway for CO_2 fixation may have an indirect effect. Because C_3 leaves become saturated with light at relatively lower intensities than do C_4 leaves, plants of the C_3 type may be more apt to succeed under shade. On the other hand, C_4 plants use water more efficiently than C_3 plants, which may make them more competitive for other growth factors. Some evidence exists that shading may disrupt the C_4 function in species with this capability (Paul and Patterson, 1980). If so, the C_4 pathway might be either an asset or a liability under competition conditions, depending upon the relative heights of the competing species.

The C_3 versus C_4 pathway may influence the relative competitiveness of species over time if the concentration of CO_2 in the atmosphere continues to increase. The burning of fossil fuels and conversion of forests to agricultural production have caused an increase in CO_2 content from the longtime constant of about 300 ppm. The level may double by the year 2025. It has been speculated that such CO_2 enrichment will make C_3 weeds more competitive with crops (such as soybeans) (Patterson and Flint, 1980).

INTERACTIONS OF GROWTH FACTORS IN COMPETITION

Light and Nitrogen

Let us first consider interactions between light and nitrogen (N). As shown in Figure 7-16, competition for light reduces roots more than leaves. Root:leaf ratio is about 1:2 for control and 1:4 for low-light plants, suggesting that ability to obtain N (and other soil

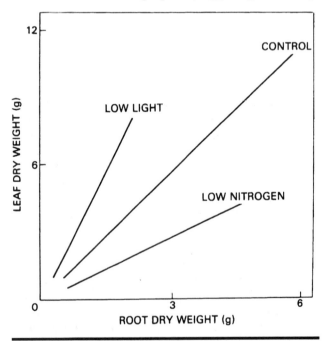

FIGURE 7-16. General effect of below-optimal supply of light and nitrogen on root and leaf weight of bean (*Phaseolus vulgaris*).

Source: Brouwer and deWit, 1969. Reproduced from Root growth, with permission of Plenum Press.

factors) is also reduced. Because weeds, especially annuals, evolved under conditions of nutrient stress, we expect them to grow relatively better than desired plants when N is in short supply. Conversely, we might expect the desired plant to be the more aggressive in utilizing added N. The latter situation, in fact, frequently does not occur. Rather, the weed appears to be the more effective in both rate and quantity of uptake of added nutrients. The explanation likely lies in the comparative time necessary to complete their respective life cycles. As was pointed out in Chapter 2 (Li, 1960), most annual weeds do so in less time than the crops with which they are competing. In order to do so, they must, in effect, grow at a faster rate. The relationship to N uptake may be a result of: (1) more rapid elaboration of the root system and/or (2) the ability to utilize the N arriving at their roots more quickly, thereby allowing more N to flow to their roots.

Plants do differ in rate of root elongation, as can be seen in Figure 7-17. Palmer amaranth, cocklebur, barnyardgrass, and crabgrass had extended their roots farther in 15 days than sorghum had in 20 days. There is also evidence that uptake of nutrients occurs over a shorter period in weeds. In quackgrass, Bandeen and Buchholtz (1967) found that by maturity in mid-July, quackgrass growing with corn had taken up 55%, 45%, and 68% of the total N, P, and K, respectively, taken up by the weed for the entire season. Although comparable data were not obtained for corn, the quantities obviously would be much less since corn, at this time, would still have the large majority of its dry matter production ahead of it (see Figures 7-10, 7-11, 7-12).

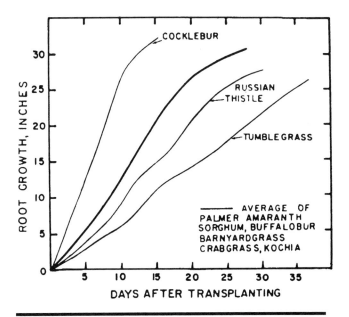

FIGURE 7-17. Differences in rate of root elongation for plants.

Source: Wiese, 1968. Reproduced with permission of the Weed Science Society of America.

Figure 7-16 also shows that shoot growth is restricted more than root growth by low N. The root:shoot ratio is about 2:1 for control plants and 4:1 for low-light plants. This relationship is further influenced by moisture supply. Since N moves passively with water to the root, it follows that competition for N is not apt to occur unless there is sufficient soil moisture for mass flow. Without mass flow, N would be left to move by diffusion, which is a slow process, thus minimizing the opportunity for differential (competitive) uptake.

Light and Water

As shown in Figure 7-18, velvetleaf shading reduced evaporative demand of cotton under wet soil conditions. This reduced cumulative total water flow in cotton by 45% (Table 7-2). Reduction in evaporative demand of velvetleaf from shading by cotton re-

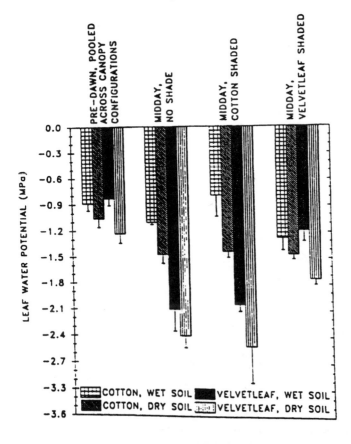

FIGURE 7-18. Leaf water potentials of cotton and velvetleaf grown in wet and dry soil under reciprocal shading.

Source: Salisbury and Chandler, 1993. Reproduced with permission of the Weed Science Society of America.

TABLE 7-2. Cumulative stem flow of water in cotton and velvetleaf as affected by shading.

Plant Canopy	Soil Moisture	Cotton	Velvetleaf
		grams per day (% of unshaded)	
Cotton	wet	55	87
shaded	dry	34	15
Velvetleaf	wet	92	31
shaded	dry	80	29

Source: Data from Salisbury and Chandler 1993.

sulted in 69% reduction in cumulative water flow in velvetleaf under wet conditions. For both species, shading reduced plant stress by reducing energy available to drive transpiration. Under dry conditions, evaporative demand of cotton was relatively unaffected by velvetleaf shading and cumulative total water flow was reduced 66%. Evaporative demand of cotton-shaded velvetleaf was reduced under dry conditions and cumulative total water flow was reduced 71%. Thus, when velvetleaf loses its competitive edge for light, demand for water is also reduced. This implies greater competitiveness for cotton under dry conditions. However, whether soil moisture is limited or ample, light competition is an interacting factor. Other studies have shown that water uptake decreases as light intensity decreases and that weeds vary (Jones and Walker, 1993). Common cocklebur water uptake was about double that of sicklepod (*Cassia obtusifolia*) at low light intensity (0.03 kw m^{-2}).

A practical question suggested by the above discussion is the merit of adding the competed-for factor to offset reduction in crop yield. For example, is it practical to add nitrogen (or other soil factors) to offset potential yield depression from a heavy weed infestation? The answer can be deduced by inspection of Figure 7-19. Where maximum crop yield (Y_1) is obtained by 1.0 unit of the growth factor Q, infestation of an equal number of weeds reduces the yield to Y_1' since the unit of growth factor must be shared (0.5 for crop, 0.5 for weed). For maximum yield to be maintained, the amount of growth factor would have to be doubled so that the crop and weed could each have 1.0 unit. Thus, although it is theoretically possible to meet the needs of both the weed and the crop, it is not practical. This picture is further complicated by interactions among growth factors. Doubling the amount of N, for example, could be expected to change the total amount of water needed for maximum crop yield. Also, aboveground growth form and size would be changed, thus influencing competition for light.

Figure 7-19 can also be used to illustrate a principle important in weed–crop relationships: Weedy crops respond more to the addition of a growth factor. For a weed-free crop at the 0.2 Q level, addition of 0.8 units of Q approximately doubles the yield (Y_2 versus Y_1). If weedy, crop yield is more than tripled (Y_2' versus Y_1') by the addition of 0.8 units of Q.

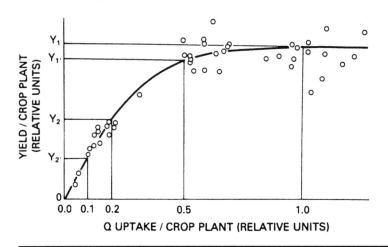

FIGURE 7-19. Effects of competition on the quantity of soil-supplied growth factor Q needed to give maximum yield. The uptake of growth factor needed to give maximum yield is taken as 1 unit. Y_1 and Y_2 are yields of weed-free crop plants for 1.0 and 0.2 units of Q. Y_1' and Y_2' are comparable crop yields where weeds are present at an equal density. The small circles are data obtained from research that served as the basis for the plotted curve.

Source: Trenbath, 1976. Reproduced from Multiple cropping, ASA Special Publication no. 27, 1976, with permission of the American Society of Agronomy, Crop Science Society of America, and Soil Science Society of America.

Carbon Dioxide and Other Growth Factors

Although, as already noted, competition for CO_2 is rare under field conditions, differences in CO_2 assimilation may lead to competition for other growth factors. Oliver and Schreiber (1974) found that pigweed net carbon exchange was at least 10 μmol CO_2 m^{-2} s^{-1} higher than birdsfoot trefoil during early stages of canopy development. More efficient CO_2 use helped it to grow more rapidly, thereby increasing its chances of competing for light.

Growth Regulators and Growth Factors

A final factor that may interact in a variety of ways with competition for growth factors is the site of production and action of endogenous plant growth regulators. As can be seen in Table 7-3, both where the growth regulators are produced in plants and where they exert their effects vary, depending upon the substance. For example, both abscisic acid and cytokinins are produced in roots, but abscisic acid promotes leaf abscission, whereas cytokinins reduce leaf senescence (and thus abscission). Auxins are produced mainly in shoots. It follows that competition for a given growth factor could differentially affect either production or action of growth substances or both, depending upon the plant part most affected by the competition. It seems likely that all competition effects are at least partially explained by the interacting influence of plant growth substances.

Furthermore, the action of growth-regulating substances may provide the explanation for some of the crop yield reductions not readily explained by competition as such. For example, competition seems inadequate as an explanation for losses sometimes caused by weed presence during only the first 2 or 3 weeks after crop emergence. Losses occasionally occur in several crops, including corn, cotton, flax, rice, soybeans, several vegetables, and spring wheat, from such brief exposure to weeds (Zimdahl, 1980). In that short time, however, it is difficult to visualize that weed growth would be sufficient to interfere with light, water, or nutrient availability to the crop in such a way that the crop would not completely recover.

On the other hand, as mentioned in the discussion of competition for light, interference with light availability during that time could affect production, transport, or action of one or more growth regulators. This interference could have a lasting effect on those plants so affected. Thus, although macroeffects of competition for water, nutrients, and light continue to be of primary importance in explaining crop losses from weeds, we must be mindful of the secondary role of interactions between availability of growth factors and growth-regulating substances.

As we have seen, competition for one growth factor will likely affect competition for others. We also learned that weeds are active partners in ecosystem development. Additionally, growth form changes in response to conditions imposed by the environment. The relationship among growth factors in competition thus may be represented schematically as follows:

Competition for Light
↓
Reduced Production of Photosynthate
↓
Lower Root:Shoot Ratio
↓
Reduced Uptake of Soil Factors
↓
Reduced Root:Shoot Ratio
↓
Competition for Light

The schematic diagram may be entered at any point, thereby establishing the probable next phase of competition. For example, if a soil factor is in short supply early in a weed–desired plant situation, the loser is started on the road to reduced shoot growth and competition for light. If competition for light occurs, the loser is started on the road to competition for a soil factor. The schematic representation of interactions provides a simple way of perceiving an answer to the question: Which comes first,

TABLE 7-3. Summary of information on endogenous plant growth substances.

	Auxins	Cytokinins
Chemical nature	Indol-3-acetic acid (IAA) and related compounds after conversion to IAA.	6-substituted anmino-purines and their ribosides and ribotides, e.g., Zeatin.
Production and occurrence	Produced mainly in meristematic and growing regions of shoots; senescent tissue has also been tissues. Found in most tissues.	Root apices appear to be a major source; found also in seeds, immature fruits, and shoots.
Transport	Move readily from shoots to roots in phloem and more slowly by cell-to-cell polar transport, basipetally in shoots and acropetally in roots.	Move in xylem from roots to shoots and weakly by cell-to-cell basipetal polar transport in shoots.
Main effects	*Promote:* elongation of stems and coleoptiles, photo- and geotropic curvature, adventitious rooting and lateral root initiation, xylem differentiation, fruit growth, cambium activity, and leaf epinasty. *Inhibit:* root elongation, leaf senescence, and fruit abscission. Maintain apical dominance of axillary buds.	*Promote:* cell division, leaf and cotyledon expansion, seed germination, coleoptile elongation, stolon and shoot initiation, translocation of assimilates and inorganic phosphorus, and transpiration. *Inhibit:* leaf senescence. Release some seeds from dormancy and axillary buds from apical dominance.
Factors affecting production, movement, and action	Production is inhibited by zinc and phosphous deficiencies and increased by gibberellins and cytokinins. Destruction is promoted by light, ethylene, and several phenolic compounds. Polar transport is stimulated by cytokinins or gibberellins and inhibited by abscisic acid or ethylene; and slow in older tissues. Action is modified by cell type, stage of development, and other hormones.	Little information available. Production in roots and movement to the shoots inhibited by flooding, drought, and high temperatures. Production in seeds increased by chilling and ethylene. Some production possible from the breakdown of RNA during senescence. Effects modified by other hormones.

Source: From Russell, 1977. Reproduced from Plant root systems: Their function and interaction with the soil, by permission of McGraw-Hill Book Company (UK), Limited.

Gibberellins	Ethylene	Abscisic Acid
A gibbane skeleton carboxylated at position 10 of the central ring. Twenty-three such compounds had been isolated from higher plants by 1973.	Olefine gas, C_2H_4 (also called ethene).	A dextrorotatory sesquiterpene.
Root apices believed to be a major source; found also in seeds, young stems, and leaves.	Produced by all parts of plants, particularly by ripening fruit, apical growing zones, and senescing tissues.	The root cap is one site of synthesis; found also in seeds, fruit, tubers, leaves, and buds.
Move from roots to shoots in xylem and from leaves in phloem and by cell-to-cell basipetal polar transport in shoots.	Little evidence available but can move from roots to shoots.	Moves in the stele from leaves, cotyledons, and roots and by cell-to-cell basipetal polar transport in shoots.
Promote: stem elongation (especially in dwarf plants) by increasing cell elongation and division, flowering in some long-day plants, seed germination, leaf expansion, abscission, and fruit growth. *Inhibit:* leaf senescence, adventitious rooting, and fruit ripening. Release buds from apical dominance and winter dormancy and root elongation from inhibition by light. Involved in photo- and geotropic curvature of stems.	*Promotes:* senescence, germination, adventitious rooting, leaf epinasty, abscission, fruit ripening and stem elongation in some water plants. *Inhibits:* stem and root elongation, cell division, stelar differentiation, geotropic bending of stems and roots and hypocotyl hook opening in legumes. Releases axillary buds from apical dominance.	*Promotes:* abscission, bud dormancy, tuber formation, adventitious rooting, leaf senescence and stomatal closure. *Inhibits:* seed germination, axillary bud growth, transpiration, stem and root elongation, ion transport and flower initiation. Involved in geotropic curvature of roots.
Production in roots and movement to shoots inhibited by flooding. Production in shoots inhibited by short days. Seed production stimulated by chilling, light, and abscisic acid. Auxin and ethylene sometimes required for full activity. Abscisic acid normally inhibits activity.	Production increased by fruit ripening, senescence of leaves and flowers, mechanical wounding, flooding, drought, and other hormones. Production affected by light and inhibited by anaerobiosis. Little known about ethylene breakdown and factors affecting transport. Action modified by auxin and other hormones and anatagonized by carbon dioxide.	Production increased by drought, flooding, nutrient deficiency, saline conditions, ripening, and senescence. Light and short days have little effect. Little known about breakdown and factors affecting transport. Inhibitory effects reversed in part at least by cytokinins, gibberellins, and auxin.

competition for light or competition for soil factors? Competition for light comes first if soil conditions are adequate, but first for soil factors if their supply is short enough to slow LAI development and light levels are high.

In summary, the complexities of weed competition with crops can best be understood and dealt with by keeping in mind that the leaf is the site of competition for light and the root the site of competition for soil factors. Any practice in crop management—selection of crop or variety, date of planting, crop stand, fertilizer management, and so forth—designed to selectively enhance the crop canopy and root volume occupied will assist the crop in its competitive struggle with weeds.

Competition for light is probably the most widespread competed-for growth factor. Competition for water and nitrogen is also common, and competition for nonmobile mineral nutrients, CO_2 and oxygen, is quite uncommon if, in fact, competition for the latter two occur at all under field conditions.

REFERENCES

Aldrich, R.J. 1987. Interference between crops and weeds. In Allelochemicals: Role in agriculture and forestry. Am. Chem. Soc. G.R. Waller, ed., pp. 300-312. Symposium Series 330. Washington, D.C.

Alkamper, J. 1976. Influences of weed infestation on effect of fertilizer dressings. Pflanzenschutz-Nachrichten 29:191-235.

Bandeen, J.D., and K.P. Buchholtz. 1967. Competitive effects of quackgrass upon corn as modified by fertilization. Weeds 15:220-224.

Begonia, G.B., and R.J. Aldrich. 1990. Changes in endogenous growth regulator levels and branching responses of soybean to light quality altered by velvetleaf (*Abutilon theophrasti* Medik.). Biotronics 19:7-18.

Begonia, G.B., R.J. Aldrich, and C.J. Nelson. 1988. Effects of simulated weed shade on soybean photosynthesis, biomass partitioning and axillary bud development. Photosynthetica 22:309-319.

Begonia, G.B., R.J. Aldrich, and C.D. Salisbury. 1991. Soybean yield and yield components as influenced by canopy heights and duration of competition of velvetleaf (*Abutilon theophrasti* Medik.) Weed Res. 31:117-124.

Black, J.N. 1958. Competition between plants of different initial seed sizes in swards of subterranean clover (*Trifolium subterraneam* L.). Aust. J. Agric. Res. 9:299-318.

Bleasdale, J.K.A. 1960. Studies on plant competition. In J.L. Harper, ed., The biology of weeds, pp. 133-143. Oxford: Blackwell Scientific.

Brouwer, R., and C.T. deWit. 1969. A simulation model of plant growth with special attention to root growth and its consequences. In W.J. Whittington, ed., Root growth, pp.224-244. London: Butterworths.

Burnside, O.C. 1972. Tolerance of soybean cultivars to weed competition and herbicides. Weed Sci. 29:294-297.

Chambers, E.E., and L.G. Hold. 1965. Phosphorus uptake as influenced by associated plants. Weeds 13:312-314.

Chiariello, N., J.C. Hickman, and H.A. Mooney. 1982. Endomycorrhizal role for interspecific transfer of phosphorus in a community of annual plants. Science 217 (4563):941-943.

Davis, R.G., W.C. Johnson, and F.O. Wood. 1967. Weed root profiles. Agron. J. 59:555-556.

Donald, C.M. 1963. Competition among crop and pasture plants. Adv. Agron. 15:1-118.

Felton, W.L. 1976. The influence of row spacing and plant population on the effect of weed competition in soybeans (*Glycine max*). Aust. J. Exp. Agric. An. Husb. 16:926-931.

Harper, J.L. 1977. Population biology of plants. New York: Academic Press.

Jones, R.E., and R.H. Walker. 1993. Effect of interspecific interference, light intensity, and soil moisture on soybean (*Glycine max*), common cocklebur (*Xanthium strumarium*), and sicklepod (*Cassia obtusifolia*) water uptake. Weed Sci. 41:534-540.

Knake, E.L., and F.W. Slife. 1962. Competition of (*Setaria faberii*) with corn and beans. Weeds 10:26-29.

_____. 1969. Effect of time of giant foxtail removal from corn and soybeans. Weed Sci. 17:281-283.

Lemon, R.R., and J.L. Wright. 1969. Photosynthesis under field conditions, XA. Assessing sources and sinks of carbon dioxide in a corn crop using a momentum balance approach. Agron. J. 61:405-411.

Li, M.Y. 1960. An evaluation of the critical period and the effects of weed competition on oats and corn, Ph.D. dissertation. Rutgers University, New Brunswick, NJ.

McWhorter, C.G., and E.E. Hartwig. 1972. Competition of johnsongrass and cocklebur with six soybean varieties. Weed Sci. 20:56-59.

Moulani, M.K., E.L. Knake, and F.W. Slife. 1964. Competition of smooth pigweed with corn and soybeans. Weeds 12:126-128.

Oliver, L.R., and M.M. Schreiber. 1974. Competition for CO_2 in a heteroculture. Weed Sci. 22:125-130.

Patterson, D.T., and E.P. Flint. 1979. Effects of chilling on cotton (*Gossypium hirsutum*), velvetleaf (*Abutilon theophrasti*), and spurred anoda (*Anoda cristata*). Weed Sci. 27:473-479.

_____. 1980. Potential effects of global CO_2 enrichment on the growth and competitiveness of C_3 and C_4 weed and crop plants. Weed Sci. 28:71-75.

Paul, R.N., and D.T. Patterson. 1980. Effects of shading on the anatomy and ultrastructure of the leaf mesophyll and vascular bundles of itchgrass (*Rottboellia exaltata*). Weed Sci. 28:215-224.

Russell, R.S. 1977. Plant root systems: Their functions and interactions with the soil. Maidenhead, Berkshire, England: McGraw-Hill.

Salisbury, C.D., and J.M. Chandler. 1993. Interaction of cotton (*Gosypium hirsutum*) and velvetleaf (*Abutilon theophrasti*) plants for water is affected by their interaction for light. Weed Sci. 41:69-74.

Smith, R.J., Jr. 1974. Competition of barnyardgrass with rice cultivars. Weed Sci. 22:423-426.

Staniforth, D.W. 1961. Responses of corn hybrids to yellow foxtail competition. Weeds 9:132-136.

Sweet, R.D. 1976. When it comes to competing with weeds, some are more equal than others. Crops and Soils 28:7-9.

_____. 1979. Influence of variety and spacing of potatoes on yield and weed suppression. Proc. NEWSS 33:110. Boston, MA.

Trenbath, R.R. 1976. Plant interactions in mixed crop communities. In Multiple cropping, ASA special publication no. 27, pp. 129-169. Madison, WI: American Society of Agronomy.

Vengris, J., W.G. Colby, and M. Drake. 1955. Plant nutrient competition between weeds and corn. Agron. J. 47:213-216.

Wiese, A.F. 1968. Rate of weed root elongation. Weed Sci. 16:11-13.

Wiese, A.F., and C.W. Vandiver. 1970. Soil moisture effects on competitive ability of weeds. Weed Sci. 18:518-519.

Zimdahl, R.L. 1980. weed–crop competition: A review. Corvallis: International Plant Protection Center, Oregon State University.

8

Allelopathy in
Weed Management

CONCEPTS TO BE UNDERSTOOD

1. Allelopathy probably influences the composition of the weed community: the influence may be especially important if perennial weeds are involved.
2. Allelochemicals are secondary plant products that inhibit both germination and growth of plants; in doing so, many metabolic processes may be affected.
3. Allelochemicals may exert their effect through the atmosphere (volatilization) and through the soil (leaching, exudation, and decomposition).
4. Evidence exists that production of allelochemicals is genetically controlled.
5. Allelopathy must be recognized as a potential compounding factor when weeds are competing with desired plants.
6. Allelochemicals are produced in all plant parts.
7. Many environmental factors influence the quantity of allelochemicals produced in plants; there appears to be a direct relationship between stress on the plant and quantity produced.
8. A large number of chemicals, widely different in their nature, cause allelopathic effects.
9. The fact that the capacity to produce allelochemicals is genetically controlled and the fact that the quantities produced are determined by a number of environmental factors offer encouragement that ways can be developed to use them in weed management.

Allelopathy has been defined as any direct or indirect, inhibitory, or stimulatory effect by one plant on another through the production of chemical compounds that escape into the environment. In general terms, *allelochemicals,* the active chemicals responsible for allelopathy, are categorized as *secondary plant products*. Plants produce many such products that do not function directly in primary biochemical activities that support growth, development, and reproduction. These products are chemically very diverse and are now widely recognized to function as an integral part of the plant's natural defense or survival mechanisms.

Based on their known chemistry, such chemicals can be expected to be harmful if present in sufficient concentration and proximity to a neighboring seed or growing plant. Further, harmful effects of one plant on another that cannot easily be explained by depletion of needed resources have long been observed. Among the examples are problems with orchard establishment on old orchard land, with reforestation, with reestablishment of alfalfa in old alfalfa fields, and with certain crop rotations and monoculture. A sizable literature has accumulated that documents the toxic effects of chemicals produced by one plant on another. In the most recent compilation, some 1,500 papers on various aspects of allelopathy were surveyed (Rice, 1995).

A feature that distinguishes allelopathy from competition is that some factor is being added to the environment in allelopathy as opposed to some factor being removed from it in competition. There are two types of allelopathy: (1) true and (2) functional. *True allelopathy* is the release into the environment of compounds that are active in the form in which they are produced by the plant. *Functional allelopathy* is the release into the environment of substances that are active as the result of transformation by microorganisms. Both types may be important for relationships between weeds and desired plants.

HISTORICAL BACKGROUND

Knowledge about allelopathy developed slowly but has accelerated rapidly in the past 10 to 20 years. In reviewing its history, Rice (1984) traces its beginnings to DeCandolle in the early 1800s, who was among the first to suggest that some plants excrete substances injurious to other plants. This suggestion was based on the observation that some crop plants grew poorly in association with certain weeds or other crops or in soil previously planted to other crops. It was nearly 50 years before the possibility again emerged with a report by Stickney and Hay in 1881 of the harmful effect of black walnut trees on the growth of other plants beneath them. Then, more than 40 years passed until, in 1925, Massey reported studies of the black walnut effect in some detail.

The results of Massey's study, illustrated in Figure 8-1, indicated that a deleterious root relationship existed between walnut and tomato plants. Tomato plants beyond the walnut roots were healthy, whereas those within the area of the walnut roots wilted and died. Over the next 40 years, the number of reports of similar effects expanded steadily. Most reports involved crop plants and fruit trees, but at least one involved weeds. In 1950, Keever observed that horseweed disappeared quickly from abandoned fields. In follow-up studies, it was found that decaying roots of this weed inhibited growth of seedlings of the same species. From about 1965 on, there has been a growing interest in possible allelopathic relationships between weeds and crops. Evidence now exists that indeed both desirable plants and weeds introduce chemicals into the environment that are toxic to themselves and to plants of other species.

Nonetheless, the concept of allelopathy is still a matter of controversy. Early observations and results of research were largely descriptive rather than analytical and therefore provided only circumstantial evidence for it, leaving room for explanations other than allelopathy for the observed results. Ideally, proof of a specific allelopathic effect requires following a sequence of steps (Putnam and Tang, 1986), including: (1) demonstrating interference by using suitable controls, describing symptomology, and quantifying growth reduction; (2) isolating, characterizing, and assaying the suspected chemical against the species that were previously affected; (3) identifying the chemical; (4) demonstrating that the isolated chemical causes the observed effect when added back to the system in which the plant is growing; and (5) monitoring release of the

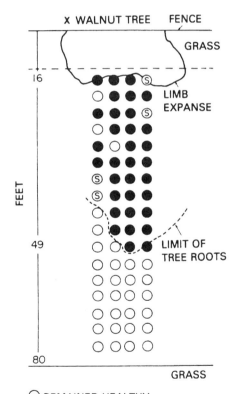

FIGURE 8-1. Allelopathic effects of a black walnut tree on tomato plants 8 weeks after transplanting.

Source: Rice, 1984. Reproduced courtesy of Academic Press, Inc.

chemical from the donor plant and detecting it in the environment around the recipient. For reasons discussed later, the research required for proof of allelopathy is extremely complex. Thus, only a few true allelochemicals are known today. The allelochemical know as juglone, produced by black walnut, was the first to have passed through the five steps required for proof.

Our concern is not so much with the proof as with the influence in relationships of weeds and desired plants. The evidence for allelopathic effects is indeed overwhelming, even though few proofs are available. Thus, it is essential we examine what is known in terms of the plant relationships.

PROBLEMS IN STUDYING

Allelopathy is a particularly difficult phenomenon to study. It is difficult to separate the effects of allelopathy from those of competition because growth and yield may be influenced by each (Qasem and Hill, 1989). For example, adverse effects of plant residues on seed germination and plant growth could be the result of immobilization of large amounts of nutrients by microorganisms involved in decomposition, by allelochemicals, or both. Care must be taken to exclude competition as a factor, which is not easy because it also often involves a complex of factors, as discussed in Chapter 7.

Further, except under desert conditions where the effect may occur through the atmosphere, the effects of allelopathy are manifested in the soil environment. The soil environment inherently provides myriad physical, chemical, and biological processes that may interact with allelochemicals and thus interfere with their study. Addition of large amounts of plant residues to soils may affect structure, moisture content, fertility, and microbial activity masking the expected allelopathic effect.

Finally, isolation of suspected allelochemicals has commonly required maceration or grinding and chemical extraction of fresh or dried tissue. Thus, altered, rather than naturally occurring, compounds are collected during extraction. Under these circumstances, it is difficult to know if the collected compounds have the same effect and activity as the native compound or if effects are due to chemicals that otherwise may not even be released under field conditions. The variety of experimental techniques used in allelopathy research impairs the ability to compare results with previous research. To confirm that allelopathy occurs, more than one test for allelopathy should be used (Qasem and Hill, 1989).

SIGNIFICANCE FOR WEED–DESIRED PLANT ECOLOGY

Allelopathy is significant for weed–desired plant ecology in three respects: (1) as another factor affecting changes in weed composition; (2) as another source of weed interference with growth of desired plants and yield; and (3) as a possible tool in reducing losses of desired plants from weeds. All three roles are considered here.

As efforts expand to document the reason for changes in weed composition, we must be alert to the possibility that allelopathy may be involved. Similarly, as we move toward a better understanding of competition, it is essential that any allelopathic relationships be fully accounted for. In particular, allelopathy must be kept in mind to help explain results not easily explained by other environmental circumstances. This is not to say that allelopathy should automatically be identified as a contributing factor when all other factors seem to have been ruled out. Rather, it is a matter of recognizing that allelopathy should not be overlooked when the results appear to be anomalous.

Effects on Weed Composition

Plant succession. As discussed in Chapter 3, there is a natural succession of plants in nature. In effect, plants change the environment, thus leading to a predictable succession, with the early colonizers being those species that rely upon large numbers of seeds and the later entrants those species that rely on their competitive ability. Rice and his co-workers (Rice, 1984) observed instances where succession seemed contradictory of these established ecological concepts. The inconsistency centered on the annual triple-awned grass (*Aristida oligantha*) in the succession shown in Table 3-4. In particular, why does such a noncompetitive species like triple-awned grass take over so quickly from the relatively robust species of the pioneer stage, and why does the annual grass and perennial bunch grass stage last so long? As we see, triple-awned grass begins to enter within 3 years after the onset of succession and is still present after 10 years.

From about the mid-1960s on, Rice and his co-workers (1984) conducted a number of different studies designed to determine if toxins produced by plants in the succession stages could be a factor in the observed anomaly. Studies were designed to evaluate the effects of additions of plant parts to soil, the effects of growing plants, and the effects of leachate from plant parts. The results of many of these studies are summarized in Table 8-1. Stage 1 species are toxic (+) toward themselves and less toxic (−)—if toxic at all—toward triple-awned grass. The exception seems to be crabgrass, which is toxic both to stage 1 species and to triple-awned grass from stage 2 systems. Because crabgrass seedlings are quite sensitive to the presence of other pioneer species, it is one of the first to be lost, thus allowing triple-awned grass to enter. Further, many pioneer species produce allelochemicals toxic to themselves that serve to shorten their persistence.

The persistence of triple-awned grass and the associated delay in takeover by the true prairie species is attributed to nitrogen relationships. It has been found that nitrogen fixation by bacteria and by bluegreen algae, as well as nodulation of legumes, may be inhibited by many species of the pioneer stage and by triple-awned grass of the second stage. It has also been shown that the nitrogen requirements of species increases

TABLE 8-1.   Toxicity of selected pioneer weed species (stage 1) toward themselves and toward triple-awned grass (a stage 2 species) in tall grass prairie old-field succession.

Stage 1 Species	Toxicity	
	Stage 1 Species	Triple-Awned Grass
Johnsongrass	+	Slight
Wild sunflower	+	−
Crabgrass	+	+
Western ragweed	+	−
Prostrate spurge	+	Slight
Flowering spurge	+	−

+ Indicates a toxic effect.
− Indicates a nontoxic effect.
Source: Data from Rice, 1984.

as succession progresses. Thus, succession from stage 2 is delayed because the nitrogen supply is inadequate for the associated species to compete successfully with triple-awned grass and its lower requirements.

Allelopathic substances have been implicated as influencing plant succession in central and southern Japan (Kobayashi et al., 1980). Goldenrod (*Solidago altissima*) and fleabane (*Erigeron* spp.), which are dominant species in stage 2, produce allelochemicals highly toxic to the perennial grass eulalia (*Miscanthus sinensis*) of stage 3 and to common ragweed of stage 1. The allelochemicals were identified as C_{10}-polyacetylenes.

These studies of the role of allelochemicals in succession have important implications for ecology of weeds and desired plants. Differential production of allelochemicals by desired plants and weeds and differential responses to them may well be involved in the shifts in weed competition associated with changes in production practices.

Patterns of perennial distribution.   Because of the characteristic of perennial species to concentrate offshoots around a parent—known as the *patch effect*—it can be reasoned that allelopathy could be especially beneficial to such species. The very fact that dense colonies of some perennials frequently occur essentially as pure stands in itself implicates allelopathy. Allelopathy has been shown to affect plant community composition in abandoned fields (Rice, 1984) and in native grassland (Muller, 1957). Curly dock affected the vegetational patterns in abandoned fields by releasing allelochemicals from fallen leaves (Einhellig and Rasmussen, 1973). The density and biomass of other plants within 2 m of curly dock plants were reduced suggesting that allelopathy directly impacted invasiveness of colonizing plants.

There is less evidence of allelopathy as a factor in the composition of weeds in land under cultivation. Work by Steenhagen and Zimdahl (1979) indicates the presence of such weed-to-weed inhibition in the perennial weed leafy spurge. Their work involved both field observations and soil evaluations in the greenhouse. The field observation utilized an infested area that had been undisturbed for 4 years. Species diversity was measured across the area and on the surrounding area. They found that quackgrass and common ragweed, although present on the perimeter, did not occur in the high-density areas of leafy spurge; some other weeds, such as kochia and crabgrass, were found associated with leafy spurge. Work in the greenhouse indicated that the effect was probably not due to competition. They point out, however, that the active agent has yet to be isolated and shown to exert its effect through the soil. Even so, it is highly probable that allelopathy in perennial weeds is frequently a factor determining the makeup of weed communities.

Weed seed longevity. The effect of allelochemicals on the longevity of weed seeds deserves special consideration in a weed–desired plant ecology context because of its potential impact on the success of attempts to alter the ecological environment to favor the desired plant. Hence, if weed seeds contain antimicrobial agents, or if such agents are produced as the result of decomposition of plant material, this characteristic must be dealt with in attempts to reduce interference from weeds over time. Rice (1984) suggested that antifungal properties of weed seeds may be the major reason for the observed longevity of some weed seeds in soils. Ecologically, this role of allelopathy may be most significant in agroecosystems.

The presence of antimicrobial agents in weed seeds was demonstrated by Kremer (1986) working with velvetleaf. Bioassays against a collection of 241 microorganisms revealed that substances were released from velvetleaf seeds that inhibited growth of 58% of the bacteria and all the fungi. Seeds with hard and water-impermeable seed coats, characteristic of those seeds persisting in soils, had greater inhibitory activity than soft, imbibed seeds. Subsequent work showed that aqueous extracts of velvetleaf seed coats contained six flavonoid compounds that not only inhibited growth and sporulation of potential seed pathogenic fungi but also inhibited seedling growth of three plant species (Paszkowski and Kremer, 1988). The compounds in velvetleaf seed coats appear to function in a defensive role against seed decomposers and competing seedlings. These results support the contention that longevity of weed seeds in soil is due in part to the involvement of allelochemicals.

Effects on Weed Interference with Desired Plants

Allelopathy may be a factor in weed interference with desired plants in two respects: (1) in inhibiting germination and seedling establishment and (2) in inhibiting growth of the desired plant. The effect on germination may be the easiest to identify, but effects on growth may be the most common under field conditions.

Crop seed germination. The stand of five crops was reduced by a heavy infestation of quackgrass in the soil previous to planting crops (Kommedahl et al., 1959). The results shown in Table 8-2 indicate alfalfa and barley to be particularly sensitive. In these studies, the heavy infestation of quackgrass was incorporated into soil preventing active growth of the quackgrass. Missed rhizomes were weeded out when the quackgrass shoots emerged. Thus, the effect could be due either to the presence of an active agent released from the living rhizomes before or when they were incorporated (true allelopathy) or to the production of toxic substances during decomposition of the rhizomes by microorganisms (functional allelopathy), or to both.

TABLE 8-2.   Effect of previously infested quackgrasss soil on the stand of 5 crops 3 months after sowing.

Crop	Percent Reduction of Stand on Infested Soil
Alfalfa	56
Flax	81
Barley	52
Oats	76
Wheat	66

Source: Data from Kommendahl et al., 1959.

Many weed seeds were reported to contain true allelochemical inhibitors to crop seed germination (Gressel and Holm, 1964). In their studies, weed seeds were ground, extracted with water, and the extract tested for effects on germination. Thus, possible reaction of microorganisms was excluded. Table 8-3 shows first that many weed seeds prepared in this way yield substances toxic to germination of a variety of crops when compared with the water control. Second, the number of hours for 50% germination show that the effect is more of delay than of prevention of germination.

Rye established and grew poorly in fields in Oklahoma where infestations of western ragweed (*Ambrosia psilostachya*) occurred prior to planting. Subsequent experiments revealed that aqueous extracts of western ragweed foliage and rhizomes inhib-

TABLE 8-3.   Inhibition of crop seed germination by water extracts of weed seeds.

	Crop Germination*				
	Percent Germination at 23 Hours		Hours for 50% Germination		
Weed Species	Alfalfa	Turnip	Pepper	Timothy	Tomato
Water control	82	64	117	68	44
Velvetleaf	5	25	>250	>250	106
Redroot pigweed	29	54	153	87	56
Common ragweed	31	12	132	92	61
Yellow rocket	5	16	>250	>250	103
Indian mustard	5	32	>250	>250	59
Lambsquarters	50	42	164	106	63
Crabgrass	76	41	152	72	54
Barnyardgrass	41	43	151	84	58
Pennsylvania smartweed	29	52	128	70	52
Common purslane	46	47	194	90	59
Yellow foxtail	20	43	182	92	57

   * Alfalfa variety California Ranger, turnip variety Purple Top, pepper variety California Wonder, tomato variety Roma.
   Source: Data from Gressel and Holm, 1964.

ited germination not only of rye but also alfalfa, switchgrass, yellow sweetclover, and tomato (Darymple and Rogers, 1983).

Thus, there is considerable evidence that weeds may inhibit germination of crop seeds. All weed parts and both true and functional allelopathy are implicated. The significance for field conditions of the observed effect with weed seeds is not clear. It is possible for weed seeds to be near enough to crop seeds for leaching of a toxic agent to occur at precisely the right time to inhibit germination of the crop seeds. The chances for this combination of proximity and timing is probably less likely than for toxic agents derived from decomposition of residues, especially since the residues represent a much larger biomass.

Crop growth. There is much evidence that allelochemicals from weeds inhibit growth of desired plants. Among the earliest reported was that of soil previously infested with quackgrass inhibiting the growth of alfalfa, flax, barley, oats, and wheat (Kommedahl et al., 1959). Subsequent work has established that the effect is probably the result of decomposition (functional allelopathy) of the rhizomes (Rice, 1984).

Many weeds have been implicated as being allelopathic toward crops with which they are growing. Separating allelopathy from competition is difficult in such situations. The following examples demonstrate different approaches for separating allelopathy from other interferences of weeds with desired plants.

Allelopathy of giant foxtail toward corn was shown by using a staircase arrangement of pots in the greenhouse (Bell and Koeppe, 1972). A diagrammatic representation of one line of the apparatus is shown in Figure 8-2. Test lines contained pots of corn alternating with pots of giant foxtail. Control lines contained only pots of corn. Solution was supplied to the uppermost pot from the quartz sand growing medium to a funnel and into the next pot in the series. The culture solution, after leaching through each pot in the series, was then pumped from a collecting reservoir back to the supply reservoir.

By using full-strength nutrient solution and monitoring for conductivity and pH, Bell and Koeppe maintained uniform growing conditions in the test and control lines. Figures 8-3 and 8-4 show the effects on corn of different kinds of exposure to giant foxtail. From Figure 8-3, we see that foxtail seedlings had no harmful effects on corn, but mature, live foxtail plants markedly reduced corn height (white bars), fresh weight (black bars), and dry weight (hatched bars). Since nutrients, water, and light were not limiting factors, it is assumed that a toxin to corn was released from the mature giant foxtail roots.

This method provides a direct way of identifying allelopathic effects but does not address the matter of relative biomass to that encountered in the field. As can be seen, mature dead plants and incorporated foxtail residue also inhibited the corn. The incorporated residue was more inhibitory than either the living or dead whole plant. It was not determined if this effect was due to microbial transformations or simply to greater release from the macerated and incorporated material. Results from evaluation of time of exposure, shown in Figure 8-4, provide at least circumstantial evidence for micro-

biological activity. Maximum effects from incorporation (Figure 8-4) showed up somewhat later than effects from the intact living plant (Figure 8-4A). Maximum reduction
in fresh weight (black bars) and dry weight (hatched bars) shown for the incorporated
material occurred at the last date (31 days), whereas at 23 days, nearly maximum reduction had occurred with the living, mature foxtail. If only physical release was involved, we might expect maceration of incorporated plant material to cause earlier expression of effects. It may be assumed that time is required for the microbial
populations to reach a fully effective level, whereas no such time element is involved
with exudation from the living plant. In addition to showing that living foxtail roots re-

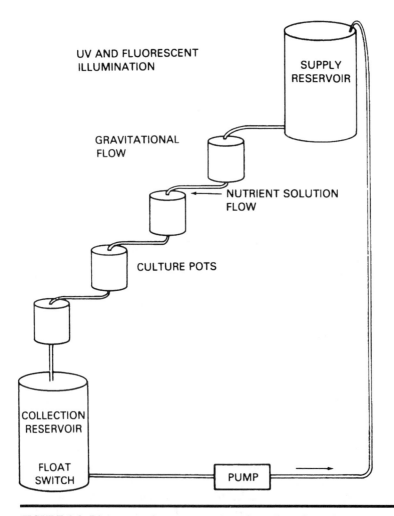

**FIGURE 8-2. Diagrammatic representation of one line of a staircase apparatus for ascer-
taining allelopathic effects.**

Source: Bell and Koepke, 1972. Reproduced from Agron. J., vol. 64, 1972, pp. 321-325, with permission of
the American Society of Agronomy.

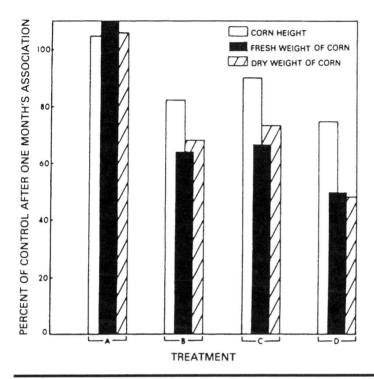

FIGURE 8-3. Allelopathic effects of giant foxtail on corn measured by the staircase apparatus shown in FIGURE 8-2. Treatments included corn seedlings (A) started together with giant foxtail seedlings; (B) growing with mature, live giant foxtail; (C) growing with dead giant foxtail plants; and (D) growing in contact with macerated giant foxtail leaf and root material incorporated into the sand culture pots.

Source: Bell and Koepke, 1972. Reproduced from Agron. J., vol. 64, 1972, pp. 321-325, with permission of the American Society of Agronomy.

lease a toxin into the soil, this research also shows that both true and functional allelopathy may be involved in explaining the full allelopathic effects of foxtail on corn. It seems likely that both types are commonly involved in other instances of allelopathy of weeds toward desired plants.

Allelopathic effects of yellow nutsedge on growth of corn and soybeans were studied by using indirect methods to relate effects to quantities of weed biomass representative of those encountered in the field (Drost and Doll, 1980). Various quantities of ground nutsedge tubers and foliage were either extracted with water or incorporated into potting media and tested for effects on the crops. It can be seen from Table 8-4 that 0.5% weight of tubers per weight of sand either incorporated or extracted with water reduced root growth of soybeans in pots in the greenhouse. This weight of tubers is representative of the quantity present in the upper 15 cm of soil in Wisconsin. Thus, the reduction due to water extracts of tubers suggests the potential inhibition that might occur from yellow nutsedge growing with soybeans. That is, it can be postulated that tox-

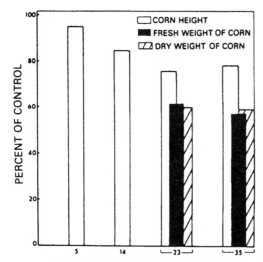

LENGTH OF ASSOCIATION (DAYS)
A. Effect of living, mature foxtail on corn

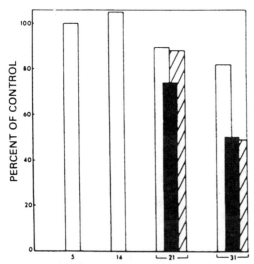

LENGTH OF ASSOCIATION (DAYS)
B. Effect of incorporated foxtail plant material on corn

FIGURE 8-4. Time needed for maximum allelopathic effects from living, mature foxtail plants vs. macerated and incorporated foxtail plants.

Source: Bell and Koepke, 1972. Reproduced from Agron. J., vol. 64, 1972, pp. 321-325, with permission of the American Society of Agronomy.

TABLE 8-4. Inhibition of soybean root growth by both yellow nutsedge tuber residue and water extracts of tuber residues.

Quantity of Nutsedge Tuber Residues (%)[a]	Soybean Root Growth (%)[b]	
	Tuber Residue Incorporated	Water Extracts of Tubers
None	100	100
0.50	45	73

[a]Weight based on 1500 grams of silica sand.
[b]Based on untreated control.
Source: Data from Drost and Doll, 1980.

ins in the tubers are released into the soil–soybean root environment under moist field conditions.

In this same study, it was also found that exposure of the crop seeds to nutsedge residues placed at the same level as the seed was relatively more inhibitory to crop growth than placement either below or above the seed (Figure 8-5). This finding suggests that the depth of incorporation of allelopathic weed residue may influence the degree of allelopathic effects.

Although the greenhouse studies with foxtail and nutsedge provide valuable information, it is the effect encountered in the field that is important in interpreting and dealing with allelopathy. It is unwise to assume that results obtained under greenhouse conditions will be the same under field conditions. In fact, results will likely not be the same. One reason is that competition and allelopathy, where it occurs, are both quite

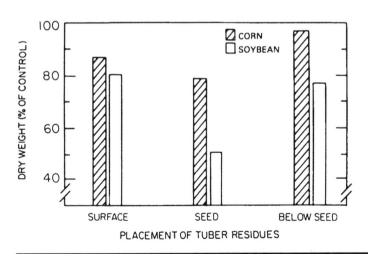

FIGURE 8-5. Allelopathic effects of yellow nutsedge residue toward corn and soybean seed at seed level, below the seed, or at the soil surface.
Source: Drost and Doll, 1980. Reproduced with permission of the Weed Science Society of America.

likely to be involved in weed and desired-plant situations in the field. Stress (competition) appears to have a marked effect on production of allelochemicals, as will be discussed later. Further, both true and functional allelopathy may well be involved. Therefore, allelopathy must ultimately be evaluated in the field to be certain of its significance in weed–desired plant relationships.

A way of dealing with these complexities has been suggested by Dekker et al. (1983). These workers suggest the design of experiments and analysis of data that *collectively* provide *inferential* measures of allelopathy. By drawing upon what is known about plant and density relationships, some divergence from that expected for competition provides an inference for the presence of allelopathy. Use of replacement experimental designs is basic to this approach. In *replacement designs*, total density is held constant and the proportions of each of two species varied. By measuring several growth parameters and subjecting the data to mathematical and graphical treatments, we can develop inference statements.

This approach was applied to velvetleaf interacting with soybeans. It was found that, in field tests, velvetleaf at 2.4 to 4.7 plants m^{-2} reduced soybean flowering node numbers, soybean dry matter, and soybean seed production (Dekker and Meggitt, 1983). Because the soybean plant population was constant and other factors such as soil fertility were not limiting, these results supported the presence of an allelopathic mechanism. Because much of the interpretation is theoretical, application to other weed–crop situations and collection of discrete data are needed. Nevertheless, the approach provides valuable background for evaluating allelopathy under field conditions.

Irrespective of the explanation, it is clear that allelopathy from weeds may be involved in the detrimental effects of weeds on crops. The effect may be expressed both through a reduction in germination and through a reduction in growth. Under field conditions, it is speculated that reduction in growth may be the effect most frequently encountered. Further, microorganisms are commonly involved, both to implement release of the toxic agent and to produce such agents. For example, allelopathic substances were exuded from living johnsongrass rhizomes or were released from decaying rhizomes into soil that reduced soybean growth (Lolas and Coble, 1982). If allelochemicals are exuded by the weed, removing it should alleviate the allelopathic effect. If, on the other hand, allelochemicals are released from weed residues during decomposition, weed growth will need to be prevented to avoid crop damage.

ALLELOCHEMICALS

There is overwhelming evidence that allelochemicals are a part of the natural environment within which plants grow. Further, we have seen that they are a factor in ecological relationships between weeds and desired plants. Thus, an understanding of the general nature of such chemicals is important in appreciating their significance for relationships between weeds and desired plants, as well as in possible application of such knowledge in weed management. This is not the place for an in-depth considera-tion of their chemistry. Other sources for such information are Duke (1986) and Rice (1984). Here we consider allelochemicals in terms of their chemical nature and signif-icance for weed and desired-plant interrelationships.

A limited number of compounds have been identified and implicated in allelo-pathic activity in plant tissues. These compounds include juglone from black walnut, scopoletin from oat, the hydroxamic acids from rye, and sorgoleone from grain sorghum. Identification of compounds suspected as allelochemicals is important in several aspects. It satisfies the key step in providing evidence for proof of allelopathy by a particular plant (Putnam and Tang, 1986). Knowledge of allelochemicals respon-sible for suppression of weeds is useful in developing plant varieties with selected al-lelopathic traits. Such varieties could be integrated in enterprise operations as compo-nents of weed management systems.

Chemical Nature

Figure 8-6 shows the major groups of organic compounds implicated as allelochemi-cals and their relationship to primary plant metabolism. It is important to note that ac-etate and amino acids are the basic components from which the secondary compounds are derived. Although no detailed discussion of the chemistry is given here, it should also be noted that the allelochemicals represent a wide spectrum of chemical classes. Within these classes, there may be many individual chemicals. Thus, it is not surpris-ing that the isolation of compounds responsible for allelopathic effects is difficult.

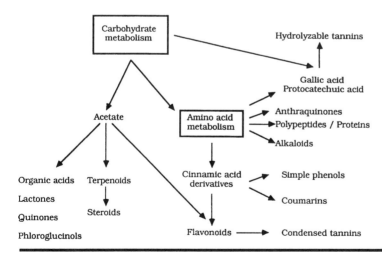

FIGURE 8-6. Metabolic relationships of the major groups of secondary compounds to pri-mary metabolism.

Source: Modified from Rice, 1984.

Such isolation—and verification of effect—is of course important to full understanding of their role in weed–desired plant and weed–weed relationships and to their use in weed management. As helpful as such information might be, the lack of it need not preclude research to identify possible species relationships in weed communities. Also, as our knowledge expands of the occurrence of such chemicals and of the factors affecting their production, possibilities for using them in weed management can also be increased. Such possibilities are considered in Chapter 15.

Sources in Plants

In discussing effects upon plants early in the chapter, reference was made to roots, seeds, and leaves as sources of allelochemicals. Rice (1984) cites literature to show that the chemicals may be produced in other major organs. Leaves may be the most consistent source of such inhibitors. Roots are considered to contain fewer and less potent toxins or smaller amounts. More complete knowledge of their location in the plant is needed in connection with efforts to use the production of such chemicals by growing plants to suppress weeds. To have maximal effect against weeds growing with a desired plant, the allelochemicals need to be concentrated in the roots, leaves, or stems, rather in the flowers and fruits. If concentrated in the flowers and fruits, it is unlikely they could be available in time to prevent interference from the weed. In instances where the plant material containing the toxin or from which the toxin is produced is to be incorporated in soil to inhibit germination and growth, location within the plant may be relatively unimportant. Rather, total biomass to be incorporated and concentration of the toxin are the important aspects.

Quantities Produced

Knowledge about quantities produced is of obvious importance to possible use of allelochemicals in weed management. As our knowledge of factors affecting quantities of allelochemicals produced expands, opportunities for manipulating crop production practices to maximize allelopathy should evolve. Considerable evidence indicates that a variety of environmental factors do indeed influence the quantity of allelochemicals produced by plants (Rice, 1984). Some of the environmental factors identified as influencing quantity follow.

First, quantities of some known allelochemicals are influenced by light quality, intensity, and duration. The most significant finding for ecology of weeds and desired plants may be that quantities produced are greatest under exposure to ultraviolet light and long-day photoperiods. Thus, understory plants might be expected to produce less quantity because of the filtering out of some of the ultraviolet rays by overstory vegetation. Also, plants during the peak of the growing season could be expected to produce more allelochemicals than those same plants earlier or later in the growing season.

Second, quantities are greater under conditions of mineral deficiency. The magnitude of the differences may be severalfold, as shown for nitrogen on sunflower in Table 8-5. Deficient older leaves and stems contained 8 to 10 times as much total chlorogenic acid, a phenolic acid known to be toxic.

Third, the amount of allelochemicals produced is greater under drought stress.

Fourth, quantities may be greater under cool than under what are considered normal growing temperatures, although the location within the plant and effects on specific allelochemicals seem to be variable.

Fifth, application of plant growth regulators, such as 2,4–D and maleic hydrazide, and of other allelochemicals apparently may greatly increase the quantities of allelochemicals produced by a plant.

Sixth, infection of some plants by pathogens and attack by insects cause considerable increases in concentrations of phenolic compounds. These compounds may enhance resistance of plants to pathogens and predators and presumably may increase allelochemical activity, although this aspect has not been investigated.

Effect of stress.   As we have seen, there are many indications that environmental conditions that restrict growth tend to increase the production of allelochemicals. It is only a short step from this general effect to the postulation that allelopathy may frequently be an accentuator of competition, even though it is not a part of competition. After all, competition between weeds and desired plants in itself implies the creation or presence of stress conditions. Thus, if stress from competition increases the quantities of allelochemicals produced, it is conceivable that the allelochemicals will inhibit growth of some species and not others, thereby further reducing the ability of the affected species to compete.

Postulation of the stress effect suggests that growth factors and growing conditions may interact to influence the quantities of allelochemicals produced. Indeed, there

TABLE 8-5. Concentrations of total chlorogenic acids and scopolin in nitrogen-deficient and control sunflower plants 5 weeks from start of treatment.

Plant Organ and Treatment	μg g^{-1} Fresh Weight	
	Total Chlorogenic Acids	Scopolin
Older leaves		
Control	1139	7.2
Deficient	8884	6.4
Younger leaves		
Control	1737	–
Deficient	873	–
Stems		
Control	383	1.8
Deficient	3275	–
Roots		
Control	303	–
Deficient	490	–

– Below amounts determinable by procedure used.
Source: Rice, 1984. Data from Lehman and Rice, 1972.

is evidence that exposure to more than one stress factor may have a compound effect on quantities of chemicals produced. This evidence is shown in Table 8-6. The combination of stress for nitrogen and moisture resulted in a 15-fold increase in chlorogenic acid and a 16-fold increase in isochlorogenic acid in sunflower plants over the control. This particular combination has special relevance for weed management since both factors may be controllable, at least to a degree, depending upon the geographic location, season, crop, and production system.

TABLE 8-6.   Effects of stress factors on concentrations of total chlorogenic acids and total isochlorogenic acids in sunflower plants.

Stress Applied	μg g^{-1} Dry Weight*	
	Total Chlorogenic Acids	Total Isochlorogenic Acids
None, control	43	135
Ultraviolet light	113	203
Water	258	320
Ultraviolet light; -water	455	512
Nitrogen	458	1065
Nitrogen; ultraviolet light	310	375
Nitrogen; -water	645	2185
Nitrogen; -water; ultraviolet light	546	979

−Indicates reduced quantity; data from del Moral, 1972.
*Weighted mean of leaf, stem, and root tissues.
Source: Rice, 1984.

It is important to remember that plants producing allelochemicals and those affected by them are a part of an ecosystem. This means that rarely, if ever, will one factor vary without changes in one or more other factors. Light can be expected to interact with temperature and indirectly with soil moisture, and numerous other environmental factors. This is one more reason allelopathy needs to be ever in mind as a possible factor in explaining relationships of weeds and desired plants.

Modes of Entry into the Environment

The fact that allelochemicals can be identified in plants does not mean they necessarily have an effect in the enterprise ecosystem. To have an effect, the chemicals must enter the environment and at a time when they can exert an effect. That is, if the allelochemical is not released from the plant that produces it, obviously it can have no effect on other plants. If the allelochemical is released to the soil environment at the end of a growing season, only to be dissipated before the next growing season, it may well have no impact. There are four ways allelochemicals can enter the environment: (1) volatilization, (2) leaching, (3) exudation, and (4) decomposition. These modes of entry are shown schematically in Figure 8-7.

Volatilization.   There are many examples of *volatilization,* or release into the atmosphere, as a way of allelochemical entry into the environment under natural conditions (Rice, 1984). This mode of entry is common under arid or semiarid conditions. However, certain temperate plants emit volatile organic compounds from their vegetative residues. The most intensely studied plants that emit volatile compounds are *Amaranthus* spp., which include several important weed species, such as the pigweeds. Volatile allelochemicals collected and identified from fresh residues not subjected to decomposition inhibited germination of carrot, tomato, and onion in bioassays (Connick et al., 1989). These volatiles consisting of low molecular weight hydrocarbons, aldehydes, and alcohols could play an important role in allelopathy immediately after the plant residues are added to soil.

A phenomenon related to volatilization with possible implications for weed man-

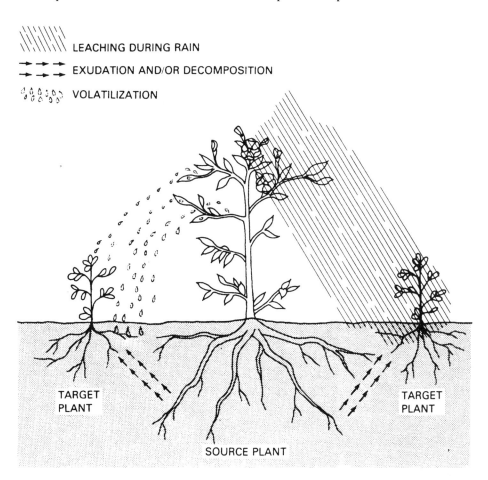

FIGURE 8-7. Schematic representation of ways allelochemicals enter an enterprise production ecosystem.

agement is the interaction between plants and other living organisms, especially insects. Many aspects of insect behavior may, at least at times, be related to plant chemistry. Some volatile compounds emitted from the growing plant serve as feeding stimulants and attract herbivorous (plant-feeding) insects (Letourneau, 1988). This suggests the possibility of interfacing allelopathy and biological control of weeds by using insects attracted to specific allelochemicals emitted by a target weed. This potential would seem to be attainable since it is likely allelochemical–herbivore interactions already occur in nature.

Leaching. Rainfall, irrigation water, or dew may transport, or *leach*, allelochemicals that are subsequently deposited on other plants or on the soil. We can visualize this mode of entry as an important factor in plant ecological relationships because it could serve to extend the time and quantities of exposure over what would occur from the other modes of entry. Here, too, there are many examples of entry into the environment in this way (Rice, 1984). Leaching of plant residues on or in the soil may also transport allelochemicals into the soil environment. Inhibition of corn and soybeans by leachates from nutsedge leaves and tubers is an example of this mode of entry (Drost and Doll, 1980).

Exudation. Exudation of allelochemicals from plant roots into the soil environment is implicated as a mode of entry by many studies. The work of Bell and Koeppe (1972) showed allelopathy is most likely the result of exudation of allelochemicals from giant foxtail roots since watering in the staircase arrangement precludes leaching from aboveground parts. However, it is possible the observed effect is simply the result of allelochemicals being leached from the roots. In this case, although decomposition cannot be ruled out, at least as a contributor, it is unlikely to be the major mechanism, due both to the short-term nature of the study and the fact that the sand medium used was likely low in microbial numbers.

Compounds exuded from roots that have been identified chemically and have undergone extensive bioassays include the hydroxamic acids of rye (Perez and Ormeno-Nunez, 1993) and the benzoquinone sorgoleone exuded from grain sorghum (Einhellig and Souza, 1992). These compounds are very potent growth inhibitors at low concentrations suggesting they strongly contribute to allelopathy by exudation as the mode of entry.

Decomposition. There is much evidence that toxic substances result from the decomposition of plant residues. Work with quackgrass and johnsongrass showed decomposition to be at least partially responsible as the mode of entry of allelochemicals (Lolas and Coble, 1982; Weston et al., 1987). It is difficult, however, to determine whether the toxic substance is contained in the residue and simply released upon decomposition or is produced instead by the microorganisms utilizing the residue as a substrate.

Identifying entry mode. As we have seen, entry of allelochemicals into the environment is indeed very complex. The toxic effects of one plant on another in the soil environment may be the result of leaching from aboveground plant parts, exudation from living roots, release as a result of breakdown of sloughed-off cells, or synthesis of toxins by organisms utilizing the plant material as a nutrient source. Separating these mechanisms is difficult at best, especially under field conditions. Further, all mechanisms could conceivably be involved at the same time in the field. However, entry by leaching and decay of litter together are more likely to have a greater effect on relationships of weeds and desired plants than entry in other ways (Putnam and Duke, 1978).

A method for collecting allelochemicals directly from roots has been developed (Tang and Young, 1982). This method, depicted schematically in Figure 8-8, allows continuous trapping of extracellular chemicals on a resin exchange column. Nutrient solution continuously circulated through the roots of donor plants transports extracellular organic compounds (exudates) from the sand culture. The hydrophobic or partially hydrophobic exudates are selectively retained on a resin column while the inorganic nutrients pass through. The trapped chemicals can then be eluted (removed from resin) with organic solvents for identification, quantification, and measurement of activity. This methodology may prove useful in separating exudation, leaching, and decomposition as modes of allelochemical entry into the soil environment.

Plant Processes Affected

Allelochemicals can affect both germination and growth of plants. These effects may be manifested through a wide variety of metabolic activities. Specific plant processes identified as being affected include: cell division and elongation, action of inherent growth regulators, mineral uptake, photosynthesis, respiration, stomatal opening, protein synthesis, lipid and organic acid metabolism, leghemoglobin synthesis, membrane permeability, and actions of specific enzymes (Einhellig, 1995; Rice, 1984). The challenge for students of weed–desired plant ecology is to identify the processes most manipulative in weed management. In this regard, those processes most directly related to competition seem to offer the best possibilities. Although all of the processes identified are, of course, involved in growth, some, especially cell division and elongation, are more directly involved than others.

POTENTIAL FOR WEED MANAGEMENT

With the mounting evidence supporting allelopathy as a phenomenon in nature, there is growing interest in the possibility of using this characteristic in desirable plants to minimize interference from weeds. Substantial progress has been made over the last 10 to 15 years in adapting allelopathy for use as a component in weed management, and results obtained thus far suggest that allelopathy can be an effective tool against weeds.

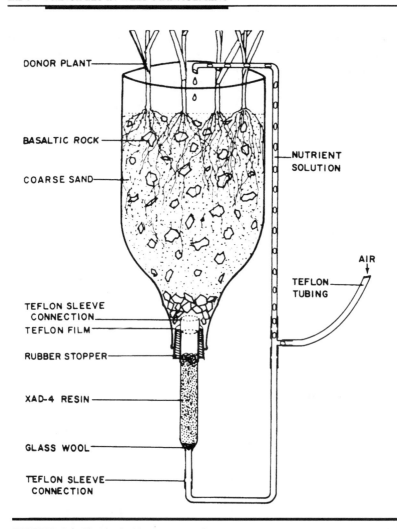

FIGURE 8-8. Hydrophobic root-trapping system.

Source: Tang and Young, 1982. Reproduced with permission of the American Society of Plant Physiologists.

Suppressing Germination and Emergence of Weeds

Allelopathy of both the growing desired plant and of its residue might be utilized to reduce weed stands. There are examples of success with each approach.

Growing crop plants. Lockerman and Putnam (1979) found that one cucumber selection of several studied markedly reduced the stand of barnyardgrass and redroot pigweed. The stand of barnyardgrass was reduced about 80% and redroot pigweed about 60% by the cucumber accession PI 169391 (Table 8-7). Cultivated sunflowers reduced the stands of both grass and broadleaf weeds by about 50% during growing seasons spanning 5 years (Leather, 1983). The reduced weed densities were attributed to allelochemicals actively exuded by sunflower roots.

TABLE 8-7.   Weed numbers in the presence and absence of selected cucumbers in the field.

Cucumber Accession or Cultivar	Barnyardgrass		Redroot Pigweed	
	Days after Planting			
	10	48	10	48
No cucumber	31	38	58	54
Pioneer	26	21	39	40
PI 169391	7	7	23	21

Source: Adapted from Lockerman and Putnam, 1979.

Obviously, such reductions in weed numbers may not be enough to prevent significant reductions in crop growth. The remaining weeds in the cucumber study caused a 25% reduction in fresh weight of the cucumber vines at maturity. Also, allelopathic activity of sunflowers in the field was variable and allelochemicals did not accumulate to the extent to provide weed suppression the year following sunflowers. These observations suggest that if an approach to weed germination and emergence suppression by growing crops is utilized, it is likely to be in conjunction with other methods of weed management. Nevertheless, it appears that allelopathy could be incorporated into crop cultivars.

Crop residues.   Substantial reductions in weed stands from crop residues have also been demonstrated. Three years of study demonstrated that grain sorghum residues greatly reduced stands of weeds in the field the year after grain sorghum was grown (Einhellig and Rasmussen, 1989). In other studies, residues on the soil surface obtained by desiccating fall-planted rye, barley and wheat with glyphosate in the following spring reduced weed densities by an average of about 90% of that in plots with no residues (Putnam et al., 1983). Rye residue reduced emergence of all weeds except yellow foxtail (Figure 8-9). Reductions as great as those shown in Figure 8-9 could be expected to prevent losses in yields of desirable plants.

Suppressing Growth of Weeds

The growing crop and its residues may also be used to suppress weed growth.

Growing crop plants.   Results of an evaluation of the worldwide collection of oat germplasm indicate the allelopathic potential of the growing plant (Fay and Duke, 1977). In this study, 3,000 accessions were screened for scopoletin content. The allelochemical scopoletin is an alkaloid compound known to be inhibitory toward plant

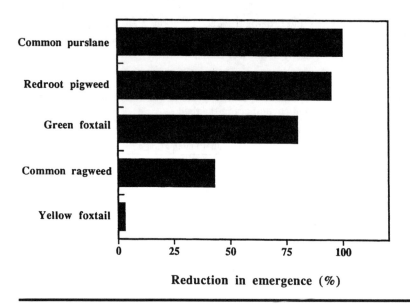

FIGURE 8-9. Reduction in emergence of weeds in rye residues. Emergence compared to control with no residues present.

Source: Modified from Putnam et al., 1983.

growth. The oat cultivar 'Garry' was used as a standard for comparison with the other accessions, and mustard was the species used to detect allelopathic activity in the bioassays. Twenty-five accessions had more scopoletin than Garry; four had 3 times as much; and accession 266281 had nearly 4 times as much. Figure 8-10 shows that accession 266281 was more inhibitory toward mustard top growth than was Garry. In fact, mustard exposed to this accession made only about one-third as much growth as mustard grown alone.

Suppression of weed growth directly in the field was demonstrated by screening rice germplasm accessions for allelopathic activity toward the aquatic weed ducksalad (*Heteranthera limosa*) under paddy conditions in Arkansas (Dilday et al., 1994). Of 10,000 accessions evaluated, 347 showed allelopathic activity by suppressing growth of ducksalad in a zone about the plant with a radius of 13 cm or greater. Weed control for these accessions ranged from 50% to 90% based on a comparison of the number of ducksalad plants within the affected zone with the number of plants around the cultivar 'Rexmont', which had no allelopathic activity. Representative accessions showing allelopathic activity are presented in Table 8-8. These germplasm accessions can be used as sources of allelopathic traits for use in varietal development programs.

The rye variety, Forrajero-Baer, selected for its ability to exude high concentrations of hydroxamic acids from roots, reduced total weed biomass in the field by 83% and 76% compared to wheat and oats, respectively (Perez and Ormeno-Nunez, 1993). The results showed that the hydroxamic acids exuded by this rye variety were a major factor in its weed-suppressing ability.

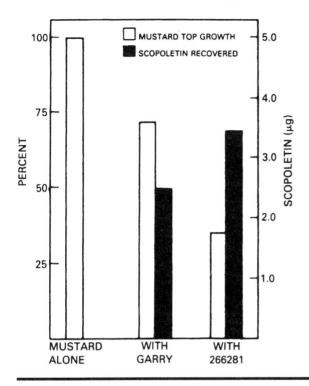

FIGURE 8-10. Varying allelopathic effects of crop germplasm on weeds.
Source: Fay and Duke, 1977. Reproduced with permission of the Weed Science Society of America.

Crop residues. Residues of rye varietal lines selected for weed-suppressing ability were shown to offer promise for suppressing weed growth in a no-till vegetable production system (Barnes and Putnam, 1983). Up to 93% control of weed biomass was obtained with rye planted in the fall, killed in spring, and vegetables planted in the residue. Yields of peas were not reduced by the residue.

From the foregoing, we see that allelopathy on the part of the crop offers several possibilities for preventing or reducing crop losses from weeds. These possibilities are examined in Chapter 15.

Novel Natural Herbicides

Identified allelochemicals offer a high potential for the development of novel, natural herbicides (Duke and Abbas, 1995). Allelochemicals may use previously unexploited molecular sites of action and provide leads to new classes of herbicides. This is particularly important with the increased incidence of weeds evolving resistance to currently

TABLE 8-8. Representative rice germplasm accessions that demonstrated allelopathic activity to ducksalad. Data from field evaluations conducted in Arkansas during 1990 and 1991.

Germplasm No.	Mean Radial Activity[a]	Weed Control[b]
	(cm)	(%)
385421	20	85
389551	18	90
400286	18	80
231649	18	70
312777	17	65
338711	15	80
385471	15	50
162176	14	85
223517	14	65
385513	13	80
Rexmont[c]	0	0

[a]Radial zone around the plant where no ducksalad growth appeared in paddy water or a reduced stand of ducksalad appeared.
[b]Weed control within affected zone based on number of ducksalad plants in check plot around control plants having no allelopathic activity.
[c]Standard rice cultivar with no allelopathic activity toward ducksalad.
Source: Data from Dilday et al., 1994.

used herbicides. Also, potential contamination of food, soil, and water may be less likely than with most synthetic herbicides used at similar rates. Development of allelochemicals as herbicides is hampered by the chemical complexity of many natural compounds making synthesis very difficult and by the problems associated with working with very small quantities available from natural sources (Duke and Abbas, 1995).

SUMMARY

Figure 8-11 summarizes, in schematic form, what is known about allelochemicals and suggests their ecological significance for relationships between weeds and desired plants. The weed and the desired plant can be affected at any time during their life cycles. The reason is the pervasiveness of allelochemicals in plant parts and in the environment, the multiplicity of factors affecting their production, and the many growth processes affected. The end effects on the ecosystem are in three broad areas: (1) weed persistence, (2) crop yield, and (3) weed composition.

The role of allelochemicals in weed persistence, and thus composition, may be most significant with perennial species. It can be postulated that the production of such chemicals would be of relatively greater value to survival of perennial than of annual weeds. As is known, most annual weeds rely upon numbers (r-strategy) for survival. Thus, their strategy is to maximize the number of different sites they can occupy. Although allelopathy toward other species conceivably might help them become estab-

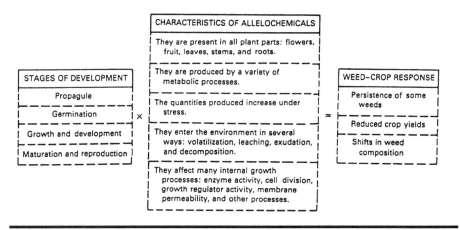

FIGURE 8-11. Schematic representation of allelopathic relationships in ecology of weeds and desired plants.

lished in those sites, it would not appear to offer any substantial advantage for their primary strategy, which is to produce many seeds. Most perennials, on the other hand, rely upon their competitive ability (K-strategy) for survival. That is, they depend upon characteristics that increase their ability to gain a disproportionate share of the resources in a given area. Production of allelochemicals that exclude other species would clearly aid a perennial in its competitive struggle. Since perennial weeds increase under reduced tillage, it can be further postulated that allelopathy will become relatively more important for all areas of weed–crop response as this practice continues to expand.

The effect on yield may well be the most significant effect for weed management. It has been established that the presence and quantity of allelochemicals is under genetic influence. Further, the amount produced by a given plant and its release into the environment are influenced by factors that can be controlled, at least to some degree, by cultural practices. Thus, the yield response offers the two-pronged approach of crop breeding and production management in utilizing allelopathy for minimizing or reducing interference from weeds.

REFERENCES

Barnes, J.P., and A.R. Putnam. 1983. Rye residues contribute weed suppression in no-tillage cropping systems. J. Chem. Ecol. 9:1045-1057.

Bell, D.T., and D.E. Koeppe. 1972. Noncompetitive effects of giant foxtail on the growth of corn. Agron. J. 64:321-325.

Connick, W.J., J.M. Bradow, and M. Legendre. 1989. Identification and bioactivity of volatile allelochemicals from amaranth residues. J. Agric. Food Chem. 37:792-796.

Darymple, R.L., and J.L. Rogers. 1983. Allelopathic effects of western ragweed on seed germination and seedling growth of selected plants. J. Chem. Ecol. 9:1073-1078.

Dekker, J., and W.F. Meggitt. 1983. Interference between velvetleaf (*Abutilon theophrasti* Medic.) and soybean (*Glycine max* (L.) Merr.): I. Growth. Weed Res. 23:91-101.

Dekker, J.H., W.F. Meggitt, and A.R. Putnam. 1983. Experimental methodologies to demonstrate allelopathic plant interactions: the *Abutilon theophrasti-Glycine max* model. J. Chem. Ecol. 9:945-981.

del Moral. 1972. On the variability of chlorogenic acid concentration. Oecologia 9:289-300.

Dilday, R.H., J. Lin, and W. Yan. 1994. Identification of allelopathy in the USDA-ARS rice germplasm collection. Aust. J. Exp. Agric. 34:907-910.

Drost, D.C., and J.D. Doll. 1980. The allelopathic effect of yellow nutsedge on corn and soybeans. Weed Sci 28:229-233.

Duke, S.O. 1986. Microbial phytotoxins as herbicides—a perspective. In A.R. Putnam and C.S. Tang, eds., The science of allelopathy, pp. 287-304. New York: John Wiley & Sons.

Duke, S.O., and H.K. Abbas. 1995. Natural products with potential use as herbicides. In K.M. Inderjit, M. Dakshini and F.A. Einhellig, eds., Allelopathy: Organisms, processes, and applications, pp. 348-362. Am. Chem. Soc., Washington, D.C.

Einhellig, F.A. 1995. Mechanisms of action of allelochemicals in allelopathy. In K.M. Inderjit, M. Dakshini and F.A. Einhellig, eds., Allelopathy: Organisms, processes and applications, pp. 96-116. Am. Chem. Soc., Washington, D.C.

Einhellig, F.A., and J.A. Rasmussen. 1973. Allelopathic effects of *Rumex crispus* on *Amaranthus retroflexus*, grain sorghum and field corn. Amer. Midl. Natural. 90:79-86.

———. 1989. Prior cropping with grain sorghum inhibits weeds. J. Chem. Ecol. 15:951-960.

Einhellig, F.A., and I.F. Souza. 1992. Phytotoxicity of sorgoleone found in grain sorghum root exudates. J. Chem. Ecol. 18:1-11.

Fay, P.K., and W.B. Duke. 1977. An assessment of allelopathic potential in *Avena* germplasm. Weed Sci. 25:224-228.

Gressel, J.B., and L.G. Holm. 1964. Chemical inhibition of crop germination by weed seeds and the nature of inhibition by *Abutilon theophrasti*. Weed Res. 4:44-53.

Kobayashi, A., S. Morimoto, Y. Shibata, K. Yamashita, and M. Numata. 1980. C_{10}-Polyacetylenes as allelopathic substances in dominants in early stages of secondary succession. J. Chem. Ecol. 6:119-131.

Kommedahl, T., J.B. Kotheimer, and J.V. Bernardini. 1959. The effects of quackgrass on germination and seedling development of certain crop plants. Weeds 7:1-12.

Kremer, R.J. 1986. Antimicrobial activity of velvetleaf (*Abutilon theophrasti*) seeds. Weed Sci. 34:617-622.

Leather, G.R. 1983. Sunflowers (*Helianthus annuus*) are allelopathic to weeds. Weed Sci. 31:37-42.

Lehman, R.H., and E.L. Rice. 1972. Effect of deficiencies of nitrogen, potassium, and sulfur on chlorogenic acids and scopolin in sunflower. Amer. Midl. Natur. 84:71-80.

Letourneau, D.K. 1988. Allelochemical interactions among plants, herbivores, and their predators. In P. Barbosa and D.K. Letourneau, eds., Novel aspects of insect–plant interactions, pp. 11-64. New York: John Wiley & Sons.

Lockerman, R.H., and A.R. Putnam. 1979. Evaluation of allelopathic cucumbers (*Cucurbita sativus*) as an aid to weed control. Weed Sci. 27:54-57.

Lolas, P.C., and H.D. Coble. 1982. Noncompetitive effects of johnsongrass (*Sorghum halepense*) on soybeans (*Glycine max*). Weed Sci. 30:589-593.

Muller, C.H. 1957. The role of chemical inhibition (allelopathy) in vegetational composition. Bull. Torrey Bot. Club 93:332-351.

Paszkowski, W.L., and R.J. Kremer. 1988. Biological activity and tentative identification of flavonoid components in velvetleaf (*Abutilon theophrasti*) seed coats. J. Chem. Ecol. 14:1573-1582.

Perez, F.J., and J. Ormeno-Nunez. 1993. Weed growth interference from temperate cereals: the effect of a hydroxamic acids-exuding rye (*Secale cereale* L.) cultivar. Weed Res. 33:115-119.

Putnam, A.R., and W.B. Duke. 1978. Allelopathy in agroecosystems. Annu. Rev. Phytopathol. 16:431-451.

Putnam, A.R., and C.-S. Tang. 1986. Allelopathy: state of the science. In A.R. Putnam and C.-S. Tang, eds., The science of allelopathy, pp. 1-19. New York: John Wiley & Sons.

Putnam, A.R., J. DeFrank, and J.P. Barnes. 1983. Exploitation of allelopathy for weed control in annual and perennial cropping systems. J. Chem. Ecol. 9:1001-1010.

Qasem, J.R., and T.A. Hill. 1989. On difficulties with allelopathy methodology. Weed Res. 29:345-347.

Rice, E.L. 1984. Allelopathy, 2nd ed. New York: Academic Press.

_____. 1995. Biological control of weeds and plant diseases: Advances in applied allelopathy. Norman: Univ. of Oklahoma Press.

Steenhagen, D.A., and R.L. Zimdahl. 1979. Allelopathy of leafy spurge (*Euphorbia esula*). Weed Sci. 27:595-598.

Tang, C., and C. Young. 1982. Collection and identification of allelopathic compounds from the undisturbed root system of bigalta limpograss (*Hemarthria altissima*). Plant Physiol. 69:155-160.

Weston, L.A., B.A. Burke, and A.R. Putnam. 1987. Isolation, characterization and activity of phytotoxic compounds from quackgrass [*Agropyron repens* (L.) Beauv.]. J. Chem. Ecol. 13:403-421.

9

Biotic Agents in Weed Management

CONCEPTS TO BE UNDERSTOOD

1. Biological control is taking place continuously in the plant world.
2. Biotic agents can be selective not only against specific weeds but also against specific plant parts.
3. Biotic agents may have greater potential value in weed management than in weed control.
4. Success with biotic agents in weed management calls for a full understanding of ecological relationships between the weed and the biotic agent and of both as a part of a larger ecosystem.
5. Evaluating effectiveness of biotic agents in weed management requires different assessments than those used in weed control.
6. Biotic agents can be easily combined with other weed control methods.
7. Biotic agents can be significant components in integrated weed management systems.

Weeds have their enemies. As discussed in Chapter 4, a significant portion of the weed seeds produced are destroyed by predators before ever reaching the soil, and many more are destroyed by predators and microorganisms in the soil. The growing weed is also preyed upon. The effects of natural enemies, or biotic agents, may go relatively unnoticed unless they are of a dramatic nature, such as the elimination of American elm in much of the eastern United States by the fungal pathogen *Ophiostoma ulmi,* the causal agent of Dutch elm disease, during the second half of the 20th century. It is reasonable, therefore, to expect some weeds to be similarly eliminated by natural enemies and logical to search for natural enemies with potential use in weed management. Recognition of the potential for using natural enemies in weed control developed from observations by early naturalists and agriculturists and was the basis for the first-attempted use of natural enemies in controlling a troublesome weed in 1902 (Harley and Forno, 1992).

Biological control of weeds is the intentional use of living organisms to reduce the vigor, reproductive capacity, density, or impact of weeds. Before considering biological control, we should again remind ourselves that we are dealing with an ecological system. Insects, plant pathogens, and other pests, along with weeds, humans, the environment, and the desired plants are all part of this system. As discussed in Chapter 3, a change in any part of this system causes changes in the other parts. Thus, as we change our practices for dealing with weeds, we must be aware of changes that will occur with the other pests. Similarly, we must take into account the potential harmful effects on biotic weed control agents of control treatments, especially pesticides, used against other pests. It was reported, for example, that fungicides used in horticultural crops may reduce control of yellow nutsedge with the rust fungus *Puccinia canaliculata*, a potential biotic agent (Phatak et al., 1983).

The strategies of biological control of weeds can be classified in two broad categories: (1) classical or inoculative, and (2) inundative or mass exposure. The *classical* strategy is based on introduction of host-specific organisms (insects, pathogens, nematodes, and so on) from the weed's native range into regions where the weed has established and become a widespread problem. The biotic agents, after quarantine to assure host specificity, are released into weed-infested sites and are allowed to adapt and flourish in their new habitat over time, eventually establishing a self-perpetuating regulation of the weed infestation at acceptable levels. Thus, classical biological control requires a time period of one to several years to achieve adequate control while the agent population builds up to effective levels.

The *inundative* strategy attempts to overwhelm a weed infestation with massive numbers of a biotic agent to attain weed control in the year of release. In contrast to classical biological control, inundation involves timing of agent release to coincide with weed susceptibility to the agent and formulation of the agent to provide rapid attack of the weed host. A development of the inundative strategy is the *bioherbicide* approach, which involves application of weed pathogens in a manner similar to herbicide applications. Since most bioherbicides have been developed by using selected fungal agents that cause such diseases on weeds as anthracnose and rust, the term *mycoherbicide* is often used in reference to these fungal preparations. A development of the inundative strategy in which native populations of natural enemies are supplemented with agents reared artificially to speed control of weed infestations is termed *augmentation*. Both the classical and inundative strategies use either *exotic* (introduced) or *indigenous* (native) natural enemies.

The objectives for discussing biotic agents are to identify the place for such agents in weed management, including their integration into current systems, and to develop an understanding of factors affecting their successful use. To do this, it is necessary to draw upon work that was designed to control a particular weed since this is the nature of most available information. The concepts appropriate to weed management are identified from research reported for weed control.

Selected examples of successful and potentially successful control with biotic agents are illustrated to identify and develop concepts and describe factors affecting implementation of biological control. No attempt is made to review available literature on the numerous biotic agents and their associated target weeds. Refer to *Biological Control of Weeds Handbook* (Watson, 1993) or *Biological Control of Weeds: A World Catalogue of Agents and Their Target Weeds* (Julien, 1992) for information on control of specific weeds.

SUCCESSFUL CONTROL OF WEEDS WITH BIOTIC AGENTS

Insects

In assessing organisms as potential biotic agents of weeds, insects have received by far the greatest attention. A listing of biotic agents and their target weeds reveals that of over 180 weed species under study for biological control, more that 75% were the target of at least one insect (Julien, 1992). Thus, it is not surprising that a majority of successful biological weed control programs have been with insects. This fact should not lead to assumptions that the greatest chance for success in biological control of weeds will occur with use of insects. Other examples exist, as discussed later, that illustrate the successful use of pathogens as potential biotic agents.

Prickly pear cactus in Australia. Control of prickly pear cactus (*Opuntia* spp.) in Australia was the first large-scale success in biological control of a weed. These cacti were introduced as ornamentals in the 1800s from North and South America, where they are native. They escaped, and by 1925, infested about 24 million ha of good grazing land; 12 million ha were infested to the extent that the land was useless. In addition, the infestations were spreading to new areas at the rate of about 400,000 ha per year.

The government established a Commonwealth Prickly Pear Board in 1920. One of the Board's first decisions was to send an entomologist to America to search for natural enemies. About 150 insects were identified as selective feeders on the cacti. Of these, 50 were sent to Australia for further study and mass rearing. Twelve insects were released and had begun to exert some pressure on the weed when the cactus moth (*Cactoblastis cactorum* Berg) was discovered to be especially effective. Larvae of this moth were collected in Argentina in 1925. Some 3,000 eggs were placed on prickly pear leaves and shipped to Australia where they were increased during the remainder of 1925, with the resulting eggs released into the field early in 1926.

Four to 6 years later (1930-1932), large areas had been freed of the pest. Following this, the insect population dropped and prickly pear regrowth occurred. The moth again increased, and by 1935 about 95% and 75% of the infested areas in Queensland and New South Wales, respectively, had been freed of the weed. The density of some infestations and effective control with the cactus moth are shown in Figures 9-1A, 9-

A. Cactus-infested range before establishment of *Cactoblastis*

B. Virtual elimination of cactus following release of *Cactoblastis*

C. Rhodesgrass (*Chloris* spp.) pasture 1 year after burning dead cactus in 9-1B

FIGURE 9-1. Biological control of prickly pear cactus by the cactus moth (*Cactoblastis cactorum*) in Queensland, Australia.

Source: Dodd, 1940. Reproduced with permission of Department of Lands, Queensland, Australia.

1B, and 9-1C. Control is actually accomplished by the larvae tunneling through the plant—including the roots—during their feeding.

St. Johnswort in western United States.   Control of St. Johnswort (*Hypericum perforatum*) was the first success with biological control in the continental United States. St. Johnswort, or klamath weed, is a perennial introduced from Europe. It became a serious problem on rangeland in western United States and Canada. By 1930, some 28,000 ha were infested with this weed in Humboldt County, California, alone.

Control of the weed by a beetle (*Chrysolina quadrigemina*) is shown in Figure 9-2. This example illustrates the importance of synchronization of the life cycles between the control agent and the target weed. The beetle adults emerge from pupae in the spring and feed on new foliage of the St. Johnswort. Both the insect and the weed go through a resting stage from mid-spring until fall. Initiation of weed growth with fall grazing is accompanied by returned feeding activity, mating, and egg-laying by the beetles. In the spring, eggs complete hatching and larvae feed on the trailing growth until shoots appear. Thus, both the adult beetle and the larvae contribute to ultimate destruction of the weed. About 3 years were required from release of 8,600 to 12,300 beetles ha^{-1} until the population built to the point where effective control of St. Johnswort was obtained.

Musk thistle in pastures in the United States.   An additional example of insect control is cited to show what can be done by focusing on seed production. Musk thistle was introduced into the United States from Europe in the early 1900s and has since become a serious weed in pastures in much of the Midwest and Appalachian regions. Biological control programs were initiated in 1969 with the introduction of two weevils (*Rhinocyllus conicus* and *Trichosirocalus horridus*) from Europe. *Rhinocyllus conicus* lays its eggs on the backs of thistle heads where larvae feed on immature seeds within the receptacle and tunnel into the stems, causing the flower heads to turn brown and die (Figure 9-3).

Weevil feeding reduces production of seed in the infested head but is partially offset by greater production in the uninfested heads (Figure 9-4). Thus, seed production per plant was not reduced until the level of infestation reached 10 weevils per head, even though production per head was reduced by 7 per head. Because even a few escapes may produce many seeds, we might conclude that a large infestation would need to be maintained for several years to cause significant reduction in the thistle population. Indeed, Surles and Kok (1976) indicated "first substantial control occurring after six years" from the 1969 to 1970 releases. Progeny from the 1969-1970 releases were reported as yielding "substantial control" after 4 years. Thus, some period of time is needed for an introduced species to become adapted to a new site. These results also reflect the characteristic of many weeds to produce more seeds than needed to fully occupy an area if all of them germinated (see Chapter 3, carrying capacity).

In other work (McCarty and Lamp, 1982), time of flowering was also shown to be a factor in the extent of larval feeding. As can be seen in Figure 9-5, weevil pupation

A. Photograph taken in 1946, with foreground showing weeds in heavy flower while rest of field has just been killed by beetles

B. Portion of the same location in 1949 when heavy cover of grass had developed

C. Photograph taken in 1966 showing degree of control that has persisted since 1949 (similar results reported throughout state)

FIGURE 9-2. Control of St. Johnswort or Klamath weed by *Chrysolina quadrigemina* at Blocksburg, California.

Source: C.B. Huffaker, 1957. Reproduced courtesy of C.B. Huffaker, University of California.

B. Larvae tunnel into bracts and into receptacles to feed on developing seeds

A. Weevils lay eggs on bracts of developing flowers

C. Larvae may tunnel into stem

D. Tunneling causes flower heads to turn brown and die

FIGURE 9-3. Control of musk thistle seed production with *Rhinocyllus conicus*.

Source: Roof et al., 1982. Reproduced courtesy of University Missouri-Columbia, Extension Division.

chambers per head (an indirect measure of weevil numbers) decreased as the season progressed. By July 17, the infestation level was down to only 0.2 chambers per head. The associated percent of well-developed seed increased from 10% on June 26 to 50% on July 17. Although not shown in Figure 9-5, seed production for the four flowering dates combined was 78%. Still left is production of a relatively large number of seeds, but the researchers indicated that proportionately more infestation will occur on late-blooming heads as weevil populations grow in subsequent years.

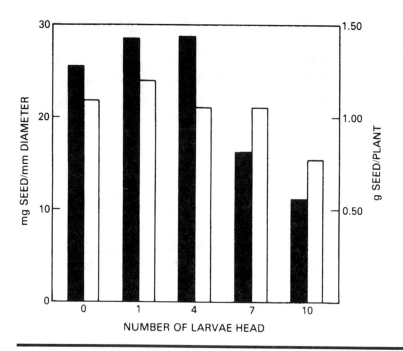

FIGURE 9-4. Effect of weevil numbers on seed production of musk thistle. Solid bars indicate terminal flower head seed weight; open bars indicate total seed weight per plant.
Source: Data from Surles and Kok, 1976.

Plant Pathogens

Use of plant pathogens as biotic agents is a more recent approach to biological control of weeds. Some striking results indicate the potential for their use in weed management. Examples of fungi, bacteria, and nematodes illustrate this potential.

Fungi

Skeleton weed in Australia and the United States. Skeleton weed (*Chondrilla juncea*) is a deep-rooted perennial introduced from Europe to Australia and to the United States. The weed is being controlled in Australia and the western United States with a rust fungus (*Puccinia chondrillina*) introduced from Europe as the major biotic agent, although selected insect agents have also been released.

The rust has been common in the field in all areas of Australia since 1972. Figure 9-6 shows changes in rosettes (weed density) that have occurred. As was true of weevils for the control of musk thistle, several years of infection (3 or more) were needed before substantial reduction occurred in the skeleton weed population. Lee (1986) provides information about the developing effectiveness of this rust with time. Where the

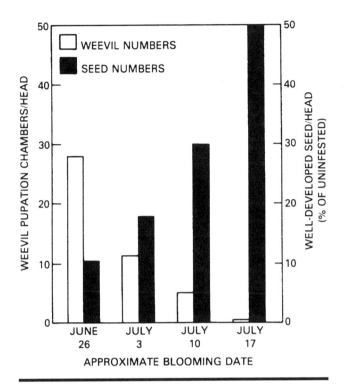

FIGURE 9-5. Effect of time of thistle flowering on extent of weevil infestation and associated seed production.

Source: Data from McCarty and Lamp, 1982.

rust had been released for less than a year, the size of the skeleton weed was reduced, but reproductive capacity was not impaired. Two years after release, the rust populations developed to the point where infection of stems and the flower calyx was severe. Such plants produced fewer flowers and seed; furthermore, the seeds had lower viability. Mortality of infected seedlings may exceed 90% under conditions of severe rust infestations. This effect on reproduction reduces the potential for spread and reinfestation, thus accounting for gradual control of a given population.

Northern jointvetch in the United States. In Arkansas, control of northern jointvetch (*Aeschynomene virginica*) by an anthracnose disease fungus (*Colletotrichum gloeosporioides* f. sp. *aeschynomene*) represented a major advance in biological control of weeds for two reasons (Templeton, 1982). First, northern jointvetch is a weed native to North America, whereas previous examples involved control of introduced weeds by pests from their native sites. Second, massive quantities of the anthracnose fungus were applied to attain control of the weed in the year of treatment rather than by introducing the agent and allowing populations adequate for control to develop over time. Treat-

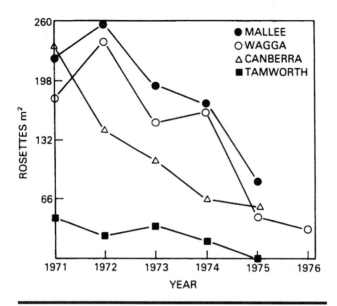

FIGURE 9-6. Reductions in skeleton weed in continuous pastures in southeastern Australia by an introduced rust plus a midge.

Source: Data from Cullen, 1976.

ment was a spray application of a spore suspension of the fungal pathogen at 188 billion spores ha[-1] in 94 L of water. In this manner, application of the biotic agent was similar to that of a herbicide. Nearly complete kill of the jointvetch was obtained on infestations in rice, as illustrated in Figure 9-7.

Bacteria

Annual bluegrass. Bacterial pathogens of weeds have only recently been identified and investigated for potential as biotic agents. *Xanthomonas campestris* pv. *poannua* is under development as a biotic agent for control of annual bluegrass, a winter annual weed in turfgrass. This pathogen is applied as a bacterial cell suspension in liquid to freshly mowed grass and enters and infects annual bluegrass tissues. The bacterium spreads through the xylem of infected leaves eventually causing wilt of the entire plant and subsequent death. Limited field trials revealed that *X. campestris* pv. *poannua* provided up to 91% control of annual bluegrass in bermudagrass turf; however, control was affected by date and rate of application and by bacterial strain (Johnson, 1994).

Composite family weeds. *Pseudomonas syringae* pv. *tagetis* produces "tagetitoxin" that causes apical leaf chlorosis in many plants of the Compositae. The strain under investigation for biological control was isolated from chlorotic terminal leaves of Canada

A. Untreated

B. Treated with a suspension of spores of
Colletotrichum gloeosporioides f. sp.
aeschynomene

FIGURE 9-7. Control of northern jointvetch in rice with the anthracnose fungal pathogen *Colletotrichum gloeosporioides* f. sp. *aeschynomene.*
Source: Reproduced courtesy of USDA, Agricultural Research Service, Stuttgart, Arkansas.

thistle (Johnson et al., 1996). Cell suspensions of this pathogen combined with a surfactant promotes rapid infection of leaves when applied as a postemergence spray. Disease symptoms include chlorosis and severe suppression of plant vigor and development. In field trials, the pathogen suppressed Canada thistle growth causing little or no competition with soybean and limited regrowth of Canada thistle in the same field during the following season. Growth of other composite weeds, including common cocklebur, sunflower, and common ragweed, is also suppressed by this pathogen, but it has no apparent effect on soybeans or corn. This pathogen shows great potential as an effective biotic agent for postemergence control of both annual and perennial composite weeds in row cropping systems.

Downy brome.   Another group of bacteria under intensive investigation for potential as biotic agents are *deleterious rhizobacteria.* Deleterious rhizobacteria differ from the pathogens just described in that they are nonparasitic bacteria colonizing plant roots and able to suppress plant growth without invading the root tissues. Deleterious rhizobacteria with potential as biological control agents were first described on downy brome (*Bromus tectorum*) occurring in winter wheat (Kennedy et al., 1991) and on several broadleaf weed seedlings (Kremer et al., 1990). Practical use for weed management involves applying inocula to establish high numbers of bacteria in the seed zone and/or rhizosphere of the target weed and initiate rapid growth-inhibitory activity. Weed management is not dependent on development of an endemic disease on established weeds, which is the basis for activity of most mycoherbicides. Rather, the rhizobacteria strategy seeks to suppress development of weeds before or coincident with emergence of desired plants, thus allowing the latter to effectively compete for growth requirements with the weakened weed seedlings (Kremer and Kennedy, 1996).

Biological control of downy brome in winter wheat by *Pseudomonas* spp. isolated from downy brome roots has been demonstrated under field conditions (Kennedy et al., 1991). A suspension of 10 million bacterial cells in 1 L water was applied to the soil surface of 1 m^2 plots infested with downy brome immediately after planting to winter wheat at three field sites in Washington. Two rhizobacterial types or isolates consistently reduced downy brome density, growth, and seed production at all three locations. Representative data from the Washtucna site (Table 9-1) showed that winter wheat densities were not affected by the bacteria and grain yield was significantly increased. The increase in wheat yields primarily was due to the growth suppressive effects of the applied bacteria on downy brome, which allowed the wheat to be more competitive. The significant reductions in seed production by the surviving downy brome plants in the rhizobacteria-inoculated plots suggested that replenishment of the seedbank was affected and could reduce future downy brome infestations.

TABLE 9-1.   Effect of selected rhizobacteria on downy brome growth and winter wheat density and yield at Washtucna, Washington, 1988.

	Downy Brome			Winter Wheat	
Treatment	Density (plants m^{-2})	Shoot Mass (g m^{-2})	Total Seeds (no. m^{-2})	Density (plants m^{-2})	Yield (kg ha^{-1})
Pseudomonas D7	164	18	9000	31	4360
Pseudomonas 2V19	188	19	17000	32	4100
Noninoculated	252	33	25000	30	3230

Source: Modified from Kennedy et al., 1991.

Viruses. Viruses have been described for some weeds, and only a few have been studied for biological control potential. A virus preparation caused leaf damage and reduced leaf growth and plant weight of milkweed vine (*Morrenia odorata*) under greenhouse conditions (Charudattan et al., 1980). The virus affects plants only in the milkweed family and was under serious consideration as a biotic agent for milkweed vine in citrus groves in Florida. A causal agent that is suspected to be a virus is responsible for rose rosette disease leading to growth suppression of multiflora rose (*Rosa multiflora*) and shows potential for biological control of this weedy shrub in pastures (Epstein and Hill, 1995).

Nematodes. Several plant parasitic nematodes exist that offer potential as biological control agents of weeds.

Silverleaf nightshade. A natural, specific host–parasite relationship described for the weed silverleaf nightshade (*Solanum elaegnifolium*) and a nematode (*Orrina phyllobia*) has been exploited as a biological control strategy (Parker, 1986). Silverleaf nightshade is a problem perennial weed in the Texas High Plains and can cause substantial yield reduction in cotton. The weed can spread faster than its parasitic nematode, resulting in silverleaf nightshade populations free of the nematode. The nematode, which incites galls on the leaves, is grown on weed hosts until the foliage senesces, at which time leaves are harvested to be used as inocula on silverleaf nightshade infestations devoid of the nematode. Field tests where galls containing the nematode biotic agent were used as inoculum (Table 9-2) have resulted in 23% and 42% reductions in silverleaf nightshade foliar biomass and density, respectively (Northam and Orr, 1982).

Russian knapweed. The nematode *Subanguina picridis* incites galls on stems, leaves, and root collars of Russian knapweed [*Acroptilon (Centaurea) repens*], a perennial weed infesting pastures and rangelands of the western United States. The nematode originally released to allow populations to build up naturally required several years before an impact on knapweed infestations was noticed. Recent developments in culturing the nematode in shoot-tips of knapweed on artificial medium allowed propa-

TABLE 9-2. **Effect of the nematode *Orrinia phyllobia* on the growth of silverleaf nightshade in Texas in 1977 and 1978.**

Treatment[a]	Plant Density			Foliar Biomass		
	1977 (no. m^{-2})	1978 (no. m^{-2})	Reduction (%)	1977 (g m^{-2})	1978 (g m^{-2})	Reduction (%)
Control	1.7	2.9	0	7.1	8.3	0
Control, disturbed	3.4	2.6	24	14.9	4.9	67
Inoculated	5.6	1.9	66	54.5	5.6	90

[a]Control = nondisturbed, shortgrass prairie; Control, disturbed = prairie with vegetation mechanically removed in 1975; Inoculated = disturbed site receiving 32 g nematode-infested galls 9 m^{-2} in April 1977.
Source: Modified from Northam and Orr, 1982.

gation of large numbers that were used for massive inoculation of the weed, resulting in infection and suppressed plant growth (Ou and Watson, 1993). Subsequent research resulted in the development of a formulation comprised of encapsulated nematodes extracted from Russian knapweed galls that extends survival during storage and maintains infectivity of the nematodes after application to the plant (Caesar-ThonThat et al., 1995).

Large Animals

Large animals can be specialized in their choice of plants to feed upon. This fact has been utilized to manage weed growth in special situations. One example is use of geese to remove weeds in strawberries and cotton. Geese prefer weed seedlings, especially grasses, to established strawberry or cotton plants and have been used successfully to prevent weed interference with the crops.

Nonagricultural Weed Control with Biotic Agents

Weed problems are not limited to agroecosystems comprised of row crops and managed pastures but are often also found in recreational areas, waterways, and wetlands. In fact, weed infestations may become so great in some recreational waters that swimming and boating are discouraged. Similarly, weeds may interfere in waterways to the extent that water flow and navigation are severely impeded and flood control and irrigation are restricted. Biotic agents have been successfully used against such problems. The grass carp (*Ctenopharyngodon idella*), a nonselective herbivorous fish from China, was first released in Arkansas and Alabama in 1963 and readily demonstrated ability to effectively control the aquatic weed hydrilla (*Hydrilla verticillata*). The fish may consume several times its own weight in vegetation daily. Grass carp have since been introduced to 29 other states for control of four aquatic weed species and continue to provide excellent weed control, including a greater than 90% reduction in hydrilla biomass in irrigation canals in southern California (Watson, 1993). The mottled water-hyacinth weevil (*Neochetina eichhorniae*) was introduced into Florida in 1972 to control waterhyacinth (*Eichhornia crassipes*) in waterways (Charudattan, 1986). This weevil and a mite (*Othogalumna terebrantis*) apparently introduced with waterhyacinth successfully reduced waterhyacinth growth over a period of years. Later work with a fungal pathogen, *Cercospora rodmanii*, combined with the insect agents yielded 99% control of waterhyacinth within 7 months. Combinations of the fungus and insects were considered the most optimum biological control approach for substantial reductions in weed populations in a single growing season.

Purple loosestrife (*Lythrum salicaria*) is a herbaceous perennial and prolific seed producer that has invaded wetlands in the north-central and eastern United States replacing native plants and adversely affecting wildlife habitats. Biological control is considered the most suitable approach for managing the weed in the wetland ecosystem. Several insect agents comprised of species feeding on rootstock, foliage, or seeds have been released in the United States and are under evaluation for their combined impact on purple loosestrife infestations (Blossey, 1995). It is expected that use of multiple species of insects rather than individuals will have the greatest success in biological control.

Biorational Approach for Biological Weed Control

The use of chemical compounds produced by microorganisms or plants in a manner similar to herbicides is a *biorational* approach to weed control. Although the use of these natural compounds is not strictly considered biological weed control since the living organism is not applied to weeds, it is relevant since the objective of both biorational and biological weed control approaches is to suppress weed growth and adversely affect weed infestations by natural means. There is considerable interest in development of natural products for use in weed management because they are highly effective against weeds, are not toxic against nontarget organisms, cause no damage to the environment, and are readily biodegradable. Thus, these compounds may serve as patterns for synthesis of new and "natural" herbicides. Currently, no less than 10 microbial herbicides have been discovered and developed for use in Japan (Okuda, 1992). Nearly all of these have been isolated from filamentous soil bacteria (actinomycetes) and are used for weed control in rice. Bialaphos, used as a nonselective preplant herbicide, is an example of a microbial herbicide. Some pathogens currently evaluated as biotic agents also produce herbicidal compounds. Tagetitoxin, from *P. syringae* pv. *tagetis* discussed above, suppressed growth of Canada thistle (Johnson et al., 1996) and maculosin, a phytotoxin specific for spotted knapweed (*Centaurea maculosa*) and isolated from the fungus *Alternaria alternata* (Strobel et al. 1990), are effective on their target weeds in the absence of their respective biotic agents. These and other natural compounds may lead to development of biorational products and offer alternatives to synthetic herbicides in the future.

ROLE OF BIOLOGICAL CONTROL IN WEED MANAGEMENT

Preventing Weed Seed Production

One role for biotic agents may be in preventing seed production of weed specimens that escape the other weed management practices. As mentioned previously, preventing seed production (renewal of the seedbank in the soil) must be accomplished for a long-term effort to reduce weed problems to be successful. In Chapter 15, we examine practices that might be embraced in attaining this goal. Here, we simply need to recognize that a producer or other land owner will probably not be able to justify more than a modest cost for practices designed to cut weed seed production.

Weed seed production in itself does not represent a yield loss for a particular crop in a particular year. But the crop must pay the cost of a treatment to reduce weed seed. Thus, a relatively inexpensive treatment is needed, but one that does not necessarily have to destroy the weed since competition may not be involved. Biotic agents may fit these requirements, especially if it is acceptable for the biotic effect to develop over a period of years and to be less than 100% effective. Both qualifications may be acceptable.

Recent work with seed-attacking organisms illustrate the potential for reducing weed seed production. The loose smut fungus, *Sphacelotheca holci*, infects johnsongrass systemically, nearly eliminating seed set as well as reducing aboveground biomass by 75% (Massion and Lindow, 1986) . Successful development of this fungus as a biotic agent could dramatically reduce johnsongrass seedling infestations by depleting the seedbank over time. Inundative releases of a seed-feeding insect, *Niesthrea louisianica,* on velvetleaf infestations at several field sites in the midwestern United States resulted in reduction in viable seeds produced by 5% to 90% (Figure 9-8), depending on the site and distance from the center of insect release (Spencer, 1988). These results suggest that seed-feeding insects released in high numbers could reduce velvetleaf seedbanks in enterprise fields where velvetleaf escaped conventional postemergence weed control although yearly re-introductions of the insect would be required.

Preventing Weed Buildup

Biotic agents may also have a place in checking weed species that have not yet reached a competitive threshold level. An example might be their use against perennial weeds that are increasing under reduced-tillage farming systems. Shifts in weed composition in response to changes in crop practices such as tillage result in development of subcompetitive populations during the early years of the shift. The general tendency is for perennial species to increase as amount of tillage decreases. For example, common milkweed infestations are increasing in the Midwest due at least partially to reduced tillage. Although common milkweed occurs in wheat, corn, and soybeans, it probably is not an important cause of reduced crop yields. Biotic agents might provide a way of keeping the weed from becoming an economic weed problem. The discovery of a bacterial disease affecting common milkweed (Flynn and Vidaver, 1995) may lead to the development of the causal pathogen *Xanthomonas campestris* pv. *asclepiadis* as a biotic agent for maintenance of common milkweed stands below economic threshold levels. The disease is a systemic blight and reduces overall plant vigor and stand density. The report of milkweed bacterial blight is significant because we have an opportunity to test the "preventive biological control approach" and because we are more optimistic regarding the likelihood for discovering similar agents on other perennial weeds.

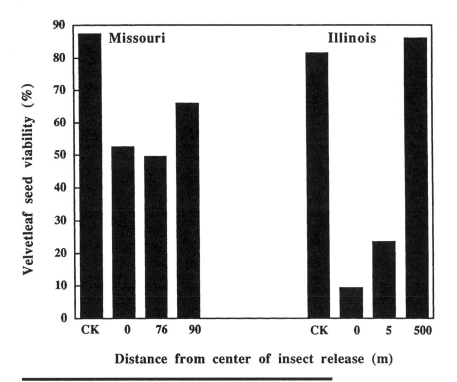

FIGURE 9-8. Reduction in seed viability of velvetleaf by the seed-feeding insect *Niesthrea louisianica* released at field sites in Missouri and Illinois in 1985. CK = check plot with no insect feeding.

Source: Modified from Spencer, 1988.

Special Weed Situations

Biotic agents may be of significant value when it is difficult to manage weeds by other means. For example, weeds in several areas cannot be controlled effectively with herbicides because of regulations severely restricting or banning herbicide use and where preservation of the environment is the primary goal. Biotic agents may provide the most effective or only alternative method of obtaining control under such conditions. These special situations include restoration of native ecosystems, wetlands, national parks, wildlife refuges, and areas bordering waterways. For example, red alder (*Alnus rubra*) is a forest weed that interferes with timber production. Red alder infestations can be suppressed by using a biocontrol fungus that is inoculated into the woody stems with a special injecting device (Dorworth, 1995). The fungus is useful for control of red alder along streams where herbicide application is prohibited and causes "slow killing" allowing slow release of nutrients from dying vegetation for use by desirable tree species as well as a gradual incursion of the crop trees.

Biological control may be useful in enterprises where the development of herbicide-resistant weed biotypes is a concern. Growth of imazaquin-resistant common cocklebur biotypes originating in soybean fields was suppressed with the mycoherbicide *Alternaria helianthi* (Abbas and Barentine, 1995). As pesticide resistance becomes more problematic with many common weeds, alternative strategies such as biocontrol can be important in maintaining adequate weed control. The use of biotic agents in special circumstances can be extended to control of parasitic weeds such as the witchweeds (*Striga* spp.), which are serious root parasites of cereal crops and considered the greatest biological constraint for food production in Africa. Herbicides and cultural methods alone have not given satisfactory suppression of the witchweeds. A soil fungus under evaluation has suppressed germination and attachment of witchweed seedlings to grain sorghum roots and increased grain sorghum yield (Ciotola et al., 1995). Development of this pathogen as a biotic agent could have a significant impact on food production in regions where *Striga* spp. are the dominant weed problem.

Weed Suppressants

Biotic agents may have a particular role as weed suppressants in fence rows, ditch banks, and other rights-of-way that serve as reservoirs for reinfestation of adjacent tilled land. Under such circumstances, some weed growth can be tolerated for the time it may take a biotic agent to reach the population level necessary to eliminate the weed as a source of reinfestation. In these situations, plant control would not be essential. Rather, control of seed production and spread by vegetative parts is all that would be needed. A practical example of limiting the spread of plants rather than elimination is the use of seed-feeding insects for destroying most of the seed crops of acacia (*Acacia* spp.) and mesquite (*Prosopis* spp.) trees in South Africa (Dennill and Donnelly, 1991; Zimmermann, 1991). The highly prolific seed production of these trees caused natural invasions of saplings into farming and grazing lands, which often resulted in impenetrable thickets making both the tree and land unusable. Since the trees have many useful purposes, biological control programs were established to use flower-, pod-, and seed-attacking insects to destroy seed production and reduce potential invasiveness while maintaining those stands that are economically productive.

INTEGRATION OF BIOLOGICAL CONTROL INTO WEED MANAGEMENT SYSTEMS

Development and rapid acceptance of biological control systems is challenged by factors limiting the spectrum of activity, efficacy, and reliability. For biological control to become a practical management option for weed management in enterprise systems, its effectiveness as a component in overall management programs must first be demonstrated rather than attempting to initiate programs based on complete biological control. Since most biological control agents are specific toward one weed and many fields contain several predominant weed species, the use of the agent for control of its target species in conjunction with herbicides selected for control of the other weeds is a logical approach. Alternatively, efforts are under way to develop genetically altered agents that are able to control several important weeds in a crop (Miller et al., 1989). This is a long-term tactic, however, as intense evaluation will be required to meet stringent regulations before approval is granted for release of such organisms in the environment.

Herbicide–Mycoherbicide Combinations

Integration with reduced rates of herbicides can successfully improve activity of mycoherbicides toward weeds. For example, the mycoherbicide *Phoma proboscis* was more effective in controlling field bindweed when combined with sublethal doses of 2,4-D than when applied alone (Heiny, 1994). Combinations of mycoherbicide applied preemergence or postemergence with trifluralin, either tank-mixed or alone as a water suspension or granular formulation, effectively controlled Texas gourd (*Cucurbita texana*), a problem weed in soybean and cotton in the southern United States (Weidemann and Templeton, 1988).

The fungus *Colletotrichum gloeosporioides* f. sp. *malvae*, endemic on round-leaved mallow, provides adequate control (about 75% kill) when applied alone as a mycoherbicide (Grant et al., 1990). Several chemical herbicides are only effective on round-leaved mallow (*Malva pusilla*) in the early seedling stage. Combinations of *C. gloeosporioides* f. sp. *malvae* with several herbicides at recommended rates were evaluated for postemergence control at the 4- to 5-leaf stage of growth. Tank mixes of the fungus with either metribuzin or imazethapyr greatly enhanced control and reduced biomass production over the fungus or the herbicides alone (Figure 9-9). These results clearly demonstrate that in some cases no single method is adequate for weed control and that combinations of methods are most effective.

Compatibility of pathogens with several herbicides suggests that they could be integrated into existing weed management strategies to broaden spectra of weed control within the crop or to enhance the activity of herbicides recommended for control of specific weeds. This can be extended to integrating mycoherbicides any time during the growing season by combining with herbicides in sequential applications for total weed control. Integrating herbicides with biological control insects may also provide satisfactory control by reducing weed density below economic thresholds more quickly than the insects alone (Messersmith and Adkins, 1995).

Combinations of Biotic Agents

Combinations of different biotic agents can enhance efficacy of control over that ex-

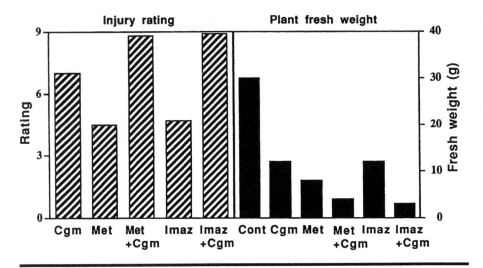

FIGURE 9-9. Effect of the mycoherbicide *Colletotrichum gloeosporioides* **f. sp.** *malvae* **and herbicides on injury rating and final plant weight of round-leaved mallow.** Rating of 0 = no effect; 9 = total kill. Cgm = *Colletotrichum gloeosporioides* f. sp. *malvae;* Met = metribuzin, Imaz = imazaquin; Cont = control.

Source: Modified from Grant et al., 1990.

hibited by either agent alone. Combinations of fungal pathogens were used to control both winged waterprimrose [*Ludwigia* (*Jussiaea*) *decurrens*] and northern jointvetch in rice (Boyette et al., 1979). Efficacy of mycoherbicides may be increased by combining selected bacteria with the pathogen (Schisler et al., 1991). Enhancement of detrimental activity on weed growth attacked by insects in association with fungi has been described (Hasan and Ayers, 1990). A seed-feeding insect combined with seed-attacking fungi significantly decreased velvetleaf seed viability and seedling emergence and increased seed infection compared to either the insect or fungus alone (Figure 9-10) (Kremer and Spencer, 1989). Pre-dispersal seed mortality on weeds escaping herbicide control may be effective in manipulating and reducing seedbanks in soil. For example, viability of insect-attacked and fungal-infected velvetleaf seeds was reduced to less than 2% after 24 months burial in soil, but nonattacked seeds maintained viability at about 80% (Kremer and Spencer, 1989), suggesting the potential of combined biotic agents for rapid depletion of the seedbank.

A practical application of soil-applied detrimental bacteria combined with insects would be in situations where the insect feeds on roots or crowns of target weeds. It has been suggested that leafy spurge control resulting from feeding by root-boring larvae of flea beetles (*Aphthona* spp.) may be enhanced due to secondary invasion of plant pathogens naturally present in soils (Rees and Spencer, 1991). Exploitation of flea beetle larvae as vectors of bacteria selective for suppression of leafy spurge could contribute an additional strategy for control of this noxious range weed.

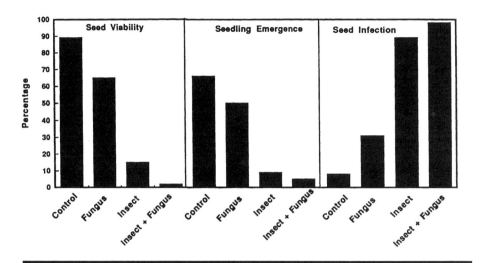

FIGURE 9-10. Effects of a seed-feeding insect (*Niesthrea louisianica*) and a seed-attacking fungus (*Fusarium oxysporum*) on production, viability, and fungal infection of velvetleaf seeds averaged over 1990 and 1991 growing seasons in central Missouri.
Source: Modified from Kremer and Spencer, 1989.

Cultural Practices

Cultural practices offer convenient application methods for integrating biological control in cropping systems. Crop rotation is a practice that may also be manipulated to encourage development of specific inhibitory bacteria on weed roots. Tillage can influence the frequency of inhibitory bacteria occurring in soil and their growth-suppressive activity. Greater proportions of indigenous rhizobacteria inhibitory to downy brome and jointed goatgrass were detected under either conventional or reduced tillage compared to no-tillage (Figure 9-11). This finding suggests that application of selected deleterious rhizobacteria during tillage may be effective in integrated weed management (Kennedy et al., 1989). Vegetative residues at or near the soil surface could serve as substrates for production of weed-suppressive agents by deleterious rhizobacteria applied directly to the residues. Previous work reported that a rotation effect in corn was due partly to certain rhizobacteria specifically associated with corn roots, and this work illustrates the potential for using such rhizobacteria to suppress weeds in crop rotation systems (Turco et al., 1990). Increasing crop interference in the field by manipulating row spacing, seeding rates, and other cultural practices to suppress early weed growth has been proposed as a viable component of integrated weed management (Jordan, 1993). Selection of highly competitive and allelopathic soybean varieties (Rose et al., 1984) matched with compatible biotic agents may provide early-season weed suppression and require only minimal subsequent postemergence weed control.

Inoculation.   A unique aspect for delivery of microbial agents to soil infested with weed seeds is either by direct inoculation of crop seeds with the agents or by promoting colonization of crop roots by the agents in formulations applied at planting. Crop roots not only may deliver microbial agents to adjacent roots of weeds but also maintain or even enhance the agent's numbers for attack of seedlings emerging later in the season.

Cover crops and mulches.   Cover crops and mulches as components of alternative management systems may be used for integrating biocontrol agents into these systems by delivering the agents on seeds and promoting their establishment in soils for attack of weed seeds and seedlings prior to planting the main crop. Previous research demonstrating that certain legume cover crops promoted populations of soilborne plant pathogens of cotton (Rothrock et al., 1995) suggests that cover crops may be useful in establishing weed-attacking microorganisms in soil well in advance of weed seedling emergence.

Integration with allelopathy.   Examples of integration of biocontrol with allelopathy have not been investigated even though augmentation of the efficacy of weed control through allelopathy seems attainable. An apparent interaction between allelopathic substances from kochia and the fungus *Rhizopus* sp. on inhibition of sugarbeet germination and seedling growth (Wiley et al., 1985) suggests the potential for discovery of

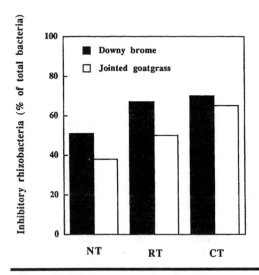

FIGURE 9-11. Proportion of indigenous rhizobacteria inhibitory toward downy brome and jointed goatgrass growing under three tillage systems in Washington. NT = no-till; RT = reduced tillage; CT = conventional tillage.

Source: A.C. Kennedy et al., 1989.

similar relationships for microbial agents and crop seeds that may be useful in weed management.

CONCLUSIONS REGARDING SUCCESSFUL USE

A number of concepts apply to biological control. The discussion that follows identifies and examines these concepts.

1. Plant pathogens may be better adapted to the bioherbicide approach than insects. They are better adapted because of their relative ease of application and storage. In any bioherbicide treatment, large numbers of the biotic agent are needed. To control jointvetch in rice, for example, 2 to 6 million spores of *C. gloeosporioides* f. sp. *aeschynomene* mL^{-1} were applied at the rate of 94 to 374 L ha^{-1}. The largest volume is somewhat higher than used for herbicides but still not so high as to make it impractical. By comparison, in augmenting natural populations of *Bactra verutana,* an insect that feeds on purple nutsedge, 60 adult pairs 4 m^2 were needed to be fully effective under caged conditions. Under field conditions, an infestation level of three larvae per shoot was needed (Frick and Chandler, 1978). This rate translates into 150,000 adult pairs ha^{-1} and possibly 10 million larvae ha^{-1}. Rearing and releasing such numbers of insects clearly pose practical obstacles for large-scale use by the inundative method. Most pathogens, in contrast, can be formulated for application as a spray or as granules and either as pre- or postemergence to weed seeds or seedlings (Daigle and Connick, 1990). Further, formulations of fungal spores and/or mycelia, bacterial cells, and nematodes offer convenient matrices in which to formulate, store, and deliver the biotic agent.

2. Eradication of the weed target is rarely, if ever, achieved. Experience in the control of prickly pear cactus by the cactus moth is an example. The cactus moth reduced prickly pear cactus infestations after 5 to 6 years to the extent that not enough weed biomass remained to sustain the cactus moth population. Cactus regrowth occurred, quickly followed by reinfestation and buildup of the cactus moth population. This type of variation in population levels of both the target weed and the biotic agent is typical of classical biological control. Fluctuations in densities of weeds and insects lead to a series of highs and lows in populations until an equilibrium population of both weed and biotic agent is established. The culmination is an ecological balance among the biotic agent, the weed, and the environment. If control is successful, this ecological balance stabilizes at a level of weed infestation below the economic threshold (Figure 9-12). In areas where the bioherbicide approach can be used, fluctuations in populations may occur but to a lesser degree than with the classical strategy. For example, with *C. gloeosporioides* f. sp. *aeschynomene* for jointvetch control, the weed may be completely controlled in the year of application, but reinfestation at a somewhat smaller

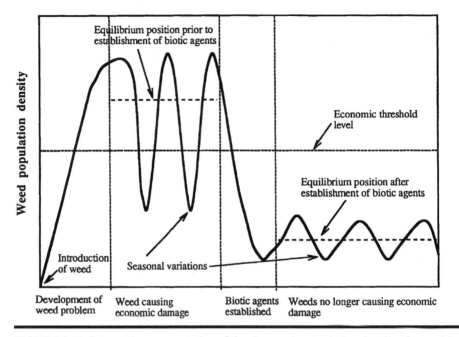

FIGURE 9-12. Schematic representation of the changes in population density of a weed before and after the establishment of a biotic agent.

Source: Modified from Harley and Forno, 1992.

population level from the soil seedbank can occur the following year, with a steady decline in reinfestations in subsequent years. The level ultimately reached will depend on the control achieved with annual applications and the length of time over which the treatments are made.

3. Control with the classical approach is usually achieved only over a period of years. Although dramatic first-year effects may occasionally occur, the more usual situation requires 3 to 10 years for the biotic agent to build to a level to provide effective control. For example, 10 years were required from the first release of the cactus moth before the infestation of prickly pear cactus was reduced to a satisfactory low level of infestation. With St. Johnswort, about 3 years was required following release of the beetle in sufficient numbers for it to reduce the weed below the economic damage level. Use of rust fungi against skeleton weed required several years before buildup was reflected in significant reduction of the weed. It should be recalled that biotic agents most likely improve in their overall efficiency with time as they become adapted to their new environment (Harley and Forno, 1992).

4. Classical biological control is more apt to be successful against introduced weeds than against native weeds. Several reasons underlie this statement. Biotic agents introduced from the native habitat can be expected to be free from constraints of predators and antagonists that might interfere with their effectiveness in the native habitat.

Also, from the standpoint of the target weed, control may be achieved before the weed has time to adjust to the presence of the biotic agent, assuming it has lost whatever tolerance it may have had for the biotic agent in its native habitat. The bioherbicide strategy stands a better chance of succeeding against native weeds, but is still constrained by the same factors affecting the inoculum approach, albeit to a lesser degree.

5. Quimby and Walker (1982) suggest that bioherbicides comprised of native pathogens may be more satisfactory than those comprised of introduced pathogens for the following reasons: (1) native biotic agents may be more readily obtained because overseas exploration and quarantine are not necessary; (2) native biotic agents may be expected to revert to original levels, thus lessening the problem of spread and adaptation to nontarget plants; (3) native biotic agents can be expected to be better adapted to the climate and environment of the area; (4) public concern may be less with native than with introduced biotic agents.

6. Success with biological control can be expected to be better against a single weed species than against a mixed population of weeds. The classical approach, almost of necessity, is restricted to single species. The problems in using a multitude of pests necessary to control a mixture of weeds simply are too great for this approach to offer much potential. The use of a single biotic agent with a broad spectrum of species upon which it is active can be expected to very much increase the chances of its attacking nontarget plants. Furthermore, control of one or a few weeds in a larger mixture of weeds can be expected to allow the uncontrolled species to pose as large a problem in competition as that posed by the entire mixture. The bioherbicide approach offers some possibility for controlling mixtures of weeds by using combinations of pathogens. However, use of "broad-spectrum" bioherbicides must address the same concerns regarding potential effects on nontarget plants as does use of classical biotic agents.

7. Perennial and biennial weeds are most likely targets for the classical strategy. Under the classical strategy, a stable and continuous food supply and shelter that the biotic agent requires of a host is best provided by a perennial plant in order for the agent to build and sustain effective control levels. Annual weeds vary greatly in numbers from year to year and are present for only part of the year. Furthermore, annual weeds may have fewer natural enemies from which to select effective biotic agents by virtue of their evolutionary heritage. As early colonizers in ecological succession, many annual weeds were continuously establishing on new sites, thus lessening the opportunities for natural enemies to become established. Finally, sexual reproduction common to annual species assures greater heterogeneity with the associated greater potential to evolve resistance than is found in species, such as many perennials, that reproduce asexually (Burdon and Marshall, 1981). These arguments do not mean that biological control does not have a place in the control of annual weeds. Bioherbicides were developed to overcome the necessity of the continuous availability of a host since high numbers of biotic agents are released at the outset, and control of the host is expected during the growing season.

8. Classical biological control has greatest potential impacts on less intensively tilled areas. The 3 or more years required for the biotic agent to attain a control level, during which some losses are occurring, cannot be tolerated in the more intensive grain

and row-crop farming systems. The early successes with biological control were all on extensive grazing land.

9. The more closely the target weed is related to the crop, the less likely the crop will be safe from activity of the biotic agent. This fact is one of the reasons why more progress has not been made in controlling weedy grasses with biotic agents. Most weedy grasses are genetically related to our major grain crops, including rice, wheat, oats, corn, and grain sorghum. Other genetic and ecological aspects are involved also. Widely distributed plant species have been shown to have more insect associates than rare species (Lawton and Schroder, 1976). It follows that the chances of finding an effective biotic agent may be better with widely distributed species. However, proportionately more of the insects associated with widely distributed plants feed on several plant species rather than on only one plant species. This relationship makes it less likely to find the necessary host specificity for practical use of the biotic agent. There is also evidence that the feeding type—chewing, mining, piercing-sucking, seed- and flower-feeding, and gall-forming—is also related to evolutionary background. That is, gall-forming and mining, which require more specialization on the part of the insect than the other three types are more common among widely distributed plants. Particularly significant for weed prevention was the finding, shown in Table 9-3, that there were many insects in the Cynareae tribe that feed on the seed, fruit, and flowers of weeds.

TABLE 9-3.   Comparison of feeding types of selected genera within the Cynareae tribe.

Chewing [no. (%)]	Mining [no. (%)]	Gall-Forming [no. (%)]	Seed-, Fruit-, and Flower-Feeding [no. (%)]	Sucking [no. (%)]
261 (30%)	184 (21%)	18 (2%)	310 (36%)	91 (11%)

Source: Data from Lawton and Schroder, 1976.

10. For biological control to be successful, the life cycle of the target weed must be synchronized with that of the biotic agent. The effective control of St. Johnswort with the *Chrysolina* beetle is a good example of this concept. The period of rapid growth of St. Johnswort in the spring and fall coincides with high activity (feeding) by beetle adults. The following spring, eggs complete hatching and the larvae feed on the shallow rootstocks until shoots appear. Additionally, best success with biotic agents can be expected where their peaks of activity coincide with those times when the target weed is most vulnerable. Thus, insects that actively feed at the time in the life cycle of a perennial weed when it is producing food for storage can be expected to be especially effective.

11. A major consideration affecting successful implementation of bioherbicides is efficacy of weed control, which is the ability to provide a satisfactory level of weed control over a short period of time with a bioherbicide that is easy to use (Charudattan, 1988). The greater the ability of bioherbicides to provide complete or nearly complete weed control, the more acceptable bioherbicides will be to enterprise operators. Enterprise operators must also be convinced that fairly quick control of weeds occurs within

2 to 6 weeks of application. Likewise, a bioherbicide should fit into standard pesticide application practices, not requiring specialized equipment or additional steps in management that require added capital and can ultimately discourage bioherbicide use. A rigorous assessment scheme for rating efficacy of bioherbicide candidates has been proposed to guide selection of those most likely to be acceptable and commercially feasible (Charudattan, 1988). However, it must be emphasized that even though many potential bioherbicides may be considerably less efficacious than chemical herbicides because weeds are not killed, weed competition is still negated through suppression of growth by the biotic agent. Growth suppression leading to elimination of competition must be demonstrated to be equivalent to weed kill in order to convince enterprise operators that bioherbicides can be an effective tool in weed management.

12. Pathogens as bioherbicides offer the best opportunity for commercialization. Three pathogens have been registered for use as bioherbicides. A liquid formulation of the fungus *Phytophthora palmivora* is marketed under the trade name DeVine® by Abbott Laboratories for the control of strangler or milkweed vine, a weed parasitic on citrus trees. *Colletotrichum gloeosporioides* f. sp. *aeschynomene* formulated as a wettable powder for use in rice and soybeans is marketed by Ecogen, Inc. under the trade name Collego®. A wettable powder formulation of *Colletotrichum gloeosporioides* f. sp. *malvae* is marketed in Canada under the trade name Biomal® by Philom Bios, Inc. for the control of round-leaved mallow in field crops. The problems of rearing, storing, shipping, and delivering insects as bioherbicides make them less likely candidates for commercialization. Commercialization is important to widespread use and rapid adoption, especially with the bioherbicide approach. It is the best way for adequate quantities of biotic agents to be made available to prospective users.

WEED MANAGEMENT VS. WEED CONTROL

It is conceivable that biotic agents will achieve a level of usage in weed management well beyond their potential in weed control. Realizing their potential in weed management will require careful articulation of the objectives and the selection of appropriate evaluation measures. With uses in management, *numbers*, used as a measure in control approaches, have little value. For example, numbers of weeds destroyed or prevented from developing by a biotic agent, although a measure of the agent's activity, do not in themselves tell us the effect on a given weed population over time. Yet, it is the latter information that will determine whether or not such a management effort should be initiated. Using biotic agents to keep a weed below an economic-loss level calls for an understanding of the nature of competition involved in order to identify the best ways to measure it. In a larger sense, uses of biotic agents in weed management require a way of measuring and assigning the costs and returns not provided by measures of control approaches. Developing satisfactory measures must be addressed for biotic technology to advance.

REFERENCES

Abbas, H.K., and W.L. Barentine. 1995. *Alternaria helianthi* and imazaquin for control of imazaquin susceptible and resistant cocklebur (*Xanthium strumarium*) biotypes. Weed Sci. 43:425-428.

Blossey, B. 1995. A comparison of various approaches for evaluating potential biological control agents using insects on *Lythrum salicaria*. Biol. Cont. 5:113-122.

Boyette, C.D., G.E. Templeton, and R.J. Smith, Jr. 1979. Control of winged waterprimrose (*Jussiaea decurrens*) and northern jointvetch (*Aeschynomene virginica*) with fungal pathogens. Weed Sci. 27:497-501.

Burdon, J.J., and D.R. Marshall. 1981. Biological control and the reproductive mode of weeds. J. Appl. Ecol. 18:649-658.

Caesar-ThonThat, T.C., W.E. Dyer, P.C. Quimby, Jr., and S.S. Rosenthal. 1995. Formulation of an endoparasitic nematode, *Subanguina picridis* Brzeski, a biological control agent for Russian knapweed, *Acroptilon repens* (L.) DC. Biol. Cont. 5:262-266.

Charudattan, R. 1988. Assessment of efficacy of mycoherbicide candidates. In E.S. Delfosse, ed., Proc. VII Int. Symp. Biol. Cont. Weeds, pp. 455-464. Ist. Sper. Patol. Veg., Rome.

———. 1986. Integrated control of waterhyacinth (*Eichhornia crassipes*) with a pathogen, insects, and herbicides. Weed Sci. 34 (Supplement 1):26-30.

Charudattan, R., F.W. Zettler, H.A. Cordo, and R.G. Christie. 1980. Partial characterization of a potyvirus infecting the milkweed vine, *Morrenia odorata*. Phytopathology 70:909-913.

Ciotola, M., A.K. Watson, and S.G. Hallet. 1995. Discovery of an isolate of *Fusarium oxysporum* with potential to control *Striga hermonthica* in Africa. Weed Res. 35:303-309.

Cullen, J.M. 1976. Evaluating the success of the programme for the biological control of *Chondrilla juncea* L. In T.E. Freeman, ed., Proc. 4th International Symposium on Biological Control of Weeds, pp. 117-121. Gainesville: Center for Environmental Programs, Institute of Food and Agricultural Sciences, University of Florida.

Daigle, D.J., and W.J. Connick, Jr. 1990. Formulation and application technology for microbial weed control. In R.E. Hoagland, ed., Microbes and microbial products as herbicides, pp. 288-304. Am. Chem. Soc., Washington, D.C.

Dennill, G.B., and D. Donnelly. 1991. Biological control of *Acacia longifolia* and related weed species (Fabaceae) in South Africa. Agric. Ecosyst. Environ. 37:115-135.

Dodd, A.P. 1940. The biological campaign against prickly pear. Commonwealth Prickly Pear Board, Brisbane, Australia.

Dorworth, C.E. 1995. Biological control of red alder (*Alnus rubra*) with the fungus *Nectria ditissima*. Weed Technol. 9:243-248.

Epstein, A.H., and J.H. Hill. 1995. The biology of rose rosette disease: a mite-associated disease of uncertain aetiology. J. Phytopathol. 144:353-360.

Flynn, P., and A.K. Vidaver. 1995. *Xanthomonas campestris* pv. *asclepiadis*, pv. *nov.*, causative agent of bacterial blight of milkweed (*Asclepias* spp.). Plant Dis. 79:1176-1180.

Frick, K.E., and J.M. Chandler. 1978. Augmenting the moth (*Bactra verutana*) in field plots for early suppression of purple nutsedge (*Cyperus rotundis*). Weed Sci. 26:703-710.

Grant, N.T., E. Prusinkiewicz, K. Mortensen, and R.M.D. Makowski. 1990. Herbicide interactions with *Colletotrichum gloeosporioides* f. sp. *malvae* a bioherbicide for round-leaved mallow (*Malva pusilla*) control. Weed Technol. 4: 716-723.

Harley, K.L.S., and I.W. Forno. 1992. Biological control of weeds: A handbook for practitioners and students. Melbourne: Inkata Press.

Hasan, S., and P.G. Ayers. 1990. The control of weeds through fungi: principles and prospects. New Phytol. 115:201-222.

Heiny, D.K. 1994. Field survival of *Phoma proboscis* and synergism with herbicides for control of field bindweed. Plant Dis. 78:1156-1164.

Huffaker, C.B. 1957. Fundamentals of biological control of weeds. Hilgardia 27:101-157.

Johnson, B.J. 1994. Biological control of annual bluegrass with *Xanthomonas campestris* pv. *poannua.* HortScience 29:659-662.

Johnson, D.R., D.L. Wyse, and K.L. Jones. 1996. Controlling weeds with phytopathogenic bacteria. Weed Technol. 10:621-624.

D.L. Wyse. 1994. Canada thistle [*Cirsium arvense* (L.)] control in soybean with *Pseudomonas syringae* pv. *tagetis.* Proc. North Cent. Weed Sci. Soc. 49:141.

Jordan, N. 1993. Prospects for weed control through crop interference. Ecol. Appl. 3:84-91.

Julien, M.H. 1992. Biological control of weeds: A world catalogue of agents and their target weeds. CAB International, Oxford, UK.

Kennedy, A.C., L.F. Elliott, F.L. Young, and C.L. Douglas. 1991. Rhizobacteria suppressive to the weed downy brome. Soil Sci. Soc. Am. J. 55:722-727.

Kennedy, A.C., T.L. Stubbs, and F.L. Young. 1989. Rhizobacterial colonization of winter wheat and grass weeds. Agron. Abstr. 53:220.

Kremer, R.J. and A.C. Kennedy. 1996. Rhizobacteria as biological control agents of weeds. Weed Technol. 10:601-609.

Kremer, R.J., and N.R. Spencer. 1989. Interaction of insects, fungi, and burial on velvetleaf (*Abutilon theophrasti*) seed viability. Weed Technol. 3:322-328.

Kremer, R.J., M.F.T. Begonia, L. Stanley, and E.T. Lanham. 1990. Characterization of rhizobacteria associated with weed seedlings. Appl. Environ. Microbiol. 56:1649-1655.

Lawton, J.H., and D. Schroder. 1976. Some observations on the structure of phytophagous insect communities: The implications for biological control. In T.E. Freeman, ed., Proc. 4th International Symposium on Biological Control of Weeds, pp. 57-73. Gainesville: Center for Environmental Programs, Institute of Food and Agricultural Sciences, University of Florida.

Lee, G.A. 1986. Integrated control of rush skeletonweed (*Chondrilla juncea*) in the Western U.S. Weed Sci. 34 (Supplement 1):2-6.

Massion, C.L., and S.E. Lindow. 1986. Effects of *Sphacelotheca holci* infection on morphology and competitiveness of johnsongrass (*Sorghum halepense*). Weed Sci. 34:883-888.

McCarty, M.K., and W.O. Lamp. 1982. Effect of a weevil (*Rhinocyllus conicus*) on musk thistle (*Carduus thoemeri*) seed production. Weed Sci. 30:136-140.

Messersmith, C.G., and S.W. Adkins. 1995. Integrating weed-feeding insects and herbicides for weed control. Weed Technol. 9:199-208.

Miller, R.V., E.J. Ford, N.J. Zidack, and D.C. Sands. 1989. A pyrimidine auxotroph of *Scerotinia sclerotium* for use in biological weed control. J. Gen. Microbiol. 135:2085-2091.

Northam, F.E., and C.C. Orr. 1982. Effects of a nematode on biomass and density of silverleaf nightshade. J. Range Manage. 35:536-537.

Okuda, S. 1992. Herbicides. In S. Omura, ed., The search for bioactive compounds from microorganisms, pp. 224-236. New York: Springer-Verlag.

Ou, X., and A.K. Watson. 1993. Mass culture of *Subanguina picridis* and its bioherbicidal efficacy on *Acroptilon repens.* J. Nematol. 25:89-94.

Parker, P.E. 1986. Nematode control of silverleaf nightshade (*Solanum elaegnifolium*): a biological control pilot project. Weed Sci. 34 (Supplement 1):33-34.

Phatak, S.C., D.R. Summer, H.D. Wells, and N.C. Glaze. 1983. Biological control of yellow nutsedge with the indigenous rust fungus *Puccinia canaliculata.* Science 219:1446-1447.

Quimby, P.C., Jr., and H.L. Walker. 1982. Pathogens as mechanisms for weed management. Weed Sci. 30 (Supplement 1):30-34.

Rees, N.E., and N.R. Spencer. 1991. Biological control of leafy spurge. In L.F. James, ed., Noxious range weeds, pp. 182-192. Boulder: Westview Press.

Roof, M.E., B. Puttler, and L.E. Anderson. 1982. Controlling musk thistle with an introduced weevil. Columbia: Science and Technology Guide No. 4867, University of Missouri, Columbia, Extension Division.

Rose, S.J., O.C. Burnside, J.E. Specht, and B.A. Swisher. 1984. Competition and allelopathy between soybeans and weeds. Agron. J. 76:523-528.

Rothrock, C.S., T.L. Kirkpatrick, R.E. Frans, and H.D. Scott. 1995. The influence of winter legume cover crops on soilborne plant pathogens and cotton seedling diseases. Plant Dis. 79:167-171.

Schisler, D.A., K.M. Howard, and R.J. Bothast. 1991. Enhancement of disease caused by *Colletotrichum truncatum* in *Sesbania exaltata* by coinoculating with epiphytic bacteria. Biol. Cont. 1:261-268.

Spencer, N.R. 1988. Inundative biological control of velvetleaf, *Abutilon theophrasti* (Malvaceae) with *Niesthrea louisianica* (Hem.: Rhopalidae). Entomophaga 33:421-429.

Strobel, G., A. Stierle, S.H. Park, and J. Cardellina. 1990. Maculosin, a host-specific phytotoxin from *Alternaria alternata* on spotted knapweed. In R.E. Hoagland, ed., Microbes and microbial products as herbicides, pp. 53-62. Am. Chem. Soc., Washington, D.C.

Surles, W.W., and L.J. Kok. 1976. Response of *Carduus nutans* L. to infestation by *Rhinocyllus conicus* Froel. (Coleoptera: Curculionidae) and mechanical damage. In T.E. Freeman, ed., Proc. 4th International Symposium on Biological Control of Weeds, pp. 105-107. Gainesville: Center for Environmental Programs, Institute of Food and Agricultural Sciences, University of Florida.

Templeton, G.E. 1982. Biological herbicides: discovery, development, deployment. Weed Sci. 30:430-433.

Turco, R.F., M. Bischoff, D.P. Breakwell, and D.R. Griffith. 1990. Contribution of soilborne bacteria to the rotation effect in corn. Plant Soil. 122:115-120.

Watson, A.K. 1993. Biological control of weeds handbook. Weed Science Society of America, Champaign, IL.

Weidemann, G.J., and G.E. Templeton. 1988. Control of Texas gourd, *Cucurbita texana,* with *Fusarium solani* f. sp. *cucurbitae.* Weed Technol. 2:271-274.

Wiley, R.B., E.E. Schweizer, and E.G. Ruppel. 1985. Interaction of kocia (*Kocia scoparia*) and *Rhizopus* sp. on sugarbeet (*Beta vulgaris*) germination. Weed Sci. 33:275-279.

Zimmermann, H.G. 1991. Biological control of mesquite, *Prosopis* spp. (Fabaceae), in South Africa. Agric. Ecosyst. Environ. 37:175-186.

10

Herbicide Use

CONCEPTS TO BE UNDERSTOOD

1. Herbicides may be applied to the soil to prevent weed emergence (pre-emergence use) or to prevent competition from growing weeds (post-emergence use).
2. The historical emphasis in herbicide usage has been on weed control.
3. Herbicides used for weed control have been responsible for sizable increases in food production.
4. Even greater contributions from herbicides are possible if they are used to minimize losses from weeds in the entire production enterprise over time.

Chemicals have long been used to control weeds. Some of those associated with the expanding application of chemistry during the later part of the 19th century introduced selective control to agriculture. However, it was the auxin-type herbicides MCPA and 2,4-D, discovered almost simultaneously in the 1940s in Great Britain and the United States, that ushered in the modern era of chemical weed control. In the United States, adoption of 2,4-D occurred rapidly from a few thousand acres in 1946 to several million acres by the end of the decade. Whereas most of the chemicals used prior to 2,4-D were inorganic, all of those introduced since are organic.

WHY HERBICIDE USAGE GREW RAPIDLY

Low Rates and Carrier Volume

There are several reasons why modern herbicides expanded so rapidly to reach the position of widespread usage they now enjoy. One important reason is that they can be applied at relatively low rates and in low volumes of water. Many of the pre-2,4-D herbicides called for the handling of considerable bulk, which was costly for transportation and cumbersome to apply. Not infrequently, the older-type herbicides were applied in large volumes of water. By contrast, early research with 2,4-D showed control to be as good when it was applied in 50 to 100 L ha^{-1} of water as when it was applied in higher volumes, provided it was evenly distributed on the weed or on the soil.

Biological Selectivity

Selectivity because of differences in internal reactions of weeds and of desired plants was another reason for the rapid expansion of herbicide use. Selectivity of the older-type herbicides was based primarily on physical differences of contact and penetration between the weed and the crop that frequently provided relatively little margin for error. With modern herbicides, on the other hand, selectivity may be due mainly to differences in the biochemistry of sensitive and tolerant species. Such biological selectivity may provide a very wide range of tolerance. For example, as shown in Figure 10-1, there is more than a 100-fold difference in tolerance between wild mustard and wheat to the herbicide chlorsulfuron applied postemergence. Such large differences in tolerance provide the necessary wide margin for complete weed kill with little, if any, injury of desired plants.

Private Industry Role

A third major reason for the rapid expansion of herbicide usage since the discovery of 2,4-D is the incentive offered private industry to carry out the necessary development and marketing of these chemicals. With all the many advantages of the newer herbicides and the spectacular results that can be achieved, their widespread usage could never have occurred without the massive effort on the part of the chemical industry to develop effective formulations and manufacturing processes and to assure their availability through effective distribution networks.

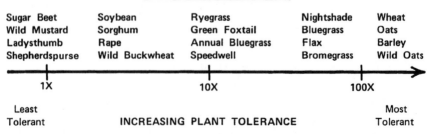

FIGURE 10-1. Variance of plant species in their biological tolerance of modern herbicides.

Source: Courtesy of E.I. duPont de Nemours and Company.

USE OF HERBICIDES IN WEED CONTROL

Postemergence and Preemergence Control

Successful use of 2,4-D to literally save a crop of corn being overrun by giant ragweed (*Ambrosia trifida*) set the tone for the use of these organic herbicides in the ensuing years. Several hundred hectares of Ohio riverbottom land in Henderson County, Kentucky, were being overrun by giant ragweed in July of 1947. Rains had prevented early cultivation, and as a result, the crop and the weeds had become too large to be cultivated. Application of 2,4-D at the rate of 0.6 kg ha⁻¹ resulted in complete kill of the giant ragweed. Although some injury symptoms occurred on the corn, a crop was harvested, whereas there otherwise would have been a total loss.

In 1948, successful selective control of weeds in corn was reported with 2,4-D applied to the soil after planting but before corn emergence (Wolf and Anderson, 1948). The use of 2,4-D to remove a weed from a growing corn crop in Henderson County, Kentucky, and to selectively control weeds by application to soil identify the two general ways herbicides are used today. The former usage is termed *postemergence* and the latter *preemergence*.

The emphasis in herbicide usage since 1947 continues to be mainly on the current season. That is, the common approach is to only consider the efficacy of a particular herbicide in controlling weeds in the current year, with relatively little attention being given to a long-term approach. This usage has been responsible for sizable increases in production of food. For example, as seen in Figure 10-2, herbicides were a major contributor to increases in yield of corn and soybeans for five Corn Belt states (Illinois, In-

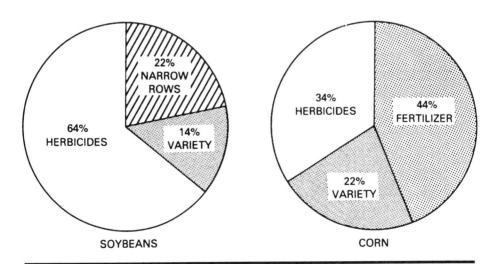

FIGURE 10-2. Contribution of technology adoption to increases in corn and soybean yields, Corn Belt region, 1964 to 1979.

Source: Data from Schroeder et al., 1981 and 1982.

diana, Iowa, Ohio, and Missouri) between 1964 and 1979 (Schroder et al., 1981, 1982). In this period, soybean yields increased 437 kg ha^{-1} in response to technology. Herbicides accounted for nearly two-thirds of this increase. Herbicides are credited with one-third of the 1,771 kg ha^{-1} increase in corn yield for the same period.

Types of post- and preemergence application. Many refinements and modifications of post- and preemergence application followed introduction of 2,4-D. As shown in Figure 10-3, herbicides may now be applied literally at any time relative to the stage of development of both the weed and the desired plant. For example, herbicides may be applied prior to planting and either prior to emergence of weeds or to growing weeds. With the expansion of reduced tillage of crops, applying herbicides to kill growing weeds and planting directly into the killed sod has become commonplace. Also, development of horizontal and wick applicators (to be discussed in Chapter 13) allows nonselective chemicals to be used selectively to remove weeds overtopping crops. Each application has its unique place in weed control. Discussion of specific control of weeds in crops can be found in several references, including Anderson (1977), Klingman et al. (1975), Crafts (1975), and Zimdahl (1993).

USE OF HERBICIDES IN WEED MANAGEMENT

As valuable as the contributions of herbicides have been in weed control, even greater contributions are possible if used in a weed management context—that is, if they are used as part of a system designed to minimize losses from weeds in the entire production enterprise over time. Using herbicides as part of a system of weed management leaves room, conceptually, for cropping and tillage practices, allelopathy, and biological agents to be used also. Furthermore, using herbicides as a tool in a weed management context may help achieve other important objectives. For example, weed suppression with herbicides, rather than their elimination, may help prevent soil erosion.

Preventing Crop Loss vs. Weed Control

The main objective with herbicide use in crop production has been and continues to be to prevent or minimize loss in crop yield. It is the competition prevented, not the degree of weed control obtained, that determines success of a herbicide treatment. However, the tendency has been to equate efficacy of herbicide use with degree of weed control obtained. This emphasis was understandable in view of the relationship between crop yield and weed density discussed in Chapter 2. But weeds that survive a herbicide treatment may, in fact, not be as competitive toward the crop as those not treated. Weeds that survived herbicides used in sugar beets were smaller and less damaging to yield than those in untreated plots (Schweizer, 1981). Giant foxtail suppressed early by sublethal rates of herbicides applied postemergence did not reduce soybean

WITH RESPECT TO CROP	PREEMERGENCE TO WEEDS		POSTEMERGENCE TO WEEDS

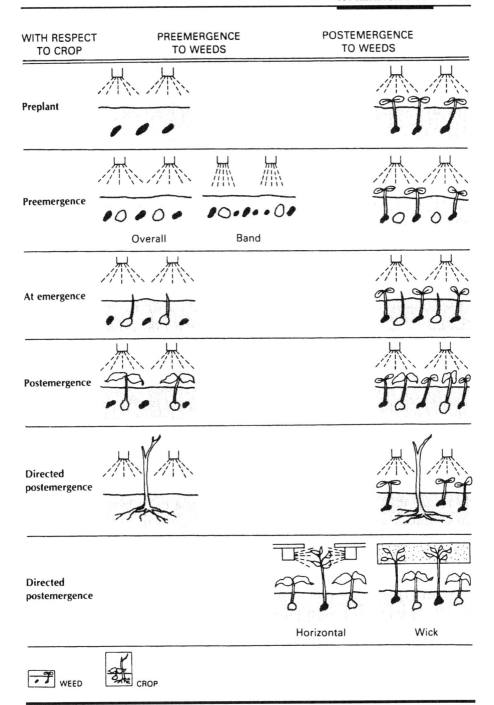

Overall Band

Horizontal Wick

FIGURE 10-3. Stages of growth of weed and crop plants when herbicides may be applied for selective weed control.

Source: After Fryer and Evans, 1968. Reproduced with permission of Blackwell Scientific Publications, Ltd.

yields even though foxtail dry weight exceeded that on untreated plots late in the season (Aldrich, 1985). For a wide variety of crops in reports of weed control in the Corn Belt, average crop yield with 70% weed control was equal to yield with nearly 100% control (Aldrich, 1987). Focusing on crop yield, rather than on degree of control, may permit less herbicide to be used, thereby reducing both the cost and the potential amount entering the environment. This role for herbicides in preventing crop loss will be expanded upon in Chapter 13.

Preventing Renewal of the Seedbank

We learned in Chapter 4 that seeds of some weeds retain viability in soil for many years, but for others, life expectancy is only a few years. Furthermore, where additions to the seedbank are prevented or are minimal, the seedbank is greatly reduced in a relatively few years of crop production (Burnside et al., 1986; Dowler et al., 1974; Roberts, 1962; Schweizer and Zimdahl, 1984). Effective use of herbicides offers a convenient and relatively inexpensive way of preventing additions to the seedbank when used as part of the weed control effort and when used to prevent seed production by escapes from control. Escapes, although they may not reduce crop yields, may well produce enough seed to restock the soil, as already discussed. Not preventing seed production by escapes from control fails to realize benefits in seedbank drawdown attained with control in production practices. Usage to prevent seed production of escapes has never been exploited, although it was discovered early in the modern era that many of the new organic herbicides could have pronounced effects on the quantity and viability of seeds produced on treated plants. Machines cannot practically be used for this purpose because the crop at this advanced stage would be physically damaged. Hand removal, although it could be effective, is simply too costly, at least in most U.S. agriculture. Also, in many cases, the weed may be the same height or shorter than the crop, precluding mechanical removal of the weed's seedheads. Giant foxtail in corn is an example.

In spite of the apparent potential value, use of herbicides for the express purpose of preventing seedbank renewal is limited. We will examine this role for herbicides when we consider herbicide application to growing plants in Chapter 13.

Drawing Down the Seedbank

Even though the seedbank can be considerably reduced when reseeding from mature weeds is prevented, the seed population that remains in the soil comprises the persistent seedbank and is the source of future weed infestations. As pointed out earlier, this persistent seedbank is composed of seeds that are in a dormant, nongerminating state. Germination, however, is the eventual fate of most seeds in soil, and stimulation of weed seeds to germinate in soil is a logical approach for depleting the seedbank. The use of herbicides expressly for reduction of the seedbank has received little attention as a practical weed management tool. Early observations of simazine, linuron, 2,4-D, or MCPA over 12- and 16-year periods suggested that long-term use of herbicides had an incidental impact on the size and composition of the seedbank (Hurle, 1974; Roberts and Neilson, 1981). The intentional use of herbicides to stimulate dormant seeds in soil as a means of eradicating weeds was first demonstrated with several carbamothioates (Fawcett and Slife, 1975). Shallow incorporation of these herbicides at less than recommended rates enhanced emergence of velvetleaf and common lambsquarters from dormant seeds. The seedlings were then eliminated by using cultural or chemical methods resulting in a net decrease in the seedbank. Subsequent approaches for germination stimulation by using chemicals known for having an effect on breaking seed dormancy, including sodium azide, cyanide, alcohols, and substituted phthalimides, were generally erratic in effectiveness and have not been widely adapted in weed management (Dyer, 1995). One exception is the successful use of ethylene to reduce witchweed seed populations in soils in North and South Carolina (Egley, 1986). Ethylene reduced dormant weed seed numbers in soil by stimulating germination. Seedlings surviving to parasitize the roots of corn were controlled by herbicides. The combined use of ethylene and herbicides greatly enhanced control and resulted in significant reduction in the witchweed-infested region of North and South Carolina.

Progress in discovery of chemicals for drawing down the seedbank will advance with establishment of protocols for screening chemicals specifically for germination-inhibiting or dormancy-breaking properties. Standard screening techniques used by industry focus on a chemical's ability to decrease emergence from nondormant seeds, which results in the selection of herbicides with little effect on seed dormancy (Taylorson, 1987). New screening methods based on attacking seed dormancy mechanisms in soil could lead to new classes of "seed-killing" compounds. Such compounds would be readily accepted for integration into weed control programs since management of seedbanks is a high priority in approaches for reducing other weed management inputs (Dyer, 1995). Application of several germination-stimulating compounds may be necessary to induce a maximum number of seeds of several weed species to germinate at a given time for subsequent elimination by cultural, chemical, or biological methods. Another potential approach for reducing seed numbers in soil is the use of chemicals to alter seed properties or "predispose" seeds to make them vulnerable to attack by soil- or seedborne microorganisms, which essentially cause the seeds to decompose or the resulting seedlings to be diseased (Kremer and Schulte, 1989).

As Part of a Total Production System

Herbicides can play a key role in achieving so-called sustainable agriculture. That is, maximum sustainable production of adequate high-quality food at reasonable prices and fair returns to the producer. Such an objective is attainable only in a production system that deals effectively with weeds. However, there will be opportunities to more effectively deal with weeds when approached from a production system perspective in which the focus is preventing crop yield loss rather than only on weed control. The variety of weeds and size of the seedbank may be reduced by a weed management program designed to accomplish these objectives over time (Schweizer and Zimdahl, 1984). With appropriate analysis of weed management effectiveness and economic returns, herbicide usage may be reduced after a few years of nearly complete control (Burnside et al., 1986; Schweizer et al., 1988). Split applications of reduced rates of herbicides with the second application aimed at both control and prevention of seed production by weed escapes may more effectively prevent crop losses than single applications with the associated benefit of minimizing seedbank renewal (Aldrich, 1985). Weed management may be chosen for crop A in year one to reduce the weed problem in crop B in year two. These and other benefits from a production system approach to weed management will be expanded upon in later chapters.

REFERENCES

Aldrich, R.J. 1985. Unpublished research.

———. 1987. Interference between crops and weeds. In G.R. Waller, ed., Allelochemicals: Role in agriculture and forestry, pp. 300-312. Am. Chem. Soc. Symposium Series 330, Washington, D.C.

Anderson, W.P. 1977. Weed Science: Principles. St. Paul, MN: West Publishing.

Burnside, O.C., R.S. Moomaw, F.W. Roeth, G.A. Wicks, and R.G. Wilson. 1986. Weed seed demise in soil in weed-free corn (*Zea mays*) production across Nebraska. Weed Sci. 34:248-251.

Crafts, A.S. 1975. Modern weed control. Berkeley: University of California Press.

Dowler, C.C., E.W. Hauser, and A.W. Johnson. 1974. Crop-herbicide sequences on a Southeastern Coastal Plain Soil. Weed Sci. 22:500-505.

Dyer, W.E. 1995. Exploiting weed seed dormancy and germination requirements through agronomic practices. Weed Sci. 43:498-503.

Egley, G.H. 1986. Stimulation of weed seed germination in soil. Rev. Weed Sci. 2:67-89.

Fawcett, R.S., and F.W. Slife. 1975. Germination stimulation properties of carbamate herbicides. Weed Sci. 23:419-424.

Fryer, T.D., and S.A. Evans. 1968. Weed control handbook, principles, vol. 1. Oxford, England: Blackwell Scientific.

Hurle, K. 1974. Effect of long-term weed control measures on viable weed seeds in the soil, pp. 1145-1152. Proc. 12th British Weed Cont. Conf.

Klingman, G.C., F.M. Ashton, and L.J. Noordhoff. 1975. Weed science: Principles and practices. New York: Wiley.

Kremer, R.J., and L.K. Schulte. 1989. Influence of chemical treatment and *Fusarium oxysporum* on velvetleaf (*Abutilon theophrasti*) seed viability. Weed Technol. 3:322-328.

Roberts, H.A. 1962. Studies on the weeds of vegetable crops. II. Effect of six years of cropping on the weed seeds in the soil. J. Ecol. 50:803-813.

Roberts, H.A., and J.E. Neilson. 1981. Changes in the soil seed bank of four long-term crop/herbicide experiments. J. Appl. Ecol. 18:661-668.

Schroder, D., J.D. Headley, and R.M. Finley. 1981. The contribution of pesticides and other technologies to soybean production in the Corn Belt region, 1964 to 1979, agricultural economics paper no. 1981-33. Columbia: University of Missouri.

_____. 1982. The contribution of pesticides and other technologies to corn production in the Corn Belt region, 1964 to 1969, agricultural economics paper no. 1982-8. Columbia: University of Missouri.

Schweizer, E.E. 1981. Broadleaf weed interference in sugarbeets (*Beta vulgaris*). Weed Sci. 29:128-133.

Schweizer, E.E., and R.L. Zimdahl. 1984. Weed seed decline in irrigated soil after six years of continuous corn (*Zea mays*) and herbicides. Weed Sci. 32:76-83.

Schweizer, E.E., D.W. Lybecker, and R.L. Zimdahl. 1988. Systems approach to weed management in irrigated crops. Weed Sci. 36:840-845.

Taylorson, R.B. 1987. Environmental and chemical manipulation of weed seed dormancy. Rev. Weed Sci. 3:135-154.

Wolf, D.E., and J.C. Anderson. 1948. Preemergence control of weeds in corn with 2,4-D. Agron. J. 40:453-458.

Zimdahl, R.L. 1993. Fundamentals of weed science. San Diego: Academic Press.

11

Herbicide Entry and Transport

CONCEPTS TO BE UNDERSTOOD

1. Herbicides, especially systemic herbicides, must commonly interact with all three tissue systems—dermal, vascular, and fundamental—to express their effects on plants.
2. The leaf, the root, and, in grasses, the coleoptile are the main points of entry of herbicides into plants.
   a. Leaf cuticle quantity and quality, pubescence, and growth form, all of which are influenced by growing conditions, affect herbicide entry.
   b. The root is a relatively nondiscriminating point of entry for herbicides. Uptake is influenced by such abiotic aspects as clay and organic matter content, rainfall, and evaporation.
   c. The presence or absence of a coleoptile and the location of the growing point within it influence the extent and affect of herbicide uptake from soil.
3. Systemic herbicides must move through the vascular system to reach the site(s) of biochemical reaction(s). Herbicides entering through the leaves must move through the living phloem. Herbicides entering via the root must usually move to the reaction vicinity in the nonliving xylem and then reenter living cells.
4. The four primary biochemical reactions affected by herbicides are: respiration and mitochondrial electron transport, photosynthesis, nucleic acid metabolism and protein synthesis, and lipid metabolism.

In the previous chapter, we discussed the place of herbicides in weed management. Here, we want to examine herbicide entry and transport in somewhat greater detail. The purpose is to develop a broad knowledge of the relationship between these aspects and plant growth as influenced by environmental conditions. This information is an aid to understanding herbicidal selectivity and factors affecting results with herbicides that are discussed in Chapters 12 and 13. Further, this background is helpful for our consideration of weed shifts in response to herbicides in Chapter 14 and the discussion in Chapter 15 of the place of herbicides in a comprehensive weed management program.

This chapter needs to be approached with such a broad, forward-looking perspective in order to visualize how what is discussed relates to the success of a weed management program under the environment of new and different production practices. For example, how might conservation tillage systems with the associated reduction in tillage and possible increases in cover crops and in plant residues influence our approaches to weed management?

Response of a plant to a herbicide involves four broad components: (1) the stage of development of the plant, (2) the genetic makeup of the plant, (3) the environment, and (4) the chemical. To understand the influence of these components on plant reaction, it is first necessary to understand something about the anatomy of a plant as it determines entry of the herbicide and movement within the plant once it enters.

For growing plants, two parts are main points of entry: the leaf and the root. A third point of entry important mainly in grass seedlings is the coleoptile node. This is not to say that herbicides cannot enter the stem and flowers, but simply recognizes that these parts represent a much smaller surface area and period of exposure for herbicide entry than do leaves and roots. Also, although dormant and nongerminating seeds may be penetrated by herbicides, such uptake is relatively unimportant as a point of entry in explaining results obtained with herbicides applied to the soil for selective control of weeds.

In general terms, but not exclusively so, leaf uptake is the primary entry in aerial applications and root uptake is the primary entry in soil applications. Shoot uptake may also be important in soil applications, however, especially in grasses and for some herbicides. Aerial applications and soil applications pose different opportunities for the environment to influence effects on the plant. As we will see, both directly and indirectly, aerial applications are relatively more affected by temperature, humidity, and rainfall as they influence plant growth. Although soil applications may also be affected by these factors, the effect is more apt to be a delayed one.

For their part, soil applications are affected by a variety of soil attributes, including organic matter content, clay content, and the soil microflora and microfauna. No matter how uniformly the herbicide may be applied to the soil surface or how uniformly it is mixed with the top 2 cm to 5 cm (two or so in.), the fact remains that individual herbicide molecules are not contiguous but are separated. With a surface spray of 0.08 m^3 ha^{-1}, the particles will be about 1.3 mm apart if perfectly spread on a flat surface (Hartley, 1960). If the spray application is perfectly mixed to a depth of 1 cm, the calculated randomly oriented distances between neighboring herbicide centers will be between 12 mm and 31 mm. Thus, soil applications offer some opportunity for the germinating and developing weed seedlings to escape the herbicide, at least for a time. This circumstance in part explains why soil applications are usually higher in rate than are aerial applications.

Where uptake from the soil is through the shoot or coleoptile node, the relatively fewer plant parts, compared with roots, decrease the chance for contacting the passive herbicide in the soil. This disadvantage is offset by the movement of the shoot and the coleoptile node through the zone in which herbicides may be located.

PLANT TISSUE SYSTEMS

The body of *vascular plants*—that is, those containing xylem and phloem—can be viewed as consisting of three systems of tissues: (1) dermal, (2) vascular, and (3) fundamental or ground. The *dermal system* provides the outer wrapping of the plant and is represented in the young plant by the epidermis. Later, with additional (secondary) growth, the epidermis may be replaced by another dermal system. The *vascular system* includes the two main conducting tissues, the phloem and the xylem. This system provides for the movement of water, minerals, photosynthate, and other chemical constituents within the plant. The *fundamental system*, or *ground system*, includes the remaining tissues, such as apical meristems and reproductive cells, that determine the overall form and function of the plant.

Even though these systems are discrete and fairly readily identifiable, as can be seen diagrammatically in Figure 11-1, the systems are nonetheless dependent upon one another. Furthermore, there is a continuity of the living matter (protoplasm) of the many, many cells that constitute tissues of these three systems.

Herbicides must commonly interact with all three systems of Figure 11-1 in manifesting their effects upon plants. There are two broad categories of herbicides based upon the nature of their killing action: (1) *contact herbicides*—that is, those that kill by acute toxicity—and (2) *systemic herbicides*—those that interfere with the weed's physiological and metabolic processes.

In theory, contact herbicides need only penetrate the dermal system to exert their effects of weakening and disorganizing the cellular membranes of the fundamental system and, thus, to destroy the plant. But it is likely that even these herbicides are moved, albeit for only very short distances, in the vascular system of the leaf, stem, or root. To better understand the entry and movement of herbicides within plants, it may be helpful to view the plant body as one in which the fundamental tissue is a sea within which the vascular system is imbedded. Viewed in this way, the fundamental tissue provides the linkage for herbicides between the dermal and vascular systems and within the plant provides the linkage between the xylem and phloem components of the vascular system.

Herbicides must reach the living protoplasm to exert their effect. As already mentioned, contact herbicides kill by contact with the living cells. Thus, entry is the main obstacle to their exerting their expected influence on the weed. Normally, it is necessary for systemic herbicides, the other broad category of herbicides, to reach the vascular system and be moved throughout the plant to the sites of action.

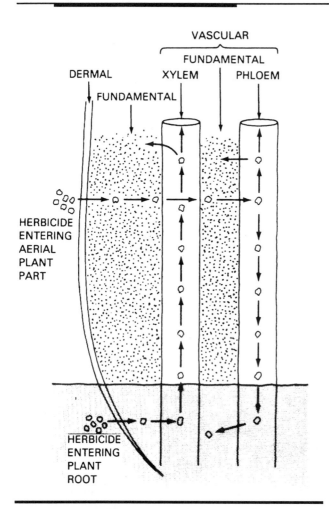

FIGURE 11-1. Simplified diagram showing the proximal relationships among the three tissue systems of vascular plants and the entry and movement of herbicides.

The relationships between some basic plant processes and herbicide entry and movement are shown in Figure 11-2. The left side of the figure pertains to photosynthesis and movement in phloem. The right side pertains to water and transport in the xylem. The ⊥ indicates points of potential interference with entry and transport. An essential point to be made by the figure is that movement in the phloem (symplast), with products of photosynthesis, may be either upward or downward to other parts of the plant. Movement in the xylem (apoplast) is only upward.

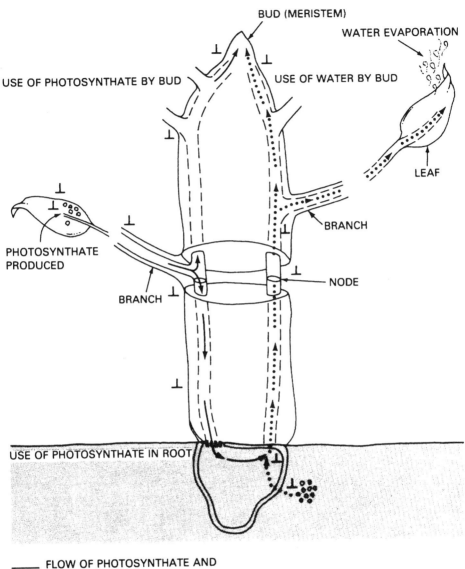

FIGURE 11-2. Schematic diagram relating metabolic processes and herbicide entry and movement in plants.

Transport Systems

Most texts treat transport in terms of two separate systems. *Apoplast* is the term for the transport system that is composed of the xylem, the nonliving cell walls of fundamental tissue, and the intercellular spaces. The term *symplast* is used for the transport system that consists of the phloem and other living (protoplasm) matter of the fundamental tissue. Although adequate for dealing with herbicide transport in mature tissue, these definitions suggest an unrealistic degree of independence between the two systems, do not adequately embrace the interrelationships of entry with movement, and leave in question the movement of herbicides in immature tissue that may contain no nonliving cells. As we shall see when we examine the xylem in greater detail, although the xylem is composed of nonliving cells in the mature tissue, it originates from cambium, which is living cell tissue. Transition from living to nonliving occurs at a very early age in the development process. Nevertheless, as already discussed, some herbicides are taken up and exert their effect during this early stage of development. *Polar transport*—that is, transport within the fundamental tissue—is sometimes identified as a third transport pathway for herbicides. Apparently, plant auxins may indeed move in this tissue system. However, it seems unlikely that this system of transport can account for herbicide movement of more than a few millimeters at most, in that it relies upon diffusion, which is a relatively slow process. Thus, transport is considered here in terms of movement in the xylem and in the phloem. Even so, we must remember that these are interrelated parts of the vascular system with ties also to the fundamental tissue. Most commonly, as depicted in Figure 11-1, herbicides interact with all tissues at some time during their time in the plant.

In order to fully understand the relationship between the plant structure and herbicide entry and action, the anatomy of plants needs to be examined in greater detail. The dermal tissue system is examined in the section where we look at the leaf as a point of herbicide entry. At this point, let us examine in somewhat greater detail the fundamental and vascular tissue systems.

Fundamental Tissue System

From the herbicide's standpoint during the entry phase, the fundamental system is an obstacle to reaching the vascular system, which is the ultimate goal of the entry phase. The fundamental tissue system, as the system from which the various parts—roots, stems, leaves, and flowers—originate, must necessarily be composed of a wide variety of cells. As already pointed out, there is a continuity of the protoplasm of living cells in the three tissue systems of plants. However, the fundamental tissue system, as well as the other tissue systems, is interlaced with intercellular spaces. These spaces may be open, or they may be filled with material excreted from the living cells surrounding them. If not filled with excreted material they may be occupied by water or by gasses, depending upon the state of hydration of the entire system. Viewed in this way, the fundamental system provides a vehicle for movement of a herbicide through either living cells or open spaces.

Vascular System

Xylem. The *xylem* is the principle conducting vehicle for water and mineral nutrients. It consists of several different types of cells, living and nonliving. The most characteristic are the tracheary parts, but fibers and living parenchyma cells are ultimately found in the xylem.

The *tracheary elements* are of primary concern in herbicide movement since they are responsible for the movement of water within the plant. There are two basic types of tracheary elements: the tracheids and the vessels. In both, as the cells mature, the secondary wall becomes lignified and contains no protoplast. The difference between the two is that the *vessels* are perforated in certain areas of contact with other vessel members. The perforations are normally on the ends, thus resulting in a long, continuous tube. The *tracheids*, on the other hand, are not perforated but do have pits in the common walls between two such walls. Thus, sap moving through these cells can do so freely from one to another in the vessels, but in the tracheids must pass through the walls, primarily the thin membranes over the pits. Vessels and tracheids of the xylem are shown in Figure 11-3A.

Phloem. The *phloem* is the food-conducting (photosynthate-conducting) tissue of vascular plants. It, like the xylem, consists of several different kinds of cells possessing different functions. The most important, from the standpoint of herbicide movement, are the sieve elements. They are of two kinds: *sieve cells* that are relatively unspecialized and the more specialized *sieve tubes*. Sieve cells are characteristic of the lower vascular plants and *gymnosperms*—that is, naked seed plants, such as conifers. Sieve tubes are characteristic of most *angiosperms*—that is, covered seed plants, such as corn. The difference between sieve cells and sieve tubes is in the degree of differentiation of the sieve areas and in their distribution on the cell walls. *Sieve areas* are areas in the cell walls with pores or holes in them through which the elements of adjacent cells are interconnected by extrusions of protoplast whose nucleus is not retained. The size of the pores may range from a fraction of a micron to 15 or more microns in some dicotyledons. A sieve cell has relatively nonspecific areas for the pores. Sieve tubes, on the other hand, have the sieve area in a more or less specified location termed *sieve plate*. Since the sieve plates occur mainly on end walls, the sieve tube members become stacked one upon another to form a conducting tube.

The phloem tube commonly includes *companion cells*. Although companion cells arise from the same meristematic cell as the sieve cells or sieve tubes, they differ from the sieve elements. They retain their nucleus and do not develop sieve areas. Rather, movement between such cells and between them and the sieve element occurs through depressed areas or *pit fields*. Gymnosperms and vascular cryptogams (which reproduce by spores) do not contain companion cells but do contain cells termed *albuminous cells* that are closely associated with their sieve cells.

The phloem also commonly contains a number of parenchyma cells besides the companion and albuminous cells. As in the xylem, these may be arranged in rings around the phloem or in rays passing through them. *Parenchyma cells* serve as a place for storage of starch, fat, and other organic foods and for the accumulation of tannins

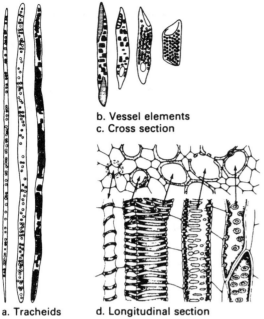

b. Vessel elements
c. Cross section

a. Tracheids   d. Longitudinal section

A. Elements usually found in xylem: (a and b) tracheids and vessel elements shown isolated from the tissue, (c) in cross section, and (d) in longitudinal section

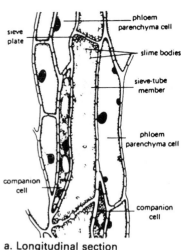

sieve plate

phloem parenchyma cell

slime bodies

sieve-tube member

phloem parenchyma cell

companion cell

companion cell

a. Longitudinal section

b. Face view of a sieve plate

B. Elements usually found in phloem: (a) longitudinal view of a mature sieve-tube member and companion cell, (b) face view of a sieve plate (black areas are actually holes in the walls)

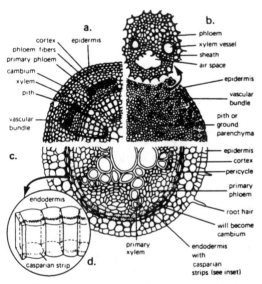

a.

cortex
phloem fibers
primary phloem
cambium
xylem
pith

vascular bundle

b.
phloem
xylem vessel
sheath
air space
epidermis
vascular bundle
pith or ground parenchyma

c.

epidermis
cortex
pericycle
primary phloem
root hair
will become cambium
endodermis with casparian strips (see inset)

endodermis

casparian strip   d.

primary xylem

C. Cross sections of (a) herbaceous dicot and (b) monocot stems and of (c) a young root; the three-dimensional drawing of the endodermal cells (d) showing position of casparian strips in the walls

FIGURE 11-3. Elements commonly found in vascular tissue and the close physical relationship among them.

Source: Salisbury and Ross, 1978. Reprinted with permission of Wadsworth Publishing Company, Belmont, CA © 1978 by Wadsworth, Inc.

and resins. Phloem may also contain fibers as a source of storage for starch. The elements in phloem are shown in Figure 11-3B. Figure 11-3C shows the close spatial relationship between the xylem and the phloem in both roots and stems.

LEAF AS POINT OF ENTRY

Figure 11-4 is a schematic diagram of a leaf showing the structural characteristics to be discussed in connection with herbicide entry. We begin our examination of this structure from the outside (cuticle) into the leaf vein. The *cuticle* is a discrete, nonliving lipophilic membrane on the outer wall of the epidermal cells. It is a major barrier to absorption of foliar-applied herbicides (Hull et al., 1982). The process of herbicide entry into the leaf is one of penetrating this nonliving exterior layer in order to reach the living, dermal (epidermal cell) layer that provides access to the fundamental and vascular (vein) systems. The cuticle contains cutin, waxes, pectin, and cellulose, all compounds with varying degrees of polarity. Lipophilic waxes and lipoidal material that predominate on the outside of the cuticle pose the first barrier to herbicide entry, termed the *lipoidal phase.* In general, nonpolar herbicides move fairly readily through this lipoidal area, while polar compounds, such as water-soluble herbicides, do not.

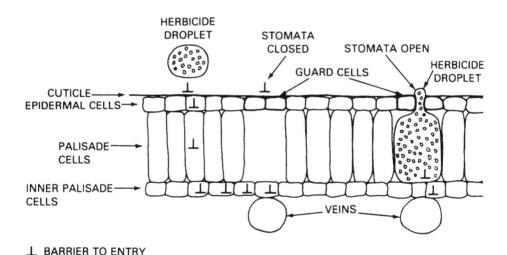

FIGURE 11-4. Cross section of the upper portion of a leaf showing barriers to entry of a herbicide to the vascular system (vein). Entry through the open stomata offers easier access to the vein than through the cuticle.

However, we know that polar herbicides do gain entry through the leaves. The leaf cuticle presents two functioning phases: the lipoidal phase just discussed and the *aqueous phase*. It is now known that the cuticle is not an uninterrupted sheet as once thought. Rather, when examined under an electron microscope, it shows layering and often has microfibrils (plasmodesmata) extending into it from the underlying epidermal cells. Thus, it is comparable in structure to a sponge with quite possibly an interconnecting network of pores. When filled with water, this network provides an aqueous route for polar herbicides to pass through the cuticle.

Factors Affecting Entry and Passage Rates

Penetration and its rate depend upon both absorption by and thickness of the cuticular layer. For the herbicide to move from the leaf surface to the interior of the leaf requires that a gradient be established. Both the concentration of the herbicide solution on the leaf surface and its rate of acceptance (translocation or metabolism) by the epidermal cell clearly influence the steepness of this gradient.

Several other factors may affect the gradient. It may be affected by purely physical aspects of the environment in which the spray droplet finds itself; it may be washed off by rain; or it may dry out, leaving the herbicide in crystalline form rather than in solution. The surface topography of the leaf itself, which varies among species (Figure 11-5), may prevent effective contact between the herbicide and the cuticle, thus preventing establishment of the gradient. For example, the gently undulating surface of nutsedge leaves allows greater contact of spray droplets than do the sharp surface features of bermudagrass and redroot pigweed. These physical factors affect entry (gradient) of the herbicide whether polar or nonpolar in character.

Since, as we shall see later, the herbicide is a passenger with photosynthate, the state of photosynthesis affects acceptance of the herbicide and, thus, the gradient. If photosynthesis is minimal or not taking place, the herbicide may accumulate in the epidermal or palisade cells and prevent a gradient from being established. Any aspect of the environment that reduces overall rate of metabolism could thus reduce herbicide uptake.

Cuticle. The thickness of the cuticle itself determines the distance the herbicide must travel. Many factors in the environment influence thickness of the cuticle. Factors that tend to cause the cuticle to be thin are shade, low temperature, and wet soil. Reduced tillage, plant residues, and cover crops may create just such conditions. Conversely, full sunlight and low moisture tend to cause the production of thicker cuticle. Also, both acid and alkaline soils result in heavier cuticle.

Age itself is an important determinant of cuticle thickness. As the plant ages, not only does the cuticle thickness increase, but in some species and under some conditions, there is a tendency for even the epidermal cells immediately beneath the cuticular layer to cutinize. A quick inspection of factors affecting thickness of the cuticle indicates that thickness may vary for leaves on the same plant. For example, those within

FIGURE 11-5. Leaf surface characteristics of various weeds.

Source: Courtesy of F.D. Hess and Wiley-Interscience.

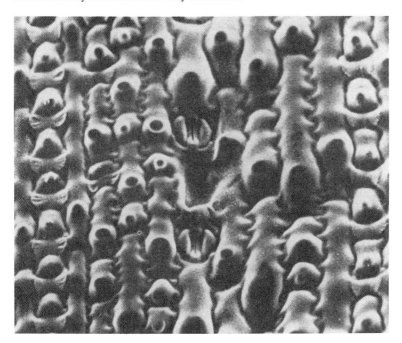

1. Bermudagrass (*Cynodon dactylon* [L.] pers.), 600×

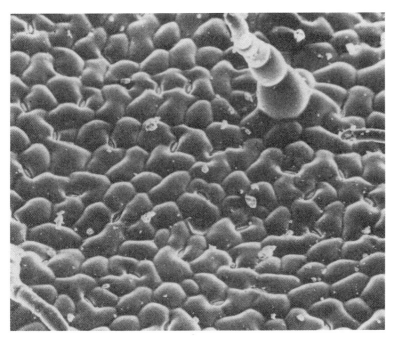

2. Redroot pigweed (*Amaranthus retroflexus* L.), 250×

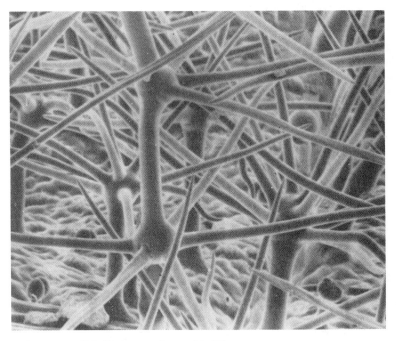

3. Common mullein (*Verbascum thapsus* L.), 250×

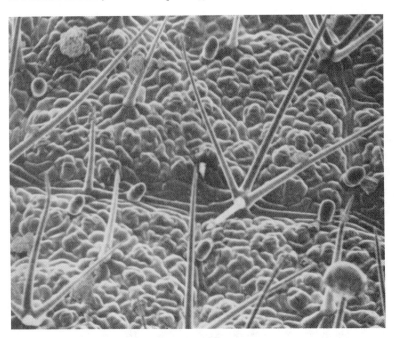

4. Velvetleaf (*Abutilon theophrasti* Medic.), 150×

5. Yellow nutsedge (*Cyperus esculentus* L.), 1,750×

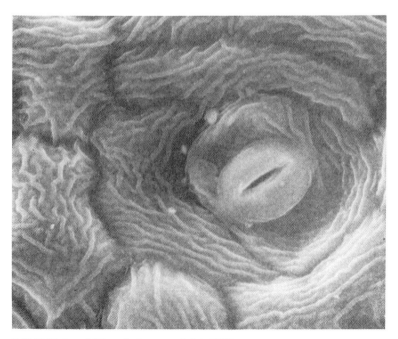

6. Field bindweed (*Convolvulus arvensis* L.), 1,250×

the canopy may be subject to rather heavy shade and may be of a different physiological age than outer leaves.

The composition of the cuticle, as well as its quantity, is also apparently affected by environmental factors, as indicated in Table 11-1. The percentage cutin is 77% under 10°C days, dropping to 50% as the daytime temperature increases to 23°C or higher. There is a corresponding change in wax percentage. These differences could affect absorption of a herbicide by the cuticle. Thus, we see that the environment can affect both the quantity and quality of cuticle present on the leaf surface. Therefore, as the point of entry for aerial applications, leaf surface can account for some of the variation in results encountered with herbicides.

TABLE 11-1. Effect of temperature on deposition of cuticle and its components, wax and cutin, in tobacco (*Nicotiana glauea*) leaves.

Day Temperature (°C)	Weight of Cuticle (g)	Wax (%)	Cutin (%)
10	0.0105	23	77
17	0.018	38	62
23	0.0132	50	50
30	0.0145	51	49

Source: Adapted from Skoss, 1955.

Stomata. Both the upper and lower surfaces of most leaves contain openings in the epidermis and cuticle called *stomata* through which an interchange of gasses occurs between the atmosphere and the subepidermal cells. Stomata with guard cells of yellow nutsedge and field bindweed can be seen in Figure 11-5. These openings are spaces between two special epidermal cells known as *guard cells.* Changes in the size and shape of these guard cells determine the size of the opening. Open stomata provide direct access of the herbicide to the living tissue and, thus, to the symplast transport system (Figure 11-4). *Palisade cells* provide a barrier to direct herbicide entry to the symplast system through the cuticle.

Turgor is the driving force for opening and closing of the stomata. An increase causes the stomata to open and a decrease causes them to close. If the stomata are closed, the cuticular barrier is continuous. When the stomata are open, penetration by herbicide sprays can be expected to be very rapid as compared with the relatively slow uptake through the cuticle.

Since many factors affect stomatal opening and stomata are present only on the lower surface of leaves of many species, stomatal entry is likely to be relatively unimportant as a point of entry for herbicides. Conditions that favor stomatal uptake also favor cuticular uptake in the aqueous phase. Because of the much larger surface area of the nonstomatal cuticle, entry through it is relatively more important. Thus, stomatal entry may be too minor to determine success or failure in control of the targeted weed species but may nevertheless, at least partially, explain the occurrence or absence of injury to the nontargeted plant.

Leaf hairs. Other leaf integuments or characteristics may influence the leaf as a point of entry of a herbicide. One is *pubescence,* or leaf hairs collectively called trichomes. These structures originate from the epidermal cells and may vary from single-celled organs to variously branched multicellular plates (Figure 11-5). Such structures clearly can affect the degree of contact between a herbicidal spray and the cuticle. Large trichomes may cause the spray droplet to shatter and bounce off (Boize et al., 1976). A thick stand of epidermal hairs could prevent contact between the herbicide droplet and the leaf surface. On the other hand, a thin stand might help to retain the spray droplets on the leaf's surface. There is some evidence also that individual hairs may serve as a point of entry for herbicides. Because species differ in the presence and amount of epidermal hairs, these structures may, therefore, be the source for some differences in effects of herbicides on plants.

Leaf shape and size. Leaf shape and size may influence entry because they determine area available for retention of the spray droplet. These differences, of course, are genetically based, but the leaf area in particular is quite plastic. That is, its size is influenced by light intensity, quality, and duration; by water stress during development; and by nutrition. Environment also affects leaf characteristics that in turn influence herbicide retention.

Leaf orientation. The orientation of leaves may increase or decrease retention of herbicide droplets depending on exposure of horizontal or vertical surfaces (Holly, 1964). Shading, drought, and low temperature caused leaves to have a more horizontal leaf orientation and greater retention of the herbicide fenoxaprop (Xie et al., 1995).

Fate of Applied Herbicide

With respect to the leaf as a point of entry, five events may occur with an applied herbicide: (1) it may be lost by runoff, by being washed off, or by volatilization; (2) it may remain on the cuticle because of evaporating down to an amorphous or crystalline deposit; (3) it may enter the cuticle but remain there in lipoidal solution; (4) it may pass through the cuticle and enter the apoplast system of cell walls and xylem; and (5) it may pass through the cuticle and cell wall system then absorb into the symplast of parenchyma or phloem cells. Obviously, unless event 5 occurs, the applied herbicide will not affect the plant.

ROOT AS POINT OF ENTRY

Factors Affecting Entry and Passage Rates

A schematic representation of a root in Figure 11-6A identifies the tissues, and something of their developmental ontogeny, important in herbicide entry. The emphasis implied in the figure is on the root during early development. This emphasis helps visualize the chronological development of two elements paramount in results with herbicides applied to the soil: the *casparian strip*—a waxy band in the cell walls of the root endodermis—and the xylem. Unlike the leaf, the root likely does not have a cuticle; or if it does, it is neither as thick nor as persistent as in the leaf. The penetration of the root through the soil, with the associated abrasive action, would tend to remove or disrupt any such layer. Furthermore, the root contains relatively more "free space" in the fundamental tissue than does the leaf. The reason is that cells in the cortex of the root commonly separate at their adjoining walls, thus adding this space to the space provided by the "poor fit" at the corners or edges of cells (Figure 11-6B). In effect, this means that a herbicide in soil, if dissolved in or dispersed in the soil solution, conceivably could be in contact with the semipermeable cell membranes of all living cells in the dermal and cortex regions, as well as rather quickly and easily reaching the vicinity of the endodermal cells that bar access to the vascular system.

Thus, the root provides a barrier to entry analogous to that provided by open stomata in the leaves. In the case of the root, however, the exposure can by visualized as continuous along the outer surface of the endodermis, as opposed to comparable exposure in the leaf only where stomata interrupt the leaf surface. Evidence supporting comparatively easy penetration of the epidermis and cortex is found in the fact that most herbicides initially are taken up rather rapidly upon reaching the soil-root interface. This fact can be explained by the filling of the intercellular space. Once this space is filled, uptake can be expected to slow down because of the need either to cross the casparian strip or cross the cell membrane shown in Figure 11-6B. As we will see in Chapter 12, however, relatively easier access of herbicides through the roots is apt to be more than offset by phenomena that can relieve the root from exposure to a herbicide (root growth beyond the herbicide zone, leaching of the herbicide, adsorption of the herbicide on soil particles, and decomposition of the herbicide).

The objective for soil-applied selective herbicides is to reach the xylem or phloem of the weed root to access sites of action throughout the plant. The two routes and combination of routes by which this objective can occur are shown in Figure 11-6B as apoplast and symplast entry. Inspection of Figures 11-4 and 11-6 suggests that there are fewer barriers to entry through the root than there are to entry through the shoot. In effect, the root presents only two barriers: (1) that posed by the semipermeable membrane of the living root cell and (2) that posed by the casparian strip in the walls of the endodermal cells.

Xylem maturity. The xylem is the tissue that provides movement of water from the soil to the aboveground parts. As shown by the grading-in of the stipling of the xylem in Figure 11-6A, xylem attains its mature status toward the juncture of the elongation

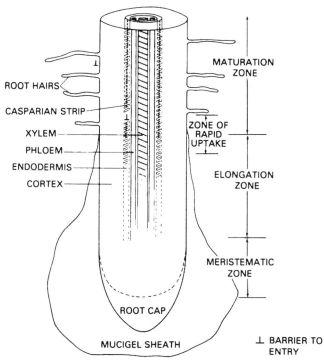

A. Three-dimensional diagram of a young root showing ontogeny and the relationship to entry

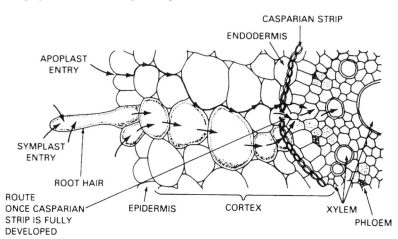

B. Transection of a root showing the two routes for entry: apoplast (xylem) and symplast (phloem). Position of casparian strip between the root exterior and the vascular vessels is also shown

---

FIGURE 11-6. Root as point of uptake for herbicides.

Source: Part A after Russell, 1977. Reproduced with permission of McGraw-Hill Book Company (UK), Ltd. Part B modified from Esau, 1965.

zone with the maturation zone. This maturation point is a relatively short distance from the root tip—in the neighborhood of 50 mm—depending on the species and growing conditions. The zone of rapid uptake of water identified in the figure includes part of the maturation zone and part of the elongation zone. This is not to say that uptake does not occur in older tissue, only that uptake is more rapid in the younger tissue. Similarly, it appears that this area is also the primary site of entry for herbicides. Thus, the age of the root as it determines maturity of the xylem has an important effect on herbicide entry.

Casparian strip. The casparian strip plays a key, and in a sense a controlling, role in herbicide entry of roots. This strip, shown in Figures 11-6A and 11-6B, is formed in the maturation zone during the early development of the endodermal cells and is an integral part of the primary wall. Chemically, it apparently consists of suberin, lignin, or a combination of both that is highly hydrophobic. Once it is fully established, the strip forces chemicals to enter via the living protoplasm in order to access the vascular tissues. Thus, this barrier provides one basis for selectivity in that it conceivably prevents access to the plants of those chemicals that cannot penetrate the cell membrane.

There is a time before the casparian strip is fully established when chemicals presumably do gain entry to the xylem by way of the intercellular and the wall spaces. Depending upon the chemical and physical nature of the herbicide, the herbicide may be restricted to entry by the apoplast, by the symplast, or by both routes.

Active and passive uptake. It is known from studies of mineral nutrient absorption by plants (roots) that such uptake may be both passive and active. *Passive uptake* may be most easily visualized as free movement with water from the soil solution into the root and upward in the xylem. *Active uptake* is uptake that depends upon expenditure of energy on the part of the plant. It is most easily visualized where the uptake occurs against a *concentration gradient*. That is, the concentration in the xylem is greater than that of the mineral nutrient at the soil–root interface. Work on uptake of herbicides indicates that both passive and active uptake are involved also and vary from herbicide to herbicide. Knowledge of the mechanism involved with candidate herbicides will help decide which herbicides would be best for the intended use and condition.

Dicots vs. monocots. The distinct difference between dicotyledonous and monocotyledonous plants in the source of their root systems may well have a bearing on the comparative importance of herbicide uptake from the soil. In dicots, the root system develops from the radical of the embryo upon its elongation and branching. In monocots, the first root derived from the root meristem of the embryo usually dies at an early age. The permanent root system then develops from stem tissue as adventitious roots. Thus, roots in monocots are closer proximally to the apical meristem of the shoot than are those in dicots.

Figure 11-7 contrasts the different anatomical structures of dicots and monocots during germination and early seedling development. The early development of adven-

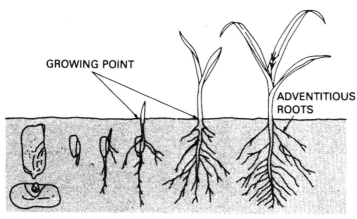

A. Seedling development of a grass plant (monocot)

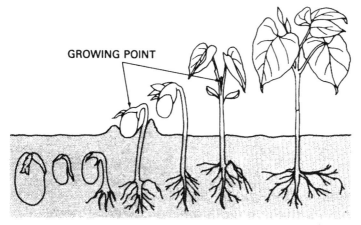

B. Seedling development of an epigeal emerging dicot

FIGURE 11-7. Differences in the location of growing points and roots relative to the soil surface of monocots and dicots as a basis for selective control of weeds with herbicides.

titious roots by monocots also means that they may well have roots closer to the soil surface than do dicots, at least during their early stage of development. Both the penetration of soil by roots and their numbers are influenced by physical factors of the soil itself, such as its density, organic matter content, and water content. All of these factors may be increased by reduced tillage. To the extent that dicots and monocots respond differently to these factors, another variable is introduced to their response to soil-applied herbicides.

COLEOPTILE AS POINT OF ENTRY

The *coleoptile* is an anatomical structure unique to the grasses. For interpretative purposes, it can be visualized as a specialized leaf that forms a cone-shaped tube surrounding the apical meristem or growing point. It has a hole at the apex through which the first foliage leaves emerge. It has stomata that, given the high humidity both internal and external to it, can be presumed to be open. Thus, the coleoptile could conceivably provide relatively easy access of a herbicide to the growing point inside. During early development of the grass seedling, this growing point may be at any place from essentially the level of the seed in the soil to the soil surface when the leaves actually emerge through the soil. In dicotyledons, by contrast, since the growing point is borne in the axis of the first true leaves that in turn is between the cotyledons, the growing point is aboveground once the cotyledons—or first true leaves—emerge through the soil. It should be noted, however, that some dicots actively take up herbicides through the hypocotyl for a time during germination and early seedling development. Legumes in the hook stage are an example. This period is the stage from the number 3 to the number 5 seedling in Figure 11-7B.

Dawson (1963) showed that uptake of herbicides from the soil could occur through the coleoptile in addition to plant roots. He found that uptake of EPTC by the primary root of barnyardgrass had relatively little effect, whereas uptake by the shoot resulted in severe injury. He concluded that the leaf is the important site of EPTC uptake in the control of barnyardgrass. Much work since then supports this finding by Dawson.

Further, it appears that the region of the so-called coleoptile node is the key, at least in determining the effect of uptake on the plant. The importance of the coleoptile was shown in work reported by Holly (1976) in which the herbicide diallate was evaluated for its effect on wheat and on wild oat when placed at different locations in the soil relative to the seed (Figure 11-8). As can be seen, both wheat (W) and wild oat (A) were very sensitive to the herbicide immediately above their seed. However, wheat was essentially unaffected where the diallate layer was close to the soil surface, whereas wild oat was nearly completely killed by this layer. The difference is explained by the fact that the coleoptile node (growing point) in wheat remains close to the seed for a time and below the herbicide layer. In wild oat, however, the coleoptile node moves upward into the treated soil at a very early stage.

BIOCHEMICAL SITE OF ACTION

Once the herbicide enters the plant's vascular system it can be moved to the site(s) of action. Identifying the primary sites of action for the large number of available herbicides is outside the scope of this text. Furthermore, such knowledge is not essential to

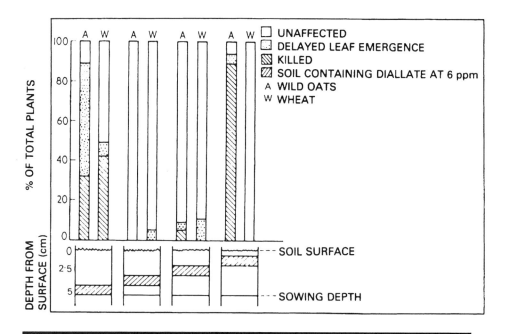

FIGURE 11-8. Differences in the location of the growing point relative to the seed in the soil of monocots as the basis for selective weed control.

Source: Holly, 1976. Reproduced with permission of Academic Press, Inc.

understanding the role of herbicides in weed management and the relationship between production practices and herbicide efficacy. Let us simply recognize that there are four primary biochemical reactions for plant responses to herbicides: (1) respiration and mitochondrial electron transport; (2) photosynthesis; (3) nucleic acid metabolism and protein synthesis; and (4) lipid metabolism (Ashton and Crafts, 1981).

SEED AS POINT OF ENTRY

Typically, seeds of many species, both weeds and crops, are not prevented from germinating in the presence of herbicides, but growth of their seedlings is arrested at an early age. This statement is not meant to imply that some herbicides may not prevent germination. However, the much more common response is for the early steps in germination to take place with further development of the seedling arrested.

REFERENCES

Ashton, F.M., and A.S. Crafts. 1981. Mode of action of herbicides, 2nd ed. New York: Wiley.

Boize, L., C. Gudin, and C. Purdue. 1976. The influence of leaf surface roughness on the spreading of oil spray drops. Ann. Appl. Biol. 84:205-211.

Dawson, J.H. 1963. Development of barnyardgrass seedlings and their response to EPTC. Weeds 11:60-67.

Esau, K. 1965. Plant anatomy. New York: Wiley.

Hartley, G.S. 1960. Physio-chemical aspects of the availability of herbicides in soils. In E.K. Woodford and C.R. Sagar, eds., Herbicides and the soil, pp. 63-78. Oxford, England: Blackwell Scientific.

Holly, K. 1964. Herbicide selectivity related to formulation and application. In The physiology and biochemistry of herbicides. L.J. Audus, ed., pp. 423-463. New York: Academic Press.

_____. 1976. Selectivity in relation to formulation and application methods. In L.J. Audus, ed., Herbicides: Physiology, biochemistry, ecology, vol. 2, pp. 249-277. New York: Academic Press.

Hull, H.M., D.G. Davis, and G.E. Stolzenberg. 1982. Action of adjuvants on plant surfaces. In Adjuvants for herbicides. R.H. Hodgson, ed., Weed Sci. Soc. Am, pp. 26-67. Champaign, IL.

Russell, R.S. 1977. Plant root systems: Their function and interaction with the soil. London: Mc-Graw-Hill.

Salisbury, F.B., and C.W. Ross. 1978. Plant physiology, 2nd ed. Belmont, Calif.: Wadsworth.

Skoss, J.D. 1955. Structure and composition of plant cuticle in relation to environmental factors and permeability. Bot. Gaz. 117:55-72.

Xie, H.S., B.C. Caldwell, A.I. Hsiao, W.A. Quick, and J.F. Chao. 1995. Spray deposition of fenoxaprop and imazamethabenz on wild oats (*Avena fatua*) as influenced by environmental factors. Weed Sci. 43:179-183.

12

Preventing Weed Emergence with Herbicides

CONCEPTS TO BE UNDERSTOOD

1. For soil-applied herbicides, selectivity may be based on both physical and biological factors.
2. Selectivity based on physical factors is the result of a separation in space between the desired plant seed and the herbicide. It may be accomplished by mechanical means, by use of adsorptive barriers, or by use of herbicide formulation.
3. Selectivity based on biological factors is the result of differences in morphology and in internal physiology and metabolism of the desired plant and weed. Safeners are biologically selective based on the use of a secondary chemical that prevents the herbicide from exerting its activity on an otherwise sensitive crop.
4. Adsorption on soil particles, leaching, volatilization, and degradation can reduce effectiveness of soil-applied herbicides in preventing weed emergence. The first three factors may also be responsible for injury of desired plants from otherwise safe herbicides.
5. Persistence in soil is important as a factor influencing weed control with a given application and of application in subsequent years, as a possible source of carryover injury to following crops, and as a potential effect on contamination of the environment.
6. Adsorption, leaching, volatilization, and degradation are primary mechanisms of herbicide removal from soil. Uptake by plants and physical removal in runoff and in soil erosion are secondary mechanisms of removal.
7. Degradation by soil microorganisms may be the result of either utilization of the herbicide as a source of carbon for growth or breakdown coincidental to utilization of some other carbon source (cometabolism). Direct utilization as a carbon source results in increased degradation with time; cometabolism results in a more constant degradation rate.

The application of herbicides to soil to prevent weed emergence is based on three well-established facts: (1) most annual weeds originate from seeds in the top 2.5 to 5.0 cm of the soil surface; (2) the germination and early seedling growth stages are generally the most vulnerable periods in the weed's life cycle; and (3) if weeds are not allowed to establish for only a few weeks, most crops can successfully compete thereafter. This method of preventing weeds from interfering with crop production offers certain practical advantages as well. One is to provide protection against conditions that may prevent later control methods, such as rainfall delaying cultivation or postemergence herbicide application until weeds either cannot be effectively controlled or have already inflicted some permanent damage on the crop. Another advantage is that herbicide application and planting the crop can be done in a single operation. Further, soil application of herbicides can be used safely on desired plants that otherwise are not tolerant after they have emerged and are growing.

HERBICIDE SELECTIVITY

The successful use of herbicides applied to the soil calls for the selective prevention of weed emergence and establishment without damage to desired plants. Figure 12-1 shows the two most commonly used methods of soil applications designed, at least in part, to ensure adequate selectivity. Application to the soil surface after or at the time of planting the crop was the earliest method used (Figure 12-1A).

For effective control to be obtained, rainfall or irrigation is needed to move the herbicide into the weed seed germination zone. Further, with most herbicides, this transfer must occur before the weed seeds germinate. For these reasons, and to reduce herbicide loss from volatilization and photodecomposition, many herbicides are now routinely incorporated mechanically into the weed seed germination zone prior to planting the crop (Figure 12-1B). Truly spectacular weed control results can be achieved, as demonstrated for preemergence applications of atrazine on corn (Figure

WEED SEED
GERMINATION
ZONE
CROP SEED ZONE

A. Application at or after planting of crops. Herbicide applied to soil surface moves downward into the weed seed germination zone with percolating water.

B. Application before planting of crops. Herbicide applied to soil surface is mechanically mixed into the weed seed germination zone.

FIGURE 12-1. Two most commonly used methods of herbicide application to soil for selective control of annual weeds.

12-2). Atrazine applied preemergence at 2.2 kg ha^{-1}, with no cultivation or other weed control, completely prevented emergence of several annual weeds that emerged in untreated plots. Effective control persisted for the entire growing season.

With any soil application method, selectivity is a function of physical or biological factors, or their combinations. This is because the herbicide, soil, plant, and climate may each influence results obtained. If the herbicide is to be effective against the weed,

A. Untreated corn

B. Corn treated with 2.2 kg ha^{-1} of atrazine applied preemergence

FIGURE 12-2. Complete and season-long prevention of emergence of annual weeds in corn.

Source: Reproduced courtesy of O. Hale Fletchall, Department of Agronomy, University of Missouri-Columbia.

it must be in contact with the germinating seed or the developing seedling, and it must enter the weed to reach the site of biochemical reaction within the plant. Simultaneously, either one or both of these processes must be avoided by the desired plant seed and seedling. Generally, physical factors are those most responsible for avoiding or assuring herbicide contact, whereas biological factors are involved in entry and biochemical reactions.

Physical Basis for Selectivity

Selectivity based on physical factors must provide differential contact of the herbicide with the root and young shoot of the weed and the desired plant in the soil. The developing root and shoot are the important points of entry for soil-applied herbicides. Although herbicides may enter the seed, the seed may be relatively unaffected if it is dormant. Further, it appears that with many herbicides, imbibition by the seed does not prevent initiation of the germination process. Herbicides otherwise equally toxic to the desired plant and the weeds may be used selectively by physically exposing the weed's points of entry and not the points of entry of the desired plant. This technique may be accomplished by physically separating the desired plant seed and the herbicide and is commonly termed *depth protection* or *placement selectivity*. In general terms, depth protection may be accomplished by mechanical means, through use of absorptive barriers, and through the use of different herbicides, all of which prevent desired plant–herbicide interaction.

Seeding desired plant below herbicide zone. The most common method for providing depth protection is by seeding the desired plant below the herbicide zone in the soil or below the zone that the herbicide may be expected to reach. Figure 12-3 demonstrates the effectiveness of this approach in early work for selective use of CDAA for annual grass control in cotton. Cotton planted 5 cm deep was reduced in stand only at the highest rate of 18 CDAA when incorporated at a depth of 3.8 cm. This rate was about 16 times the amount required for effective grass control; thus, there was a sizable safety margin for the crop. Where CDAA was incorporated to a depth of 10 cm, the stand was reduced by one-third with rate of only 2.2 kg ha^{-1}. It should be noted also that weed control was less effective with the 2.2 kg rate incorporated at 10 cm than where the herbicide was incorporated only 3.8 cm. With deeper incorporation, relatively less herbicide was concentrated in the surface soil zone in which the weeds germinated.

Other adaptations of this concept of separation have been successfully used. Nalewaja et al. (1987) were able to use trifluralin for grass control in flax by incorporating the herbicide to a depth of 10 to 12 cm below the soil surface and seeding flax at a depth of 2 to 4 cm. Rapid emergence of flax likely avoided severe injury by the herbicide, leading to minimal stand reductions and no significant decreases in flaxseed yields. Holstun and Wooten (1964) used a combination of subsurface and surface spraying to improve selective control of weeds in cotton with EPTC. They used herbi-

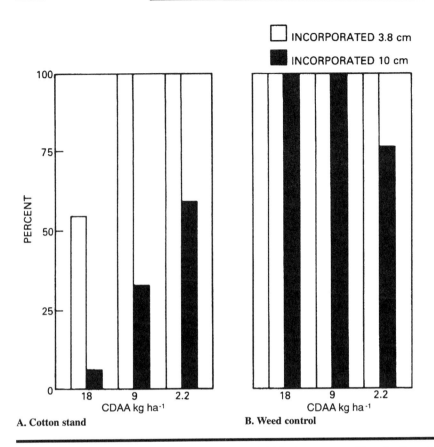

FIGURE 12-3. Seeding the crop below the herbicide layer to avoid reductions in crop stands with herbicides incorporated in the soil. Here cotton was planted 5 cm deep.

Source: Data from Kemper et al., 1963.

cides tolerated by cotton in a band over the cotton row with adjacent row–shoulder subsurface treatments of EPTC. The band application controlled the annual weeds in the row, and the shoulder applications controlled the nutsedge as well as the annual weeds beyond the row.

Absorptive barriers. A method to physically separate the herbicide and crop seed is the use of *absorptive barriers.* The technique, diagrammed in Figure 12-4, can be accomplished in one operation by the use of an implement that seeded the crop in a furrow, covered the seed by partially filling in the furrow, applied a band of absorbent (activated carbon) into the furrow, filled in the furrow, and then sprayed preemergence herbicide over the soil surface. In this manner, herbicide contact with the seed due to downward movement was avoided because the activated charcoal absorbed the herbicide before reaching the seed. Adaptations of the absorptive barrier concept include coating seeds and dipping seedlings of transplants into suspensions of activated carbon. This approach protects germinating seeds and roots of transplants from herbicide incorporated in soil because it is absorbed by the activated carbon on the seed and root surfaces. Finely ground activated carbon can also be spread on sites with herbicide and other organic chemical spills to avoid or reduce contamination of the surrounding environment.

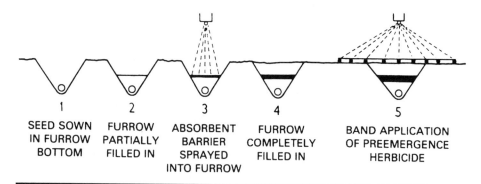

FIGURE 12-4. Reductions in crop stand from residual preemergence herbicides avoided by placing a barrier of activated carbon above the seed.

Source: Holly, 1976. Reproduced with permission of Academic Press, Inc.

Herbicide differences.   As will be discussed, herbicides vary greatly in their mobility in soil. Some move very freely with percolating water. Others are held to varying degrees on soil particles and organic constituents and move relatively little with percolating water. Thus, the use of a relatively immobile herbicide may provide a physical separation of the herbicide from the germinating desired plant and weed seed. This technique was demonstrated with 2,4-D formulations in very early research (Aldrich and Willard, 1951). As shown in Table 12-1, the butyl ester formulation caused little or no reduction in corn stands, whereas stands were reduced 23% by the triethylamine salt formulation. Sufficient irrigation water was applied immediately after 2,4-D application to percolate down to the corn seeds. Laboratory studies showed that the butyl ester moved downward less freely, which explains the field results in Table 12-1.

TABLE 12-1.   Effect of 2,4-D formulation applied preemergence on stands of corn.

4 lb 2,4-D per Acre	Corn Plants per Plot
Butyl ester	45.7
Amine salt	37.4
Untreated	48.8

Source: Data from Aldrich and Willard, 1951.

Biological Basis for Selectivity

Differences in morphology and in internal physiology and metabolism of the desired plant and the weeds may provide the basis for selectivity. Since selectivity is commonly only a matter of degree, it is often desirable to utilize physical selectivity, if possible, in addition to biological selectivity.

Morphological differences. Differences in morphology as a basis for selectivity depend upon biological factors to accomplish a separation between the desired plant and the weed and point of uptake of the herbicide. In other words, the plant is the source of separation comparable to that provided by physical factors in physical selectivity. The difference in root systems, or more appropriately in their origin within the plant, is one difference in morphology that can be the basis for selectivity. Most dicotyledonous plants have a taproot system that develops from the radicle in the embryo, whereas in most monocotyledonous plants, the primary root usually does not survive very long but is replaced by adventitious roots that develop from the coleoptile nodes (see Figure 2-10). This characteristic may be utilized with some dicotyledonous crops to provide some measure of protection from a soil-applied herbicide. Where uptake is through the root, development of the radicle and the taproot moves the zone of potential uptake away from the herbicide zone, even if the latter should advance downward. Such separation can be visualized for cotton in the research described in Figure 12-3 in that shortly after germination of cotton, its zone of potential herbicide uptake is a few millimeters deeper than the 5 cm at which the cotton seed was planted. For its part, barnyardgrass initiates roots in the surface 0.5 cm of soil and continues to have absorbing roots in this zone because of their development from the coleoptile node.

The coleoptile in grasses is itself an important point of entry for some soil-applied herbicides. The shoot apex is just above the first node of this coleoptile. Grasses differ in where the first node, and thus the growing point, is located in the soil (Figure 12-5). The difference in location of the growing point of wild oat and wheat was the basis of selective control of the weed with diallate shown in Figure 11-7. The actual location of the coleoptilar node in some species appears to be in relation to the seed, and in others, in relation to the soil surface. In barnyardgrass, for example, the node is in the upper 1.2 cm, regardless of the depth at which the seed is located. In barley, it is within about 1.2 cm of the seed, so depth of planting determines its location relative to the soil surface. This morphological difference may also provide some degree of selectivity.

Physiological and metabolic differences. With soil-applied herbicides, selectivity based on physiological and metabolic differences is primarily a matter of what happens after the herbicide enters the plant since most are readily absorbed from the soil. This uptake is apparently less specific than is uptake through the leaves. Although herbicides enter mainly in solution, some, such as trifluralin, can gain entry in the vapor phase. Species of plants differ in the quantity of herbicides (herbicide dosage) they can tolerate in the soil environment, which illustrates a type of selectivity based on the amount of herbicide applied to the soil. Herbicidal activity based on dosage depends,

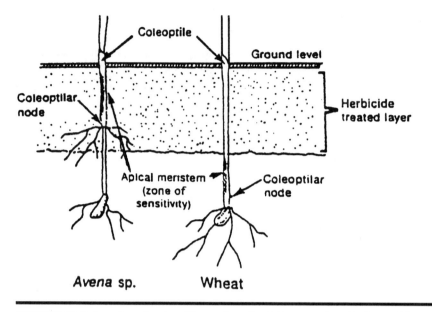

FIGURE 12-5. Location of coleoptilar nodes of oat (*Avena* sp.) and wheat with respect to their seeds, depth of planting, and herbicide placement.

Source: Moyer, 1987. Reproduced with permission of the Weed Science Society of America.

of course, on the herbicide chemical properties and soil type. A detailed discussion of herbicide physiology and metabolism is beyond the scope of this text; however, other books, including *Physiology of Herbicide Action* (Devine et al., 1993), *Metabolism of Herbicides in Higher Plants* (Hatzios and Penner, 1982), and *Mode of Action of Herbicides* (Ashton and Crafts, 1981) provide thorough treatment of this subject.

For our purposes, we need only recognize that tolerance on the part of one species to a herbicide that kills another is the result of failure of the herbicide to reach and accumulate lethal concentrations at the sites of action in the tolerant species. This tolerance may be due to differences in translocation, differences in absorption at sites other than the biochemical reaction sites for the herbicides, differences in breakdown within the plant, and differences in the complexing of the herbicide with the cell constituents.

Translocation of herbicide.   Differences between weeds and desired plants in the ease and extent of movement of herbicides within them provide a basis for selectivity. To reach the sites of action, herbicides must move within the plant, mainly in the xylem. With respect to translocation, it appears that amiben, prometryn, pyrazone, and terbacil are readily transported in the xylem of susceptible weeds to sites of activity in the shoots or leaves; whereas, with amiben and pyrazon at least, they are not readily translocated in wheat and sugarbeets, respectively. Tolerance of wheat to amiben is augmented also by complexation of the herbicide with chemical constituents within the wheat root cells to form an immobile compound identified as glycosylamiben. This immobile conjugate is not formed in barnyardgrass, which is susceptible to amiben.

Absorption of herbicide at ineffective sites.   If a herbicide is absorbed at a site other than its site of biochemical reaction, its killing action is prevented. With the phenoxy acid herbicides, it has been found that plants differ in the extent to which they are absorbed on the plant cell membrane phospholipids. Species with the ability to absorb these herbicides on the membrane phospholipids may thus be protected from injury simply because the herbicide does not reach the sites of biochemical reaction within the cell.

Detoxification of herbicide.   The ability of tolerant species to detoxify a potential herbicide appears to be the most important single factor contributing to selectivity. The herbicide within the plant represents an alien substance and thus is subject to attack by the enzymes inherent to that species or simply by interactions with chemical constituents of the cell without the aid of enzyme catalysts. Plants contain a large number of chemicals, including enzymes, and differ widely in their biochemical makeup. It is not surprising, therefore, that selectivity frequently is based on the differing ability of the weed and the crop to detoxify the potential herbicide taken up from the soil. Many chemical reactions are involved in herbicide metabolism in plants (Hatzios and Penner, 1982). The most important reactions with respect to selectivity are: (1) oxidation, (2) reduction, (3) hydrolysis, and (4) conjugation.

Oxidation is the loss of electrons. An example in herbicide activity is provided by 2,4-DB. This chemical is not toxic to plants. However, some plants have the capacity to remove two carbon atoms from the butyric acid side chain (via beta oxidation), thus creating the highly toxic 2,4-D. Many small-seeded legumes, including alfalfa, are tolerant of 2,4-DB in part because these plants are metabolically unable to perform beta oxidation.

The acceptance of electrons—*reduction*—of the nitro group of trifluralin by soybean roots probably accounts in part for this crop's tolerance of trifluralin. Susceptible species, such as corn, apparently do not reduce the nitro group.

The addition of water to s-triazine herbicides via *hydrolysis* provides part of the basis for selectivity of these herbicides. In corn, which is tolerant, hydrolysis of atrazine at the number two position on the ring to the inactive compound 2-hydroxyatrazine, appears to be mainly responsible for tolerance. Susceptible grass species, such as barnyardgrass, do not possess the biochemical mechanism for atrazine hydrolysis.

The formation of a complex compound by joining a herbicide molecule with another chemical such as a sugar, amino acid, or peptide (*conjugation*) is a source of selectivity with many herbicides. In corn, for example, tolerance to EPTC is associated with linkage of the herbicide molecule and the peptide glutathione within the corn plant to form an inactive conjugate. In susceptible plants, such as oats, EPTC is not inactivated by conjugation due to very low levels of glutathione within the root tissues.

Safeners for crop tolerance of herbicide.   Chemical compounds that have limited phytotoxicity by themselves and selectively protect crop plants against herbicide injury without protecting weeds are termed *safeners* and offer another form of selectivity. These compounds were traditionally referred to as herbicide antidotes, antagonists, and protectants, all of which appear in the weed science literature. A thorough review of the chemistry, mechanisms of action, uses, and impact of herbicide safeners is presented in *Crop Safeners for Herbicides* (Hatzios and Hoagland, 1988). The possibility of using chemicals for protecting crops against herbicides was first reported for barban in wheat. Barban is effective for control of several weeds in small grains and other crops but occasionally causes some crop injury. It was found that about 75% of the inhibition of wheat by foliar-applied barban was prevented by a chemical ("S-449") dusted on the wheat seed prior to planting (Hoffman, 1962). Later, chemicals were found that protected the crop from herbicides applied to the soil (Burnside et al., 1971), but the quantities needed necessitated seed treatment. Soon thereafter, chemicals were introduced that were sufficiently active and specific that they could be applied to soil (Chang et al., 1973). The protection was achieved without sacrificing weed control.

The degree of protection provided by herbicide safeners can be very pronounced (Figure 12-6). Corn treated with a safener was not damaged by rates of EPTC higher than needed for complete weed control. Without the safener, corn yields were cut about 10% by the lowest rate and by about 50% by the highest rate. This form of selectivity offers a way of extending the use of existing herbicides to the control of some problem weeds, such as the nutsedges, in selected crops where the margin of selectivity is very narrow. Since the protection is a result of chemical reactions within the plant, safeners provide another type of biological selectivity.

ENVIRONMENTAL FACTORS AFFECTING SUCCESS

Several factors can prevent fully satisfactory weed control with a soil-applied herbicide. The major ones to be examined in some depth here include: (1) adsorption, (2) leaching, (3) volatilization, and (4) degradation. These are all factors that under certain circumstances can prevent the herbicide from being in contact with the germinating weed seed and developing weed seedling. Such contact is important since this is the stage when many weeds are most vulnerable to effects from herbicides. Further, if we are to attain the objective of preventing emergence of the weed with the desired plant, the herbicide must be maximally effective to weeds in this stage of development.

Our major concern here is the influence of each process on results obtained, not the nature of the process itself. Thorough discussions of the processes themselves may be found in Cheng (1990) and Hance (1980).

Adsorption

Adsorption is the retention of a dissolved or gaseous substance on the surface or within

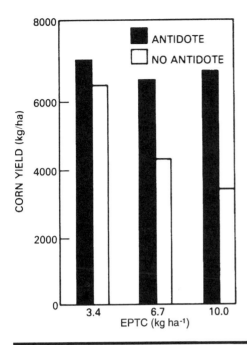

FIGURE 12-6. Protection of otherwise sensitive crops from herbicide injury by use of safeners. Here corn is protected from EPTC injury by a safener applied to the corn seed.

Source: Data from Burnside et al., 1971.

the structure of soil particles. As the definition implies, the extent of adsorption is a reflection of surface area. Surface area in turn, is determined largely by the size of the soil particle. The three broad categories of soil particle size—that is, sand, silt, and clay—expose vastly different surface areas. For any given quantity, clay particles (less than 2 µm in diameter) have 2,000 times more surface area than silt (2 to 50 µm in diameter) and 100,000 times more surface area than that of sand (greater than 50 µm in diameter). The practical effect of adsorption on herbicide use in preventing weed emergence with crops is to require somewhat higher rates on soils high in organic matter and somewhat lower rates on very sandy soils.

Clay and organic colloids. Most soils contain some clay (inorganic colloids) and organic matter, although quantities may be quite small. Because of its much greater surface area, the clay fraction of the inorganic material is much more important than the other inorganic fractions in adsorption of herbicides by soil. Further, clay particles 1 µm in diameter or smaller are chemically the most reactive. Particles in this size range can be suspended in water and remain suspended for considerable periods of time. These particles are insoluble and have properties similar to glues and other plastic materials, which accounts for the term *colloid*, from *kolla*, the Greek word for glue. Clays themselves can be differentiated based on the strength and ionic composition of the molecular lattices comprising their physical structure. For example, the lattices of kaolinite are held so tightly together by electrostatic forces that water molecules or chemical ions cannot penetrate between the lattices, and it is thus referred to as a non-expanding clay. Montmorillonite is composed of lattices that are held together by weak forces, and it is capable of expanding to accept water and ions within the clay structure. Thus, in expanding clays such as montmorillonite, herbicide molecules are not only adsorbed on external surfaces but also on internal surfaces as well. Organic matter at advanced stages of decomposition contains particles colloidal in size. The colloidal fraction of organic matter most responsible for adsorption is the humic matter derived from organic materials most resistant to decomposition in soil. This material also has charged external and internal surfaces capable of adsorbing herbicides to the same degree or more as the expanding clays. Inorganic clays and organic matter together provide soil colloids that are most important in adsorption of herbicides. Figure 12-7 is a diagrammatic representation of the adsorbing characteristics of soil particles. Particles have primarily negative charges imparted by unsatisfied oxygen charges (bonds). These unsatisfied charges are points of adsorption for positively charged sites of a herbicide molecule.

Early in the modern herbicide era, it was shown that organic matter and clays adsorbed herbicides and affected their final toxicity. In Table 12-2, we see that large differences exist among representative herbicides with respect to their adsorption to soil constituents. Adsorption is generally quantitatively expressed as a sorption coefficient, K_d, which represents the ratio of herbicide attached to soil particles (including organic matter) compared to that amount in the water phase. Thus, the K_d of 8,000 for trifluralin (Table 12-2) implies that 8,000 parts of herbicide is adsorbed for every one part of herbicide in solution. It is readily apparent, then, that trifluralin and pendimethalin are strongly adsorbed whereas dicamba and dalapon are very weakly adsorbed. We must remember that the K_d values presented are average values and vary for each herbicide depending on the properties of specific soils. This influence of soil texture, or the relative proportion of sand, silt, and clay contents, and soil organic matter content on adsorption is shown for EPTC in Table 12-3. The degree of adsorption, in turn, influences toxicity toward oats, used as a bioassay species. We also see that adsorption by dry soil is relatively higher than by moist soil. Other factors also affecting degree of adsorption include soil pH, soil drying–rewetting cycles, presence of crop residues, temperature, herbicide solubility, and degree of ionization.

Reversibility. Adsorption is reversible in that it represents an equilibrium relationship between the concentration of the substance adsorbed on surfaces and the concentration of the substance in solution. It follows that uptake of a herbicide by the weed root or seedling reduces its concentration in the soil solution, thereby allowing an amount of the herbicide adsorbed to soil particles to be released into the soil solution and become available for plant uptake. It has been found that many herbicides are affected by simple physical adsorption. Since herbicide in solution is directly available for uptake, it appears that adsorption by itself affects mainly the time required for uptake of a given herbicide rather that the total quantity taken up. Because seedlings of

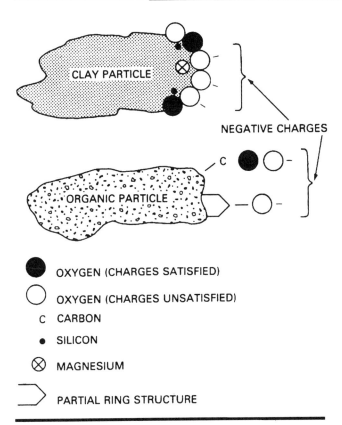

OXYGEN (CHARGES SATISFIED)

OXYGEN (CHARGES UNSATISFIED)

C   CARBON

•   SILICON

⊗   MAGNESIUM

PARTIAL RING STRUCTURE

FIGURE 12-7. Diagram of clay and organic matter colloidal particles showing the typical negative surface charge that gives the colloid its adsorptive ability.

TABLE 12-2.   Relative adsorption values of some representative herbicides.

Herbicide	K_d
Trifluralin	8000
Pendimethalin	5000
Triallate	2400
Oryzalin	600
Prometryn	400
EPTC	200
Metolachlor	200
Cyanazine	190
Atrazine	100
Metribuzin	60
2,4-D	20
Imazaquin	20
Picloram	16
Dicamba	2
Dalapon	1

Source: Data based on Wauchope et al., 1992.

TABLE 12-3. **Effect of soil type on the toxicity of EPTC to oats and on the soil adsorption of EPTC vapor.**

Soil Type	Organic Matter (%)	Clay (%)	Oats Injury (ED_{50})[1]	Relative Soil Adsorption[2] Soil at Field Capacity	Air-Dry Soil
Egbert series	4.8	79.5	1.55	20	100
Yolo silty clay	2.4	45.0	1.20	14	75
Yolo clay loam	2.4	38.7	0.88	15	70
Yolo sandy clay loam	1.4	22.5	0.44	10	70
Hesperia sandy loam	0.3	16.0	0.24	8	56

[1] The ED_{50} is the concentration in ppm that reduced the fresh weight of oats 50%.
[2] Calculated using adsorption on air-dry Egbert soil as 100, from the original data that were reported in counts per minute of radioactivated EPTC.
Source: Adapted from Ashton and Sheets, 1959. Reproduced with permission of the Weed Science Society of America.

many weeds become progressively less vulnerable to many herbicides as germination and emergence proceed, a delay in uptake due to adsorption by itself may affect success, but likely only to a minor degree. Exceptions may be: (1) on very sandy soil where all of the applied herbicide may be assumed to be completely available for uptake; (2) on soils very high in organic matter where adsorption is not of the simple physical type but rather involves penetration of the herbicide into and retention by the organic particle itself; and (3) where there is a specific adsorption within negatively charged clay particles, such as that which occurs with the dicationic (possessing two positive charges) herbicides diquat and paraquat. With diquat and paraquat, the attachment is so strong that doses normally used in crop production are completely inactivated in the soil.

A different type of adsorption is involved for the interaction of glyphosate with clays in soil. Glyphosate adsorption likely occurs through binding of the negatively charged phosphonic acid group to clays in the presence of iron, aluminum, and other metallic cations to form insoluble complexes. Although this complexation may be partly reversed by addition of inorganic phosphate, it nonetheless represents a type of adsorption less readily desorbed than where simple adsorption is involved.

Leaching

Leaching is the term applied to movement of herbicides with water in soil. Such movement can be downward, lateral, or upward. It is most commonly thought of in terms of downward movement because the volume that may move with percolating rainwater or irrigation water is much greater than the volume that moves upward or laterally. Ideally, a herbicide applied to the soil to prevent weed emergence with the desired plant would be placed in the soil zone from which the weeds originate and remain there until all of the nondormant seed had germinated. Several factors can prevent this ideal situation from being achieved. First, we need to realize that herbicides that are mainly effective when absorbed by the weed root or young shoot usually need to penetrate the soil some distance to be in contact with these organs. In fact, if the herbicide is mainly effective when absorbed by the root, it commonly needs to be distributed through 5 cm or more of the surface layer of soils to be most effective. On the other hand, with soil-applied herbicides that are mainly absorbed by the coleoptile or shoot, free movement with percolating rainwater can move the herbicide beyond its required soil zone and thereby render the herbicide ineffective. Poor control because of inadequate herbicide movement is much more common reason for failure than movement of the herbicide beyond the effective zone.

Effect of adsorption.   Although, as we have seen, adsorption plays only a minor role in total uptake, it plays a major role in leaching. Retention against leaching where adsorption occurs is due to the equilibrium relationship between the herbicide in solution and that adsorbed and absorbed. In the simplest situation of physical adsorption, replacement of the soil solution containing a herbicide with new water results in some movement of the herbicide from the soil particle into the new solution. The movement continues until the same relative proportion of dissolved herbicide to adsorbed herbicide is established—that is, until the equilibrium balance for the particular herbicide and soil colloids is attained.

For example, if 98% of trifluralin applied to a given soil is adsorbed and the soil solution containing 2% of soluble herbicide is replaced with new water, the same 98:2 relationship will be reestablished. That is, only 2% of the 98% adsorbed is released into the new water. Following this flushing, the soil particles would then have 96.04% of the original application, 98% − (2 × 98%). If this process of flushing is repeated three more times, the resulting percentages on the soil particles will be 94.12%, 92.23%, and 90.39%. If the flushing is assumed to be the result of a rain sufficient to wet or pass completely through the soil zone containing the adsorbed herbicides, even after four such rains, only about 10% of the original application of trifluralin will have moved out of the zone in which it was adsorbed. In most agricultural usage situations, the herbicide accomplishes its intended purpose long before this much percolation has occurred.

Effect of solubility.   Contrary to early belief, solubility is not a major factor in leaching, even though it is true that the portion of the herbicide in solution is the part available to be leached. The reason is found in the relative mass of the soil as opposed to that of the applied herbicide. That is, the amount of applied herbicide is normally extremely small compared to the amount of soil as a potential adsorbing surface and the amount of water as a potential dissolving medium.

This fact can be illustrated by a simple comparison of weights. The weight of the top 15 cm of soil is often estimated to be about 2,242,760 kg ha^{-1}. Thus, a 2.2 kg ha^{-1} application of a herbicide represents about one part of herbicide to one million parts of soil (the potential adsorbent). The plow layer is occupied roughly equally by solid material and by open space, commonly termed *pores*. Thus, there would be roughly 1,500 m^3 of such open space in the top 15 cm of a hectare of soil. If all of this space was occupied by water, it would amount to 1,500,000 kg. On a weight-to-weight basis, the water would need only the capacity to dissolve one part herbicide in about 680,000 parts water to completely dissolve all the herbicide in a 2.2 kg ha^{-1} application if none of it is adsorbed. Thus, it is adsorption, as influenced by type of clay and organic colloids present and interacting with the physicochemistry of the herbicide, that largely determines leachability.

The effect of solubility is well illustrated in Figure 12-8 by the restricted downward movement of the completely soluble amine salt formulation of silvex; 2,4-D; and 2,4,5-T. In this study, 500 mL was enough water to moisten dry soil to a depth of about 56 cm. Application of 500 mL to wet soil resulted in 450 mL percolating from the bottom of the tubes. Even where 500 mL was added to wet soil, these completely soluble herbicides failed to reach the bottom of the tubes. Retention against downward movement is assumed to be due to adsorption on the soil colloids.

Effect of herbicide formulation.   Figure 12-8 also clearly demonstrates that formulation affects herbicide leaching. The ester formulations of 2,4,5-T; 2,4-D; and silvex all penetrated the soil columns less than did the amine formulations. Part of the reason may be differences in solubility since the solubility of the esters is about 16 ppm, whereas the amines are completely soluble. However, solubility by itself is likely to be less important as the explanation than differences in adsorption for the reasons discussed in the previous section.

Differences between the amine salt and ester formulations of silvex; 2,4-D; and 2,4,5-T may also be due to a difference in their amphoteric properties. *Amphoteric* is a term often applied to surfactants and means that they have both a lipophilic and hydrophilic portion on the same molecule. The phenoxyacetic acid herbicides have this property in a weak sense. The esters could be expected to have this to a greater extent than the amine salts, thus increasing their adsorption. Regardless of the explanation, as we saw in Table 12-1, differences in downward movement of 2,4-D formulations can be sufficiently great that amine formulations reduce corn stands significantly, whereas 2,4-D ester does not.

Impact of leaching.   Although the likelihood of leaching causing a readily leached herbicide to be moved out of the zone from which most of the weeds originate is minimal, some evidence suggests that this situation may occur. As can be seen in Figure 12-8, for all herbicides except the emulsifiable formulation of fenac, flushing with 1,000 mL of water reduced the amount of herbicides in the surface 2.5 cm to the point that soybeans were only moderately affected, if at all. Indirect evidence supporting this possibility under field conditions is found in earlier work by Aldrich (1950), reported in Table 12-4. Aldrich found that corn stands were reduced more in a planting 9 cm deep than in one 3.8 cm deep when rain immediately followed the application, thus suggesting that rain reduced the 2,4-D concentration in soil closer to the surface.

There is also evidence that herbicides may move upward in the soil and in sufficient quantity to be toxic to plants germinating or growing in that area (Figure 12-9). The horizontal bars in the figure show the fresh weight of oats, a species sensitive to

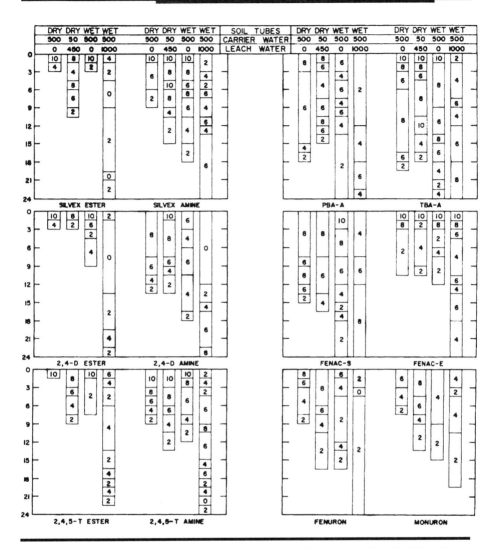

FIGURE 12-8. Differences in herbicides in their downward movement in soil. Soil moisture at time of application also influences the extent of herbicide penetration. The herbicides shown were applied to 7.6 cm (3 in.) soil columns 61 cm (24 in.) deep. Applications were made in the quantities of water shown opposite Carrier water and the columns flushed with quantities of water shown opposite Leach water. Numbers in bars are injury scores of soybeans grown in soil from the indicated depths; a score of 0 = no injury and 10 = plant dead.

Source: Wiese and Davis, 1964. Reproduced with permission of the Weed Science Society of America.

TABLE 12-4. Effect of depth of planting on reduction in corn stands from 2,4-D amine applied 4 days after planting. A 0.6 in. rain occurred immediately after application that was sufficient to wet at least to the seed level.

Pounds of 2,4-D per Acre	Corn Plants per Plot*	
	Planted 1½ in.	Planted 3½ in.
0	65.4	66.1
1	61.2	62.5
2	61.5	58.6
4	56.0	49.9

*Average of 3 replications.
Source: Aldrich, 1950.

dicamba and diphenamid, as a percent of untreated when grown in the greenhouse on soil collected from each depth. Covering the soil after watering provided for downward but not upward movement of herbicides. Leaving open after watering allowed evaporation to occur, thus providing for upward movement. The dry-open treatment provided a check against the importance of water as a vehicle for movement. As can be seen, growth of oats was only slightly reduced above or below the herbicide zone in the dry soil, indicating little herbicide movement. Oat growth in the top 2.5 cm soil layer was reduced more than 75% by dicamba in wet-open treatment, showing it is readily moved upward by water evaporation. By contrast, oat growth exceeded 50% of untreated for all diphenamid treatments except the applied level, which shows diphenamid is relatively immobile.

Figure 12-10 suggests that both the total amount of percolating water and its intensity influence extent of downward movement. Further, adsorption has a marked modifying effect. Dicamba, which apparently is adsorbed weakly (Table 12-2), is removed from the top 7.5 cm of a soil column with an added surface 12.5 cm of water, providing the water is added in small increments. If water is added in 2.5 cm increments, a total application of 25 cm does not remove the dicamba. Just the opposite occurs with diphenamid, which is adsorbed to a greater degree than dicamba. That is, more downward movement occurs when water is added in large rather than in small increments. The results with diphenamid are understandable from what we learned about adsorption and desorption. Dicamba, on the other hand, is dependent for its movement on being dissolved in water passing through the surface of the soil. In this situation, time becomes a factor. Where 12.5 cm of water was applied in 0.6 cm increments every 30 minutes, 10 hours was required to make the necessary 20 applications. Therefore, dicamba on the soil surface under this water treatment was available for transport in the passing water for 10 hours. Where the 12.5 cm was applied in 2.5 cm increments, the time for transport was only 2 1/2 hours.

In summary, leaching, which is controlled by adsorption and by the amount and duration of water moving through the herbicide zone, may affect results in a number of ways. It may help move the herbicide into the soil zone critical for optimum weed

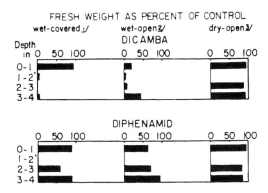

FRESH WEIGHT AS PERCENT OF CONTROL

¹Received 1 surface inch of water and covered
to prevent evaporation
²As treatment 1, but water was allowed to
evaporate freely from the soil surface.
³Soil left air-dry and uncovered.
*Herbicides were applied at this level.

FIGURE 12-9. Toxicity to oats from upward movement of the mobile herbicide dicamba in soil. Enough dicamba, which is weakly adsorbed on soil colloids, if at all, moves to the top 2.5 cm (1 in.) soil layer when provided conditions for evaporation (wet-open) to reduce oat fresh weight more than 75%. Diphenamid, which is adsorbed, does not move away from the point of application enough to reduce fresh weight even 50% under any treatment.

Source: Harris and Warren, 1964. Reproduced with permission of the Weed Science Society of America.

control. It may move the herbicide out of this critical zone for weed control, thus detracting from satisfactory prevention of emergence. Or, it may either reduce or increase the hazard of damage to the germinating desired plant seed and developing seedling.

Volatilization

Volatilization is a third major source of variability in success of herbicides used to prevent weed emergence. Volatilization is a two-stage process involving evaporation of the herbicide molecule into the air from residues present on soil surfaces and dispersion of the resulting vapor into the overlying atmosphere by diffusion and turbulent activity. For many herbicides, volatilization can be the most important source of loss from the soil during the first few weeks following application. The extent of such loss under certain circumstances can be substantial. It was calculated that EPTC, one of the most volatile herbicides, could be lost from a glass surface at a rate of 76 kg ha^{-1} day^{-1} (Hartley, 1960). Since EPTC and other volatile herbicides are used successfully

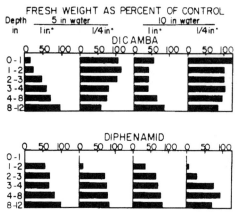

*Increments of 0.25 and of 1 surface inch
were poured directly onto the column surfaces at 30-minute intervals for the 5 in.
treatment. The 10 in. treatment was exactly
the same as two 5 in. treatments on two
successive days.

**FIGURE 12-10. Effects of duration and intensity of leaching
on movement of unadsorbed or weakly adsorbed diphenamid
and readily adsorbed dicamba in soil.** Bars showing growth of
the test plant, oats, in soil from several depths are a measure of
herbicide amount at the depth indicated.

Source: Harris and Warren, 1964. Reproduced with permission of the
Weed Science Society of America.

as soil-applied herbicides, there must be factors in the soil that prevent rates of volatile
loss such as calculated by Hartley. Adsorption and soil moisture are two such factors
that greatly influence evaporative loss.

Effect of adsorption.   Adsorption, especially upon dry soil (Table 12-3), is a major
reason even the volatile herbicides are not lost from the surface more quickly. Gaseous
substances tend to condense on solids—especially those solids having a collection of
molecules with residual attraction, such as clay and organic particles in soil—rather
than remain associated with one another. This attraction is especially strong in dry soil
because of limited competition with water for which soil particles have a stronger attraction. The soil surface available for adsorption of a gas is very large; roughly one-
half of the soil bulk is open space available to gas penetration if not filled with water.
It is also segregated into variable-sized clumps that frequently contain clay and organic
matter particles with very large surface areas. Recognizing that the soil surface is frequently dry when the herbicide is applied, we can therefore understand why even a
very volatile herbicide does not rapidly dissipate into the atmosphere.

The clay and organic matter in the soil effectively hold volatile herbicides, just as they do those in solution. Table 12-5 shows the combined relationships between these soil components and loss of the volatile herbicide EPTC from soil. Loss from the higher organic matter and clay soil (heavy texture) clearly is less than from the lower organic matter and clay soil (light texture). This work was done under laboratory conditions where EPTC was added to soil wetted to different moisture levels then allowed to dry at room temperature. Since the quantity of soil was relatively small, we might expect that nearly all of the soil particles offered an opportunity for evaporation of the adsorbed EPTC. Thus, evaporation under these conditions might be greater than would be expected under field conditions, where some of the herbicides might be at a depth at which access to the atmosphere via evaporation is restricted.

TABLE 12-5. **Composition of soils and its effect on the loss of EPTC-S[35] applied at 10 ppm during the evaporation of various amounts of soil water from wet soil.**

| | | Percent of | | Percent Loss of EPTC during Drying of Soil Containing[2] | | |
| | | Organic | | 33% | 50% | 60% |
Soil	pH	Matter	Clay	Water	Water	Water
Light texture						
Loamly sand from Klamath[1]	7.6	0.9	5	64	78	80
Ritzville very fine sandy loam	6.9	1.7	15	66	76	78
Newberg sandy loam	6.3	1.5	19	74	82	84
Heavy texture						
Chehalis loam	5.8	2.8	24	48	53	67
Loam from Eastern Oregon[1]	7.5	2.1	25	59	64	80
Willamette silty clay loam	6.8	4.0	34	52	61	66
Melbourne clay	5.5	4.9	49	40	43	61
Peat	5.6	83.3	...	14	15	25

[1] Soil series unknown.
[2] Average of triplicate determinations.
Source: Fang et al., 1961. Reproduced with permisstion of the Weed Science Society of America.

Figure 12-11 shows losses of the several s-triazine herbicides from a metal surface (Figure 12-11A) and of prometon from five different soils (Figure 12-11B). Simazine (number 1 in Figure 12-11A), among the least volatile of the herbicides, did have some loss from the metal. The most volatile of these seven s-triazines, prometon (number 4 in Figure 12-11A), by contrast lost only about 10% of the amount originally applied in 24 hours in all soils except the Tifton loamy sand. The latter soil had the lowest combined content of clay and organic matter, again suggesting the importance of these components in reducing evaporative loss. Nearly 80% of the prometon was lost in 2 hours from the metal surface.

Effect of soil moisture.   Soil moisture plays a key role in herbicide loss through evaporation. It does so through its effect on retention of the herbicide on soil particles

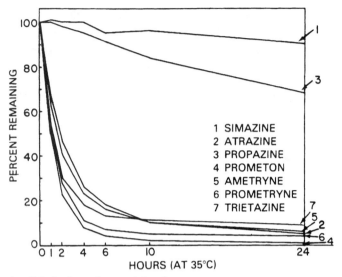

A. s-Triazine losses from metal surface

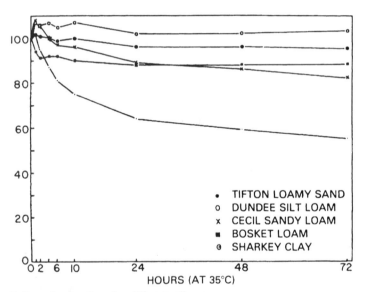

B. Prometon loss from five different soils

FIGURE 12-11. Effect of surface type on herbicide evaporation.

Source: Kearney et al., 1964. Reproduced with permission of the Weed Science Society of America.

and on movement of the herbicide with water during evaporation. As discussed earlier, adsorption on dry soil can be especially strong. No loss of even the most volatile of herbicides—in this case, EPTC—occurred from dry soil after 42 days in the open (Fang et al., 1961). This situation, of course, is not usually encountered in the agricultural use of herbicides.

The more usual situation is one in which the soil is moist, at least just below the soil surface, and rains occur at intervals following application. These conditions present a much different situation relative to evaporative loss of herbicide. As we have already seen, gases (volatile herbicides) tend to condense on soil particles. However, many clays and organic matter are strongly hydrophilic. That is, they readily displace other substances, such as an adsorbed herbicide, with water. The displaced herbicide, depending upon its volatility, may then evaporate or be carried off with the evaporating water, depending upon the moisture conditions. Referring back to Table 12-5, we see that the amount of moisture in the soil does affect loss of EPTC. If we consider only the soils identified as having heavy texture, the average relative loss is 1.0:1.1:1.4 for the three increasing moisture contents, respectively. Again, these results were obtained under laboratory conditions where all soil was provided the opportunity to lose adsorbed EPTC. Nonetheless, this study indicated that since moisture is commonly available immediately beneath the surface or at the surface following rainfall, soil moisture is commonly a factor in evaporative loss. If sufficient opportunity for evaporative loss occurs during the period immediately following application, preventive weed control could be poor. On the other hand, if an extended dry period is followed by a rainy period, herbicidal effects may appear due to the displacement of the adsorbed herbicide by water. In this case, whether or not the treatment is effective depends upon the point of entry for the particular herbicide and stage of growth of the weeds at the time.

Effect of incorporation.   Mechanically mixing the herbicide with the surface few centimeters of soil is called *incorporation*. Now a common practice with many soil-applied herbicides, incorporation is done to improve control by distributing the herbicide throughout the zone from which a majority of the weeds originate and to reduce loss from volatility.

Figure 12-12 is representative of the improvement in weed control that can be obtained by this practice. Incorporating EPTC reduced the amount needed by about one-half, as can be seen by comparing 2.2 kg ha[-1] incorporated with 4.4 kg ha[-1] not incorporated. The barnyardgrass in this study was actually planted at 1.2 cm below the soil surface. Two important points about incorporated treatments can be made from the data obtained.

First, where the herbicide involved is taken up both by the shoot and root—as is the case with EPTC—best results are obtained if the herbicide is distributed deeply enough for the root system to be exposed. Thus, even though the barnyardgrass was seeded at 1.2 cm, best control with the lowest two rates used was obtained where the EPTC was incorporated to a depth of 2.5 cm. With the herbicide incorporated at this depth, the emerging root of the barnyardgrass is exposed to it for the first 1.3 cm growth made beyond the seed.

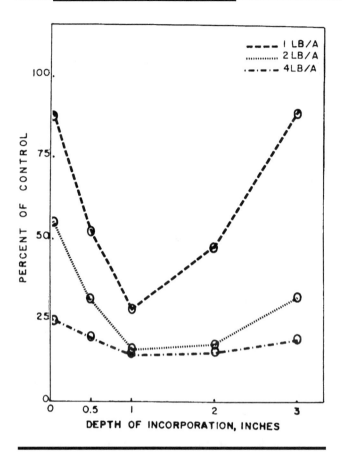

FIGURE 12-12. Soil incorporation as a means to greatly improve weed control with volatile herbicides applied to soil. Here, 2.2 kg EPTC ha⁻¹ (2 lb per acre) incorporated was as effective in controlling barnyardgrass as 4.4 kg (4 lb) not incorporated.

Source: Ashton and Dunster, 1961. Reproduced with permission of the Weed Science Society of America.

Second, control is less effective if the herbicide is distributed throughout too great a depth unless an excessive rate is used. Poorer control would appear to be simply a result of the dilution effect. Diluting the 454 g (1 lb) application by mixing it in the top 7.6 cm (3 in.) reduced the amount in the vicinity of the barnyardgrass below that needed for control.

Even though incorporation can substantially reduce volatilization of herbicide, losses due to volatilization can still occur. This is mainly a result of mass flow of water to the surface due to evaporation from the soil surface. As long as the surface of the soil remains moist, providing for a continuity of the capillary pores in the soil within which the water must move, upward movement is initiated as evaporation occurs. Because the soil is a relatively poor wick—that is, the capillaries are not continuous vertical tubes but are more a network of open pore spaces with many contortions—upward movement may be only a few centimeters. In view of the fact that the relatively shallow surface zone of soil is the source of most annual weeds in a given year, this movement may nonetheless be sufficient to reduce weed control. This source of loss may be significant even for relatively nonvolatile herbicides because the movement in mass flow with water to the surface may result in accumulation of the herbicide at that point until it reaches the evaporative level. The frequency and amount of rainfall determines the degree of such loss.

Effect of temperature.   Temperature acts as a modifier of all types of evaporative loss. In quite general terms, the temperature effect roughly doubles evaporation for each 10°C to 15°C increase. This effect is clearly demonstrated in Figure 12-13. Loss of atrazine ranged from about 20% to 40% at 35°C for the five soils (Figure 12-13A) and from about 30% to 60% at 45°C (Figure 12-13B).

Degradation

Microbial degradation or biodegradation.   The fourth major factor that may influence weed control efficacy of soil applications is *degradation,* the conversion or breakdown of the herbicide compound to simpler and, most frequently, nontoxic components. Microorganisms in soil either have the inherent ability or can be induced to attain the ability after exposure to an applied compound, to degrade literally any compound introduced into soil. It is well established that microbial degradation or biodegradation is the principle mechanism for the ultimate dissipation of most herbicides in soils. However, the brief time frame (only a few weeks) within which the herbicide needs to be available in soil for acceptable weed control efficacy causes biodegradation to be a relatively insignificant factor in determining the success of a given soil application for weed control. The exception may be the situation in which a particular herbicide is applied for 2 or more consecutive years to the same field. Very early work with 2,4-D demonstrated that biodegradation was more rapid in soil pretreated with 2,4-D than in previously untreated soil (Racke, 1990). Soil microorganisms capable of degrading 2,4-D were subsequently isolated and identified thereby verifying the microbial involvement in the rapid degradation. Subsequent studies with the carbamothioate class of herbicides developed for "reduced persistence" in the environment revealed that these soil-applied chemicals were particularly susceptible to rapid degradation in fields with historic use of the chemicals, which was the major factor involved in many of the weed control failures reported in the late 1970s (Racke, 1990). The rapid rate of herbicide degradation due to a buildup of soil microorganisms adapted for degradation of the herbicides became known as *enhanced degradation.* The development of enhanced degradation is illustrated in Figure 12-14 by the rapid breakdown of EPTC in soil previously treated (solid line) compared to slower breakdown in previously untreated soil (dashed line). Although not shown in Figure 12-14, the half-life of EPTC in soil where it was applied annually for 8 years was only one-

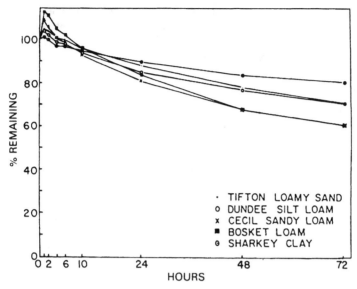

A. 35°C

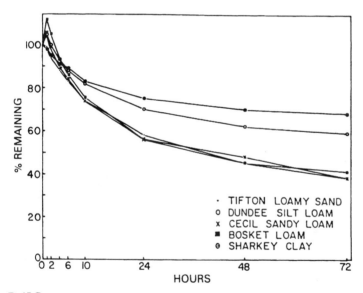

B. 45°C

FIGURE 12-13. Pronounced effect of temperature on evaporation of herbicides from soil as shown by the loss of atrazine from five soils.

Source: Kearney et al., 1964. Reproduced with permission of the Weed Science Society of America.

half (9 days) that of EPTC in previously untreated soil (18 days). Enhanced degradation may be an especially important factor critical to the success of soil-applied herbicides in monoculture crop systems.

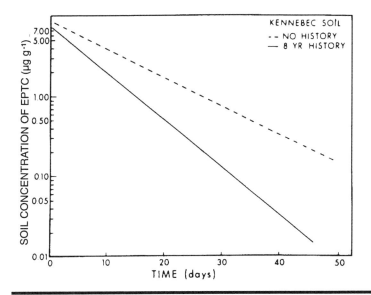

FIGURE 12-14. Effect of previous application on enhancing degradation of a soil-applied herbicide.

Source: Obrigawitch et al., 1982. Reproduced with permission of the Weed Science Society of America.

Because the carbamothioate herbicides, including butylate and EPTC, continue to be useful in controlling grass weeds such as johnsongrass, wild proso millet, and the foxtails in corn, weed management systems have been refined to circumvent enhanced degradation of these herbicides in the field. These adjustments include rotation for 1 year to no herbicide or to another class of herbicide such as atrazine or metolachlor; rotation to a different crop such as soybeans; use of "extenders" in the carbamothioate herbicide formulations that increase persistence in soil, thereby assuring weed control efficacy; and immediate, uniform incorporation in soil prior to corn planting.

Nonbiological degradation. Nonbiological degradation occurs in soils in the absence of biological mediation and is also referred to as abiotic degradation. Soil is a complex of surfaces and chemicals with which reactive herbicides may interact. For example, the negative charge on clays makes them cationic exchangers. Organic matter itself provides a wide variety of reactive chemicals; numerous reactive metal ions are typically associated with clays (magnesium, iron, copper, aluminum, and so on); and the clays have acidic and basic sites. Finally, water is nearly always present to act as a solvent, to permit formation of acid and amine salts, and to serve as a medium for many chemical reactions. Even though reactions involving the soil's dynamic nature have been demonstrated in numerous studies under laboratory conditions, few examples exist where such reactions have been a major factor in poor results from soil applications except in very unusual conditions for selected herbicides.

Photochemical degradation. *Photodegradation* may have pronounced effects on the success of many soil-applied herbicides. Light in the range from near 290 to 450 nm is a highly effective energizer of several common reactions involving organic compounds (including herbicides), such as oxidation, reduction, elimination, hydrolysis, substitution, and isomerization. We now know that many herbicides may have their activity reduced if exposed to light while on the soil surface.

Figure 12-15 shows that the detoxification effect of sunlight may be very pronounced with even a relatively brief exposure. In this case, trifluralin was completely

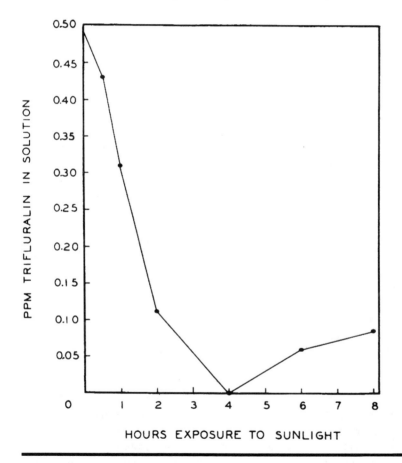

FIGURE 12-15. Rapid detoxification effect of sunlight on herbicides. Here trifluralin activity was measured by a grain sorghum bioassay.

Source: Messersmith et al., 1971. Reproduced with permission of the Weed Science Society of America.

degraded in 4 hours. Among the other herbicides found to be sensitive to photodegradation are amiben (Sheets, 1963), diruron and monuron (Harris and Warren, 1964), the s-triazines (Jordan et al., 1963), and paraquat and diquat (Crosby and Tang, 1969).

Anything that prevents exposure to direct sunlight reduces herbicide loss. Under field conditions herbicide loss is commonly less than that observed under controlled laboratory or greenhouse conditions. In the field, the soil, at least from the vantage point of the herbicide it contains, does not expose a smooth surface to the sun. Rather, the soil contains clumps of varying size and an uneven topography that protects some parts from direct sunlight. In addition, adsorption, which as we have already seen may be substantial for many herbicides, also protects against photodegradation. Under field usage, of course, loss from photodegradation is essentially prevented by incorporating the herbicides for which this may be a problem.

Weed–Crop Density

Since removal of any part of the applied herbicide from the soil reduces that available for weed control, it follows that the number of weed and crop seeds present may influence control. Indeed, this effect has been found with atrazine (Hoffman and Lavy, 1978; Winkle et al., 1981), diuron (Burrill and Appleby, 1978), and alachlor (Winkle et al., 1981).

The relationship between alachlor rate and foxtail millet (*Setaria italica*) seeding rate shown in Figure 12-16 indicates that the effect may be quite pronounced. At the densest seeding rate, even the highest rate of alachlor (1.0 ppm) only reduced growth of the test species about 60%; whereas, at the lowest density, the same reduction of growth was accomplished with only one-fifth as much alachlor (0.2 ppm). Table 12-6

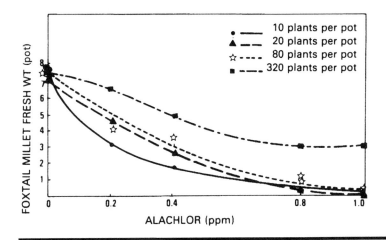

FIGURE 12-16. Influence of seed and seedling density on effectiveness of alachlor applied to soil.

Source: Winkle et al., 1981. Reproduced with permission of the Weed Science Society of America.

provides the explanation. As density increases, the amount of alachlor absorbed per plant decreases. Thus, at the high density, less herbicide is available in the plant to exert herbicidal effects.

TABLE 12-6. **Absorption of alachlor by foxtail millet as affected by plant density. Plants grown in pots containing 170 g of Sharpsburg silty clay loam soil were harvested 21 days after treatment.**

Foxtail Millet (plants per pot)	Alachlor Rate (ppmw)	Alachlor Absorbed	
		(ng per seedling)	(μg g^{-1} fresh weight)
20	0.4	42	0.7
80	0.4	20	0.4
20	0.8	96	64.0
80	0.8	40	0.7

Note: Data represents average of 2 experiments with 2 replications.
Source: Winkle et al., 1981. Reproduced with permission of the Weed Science Society of America.

HERBICIDE PERSISTENCE IN THE SOIL

The persistence of a herbicide following application to soil is important in three respects. First, it is a factor influencing weed control with a given application and of applications in subsequent years. As discussed in the previous section, some herbicides may not persist long enough to provide the needed weed control. With others, persistence in subsequent years may be reduced to the point that control is inadequate. Second, persistence also is important through its potential effect on contamination of the environment. For example, the longer a herbicide remains on or in the soil, the greater the chance for transport to adjacent areas or water supplies in surface runoff and percolation through the soil profile. Third, persistence is important as a possible source of injury to other crops grown on the field the following year (herbicide carryover). Notable examples include injury to soybeans the year following weed control with atrazine in corn or injury to corn the year following weed control with imazaquin in soybeans.

Factors Affecting Rate of Loss

The relative persistence of some common soil-applied herbicides expressed in terms of "half-life" is shown in Figure 12-17. The *half-life* is defined as the time required for a herbicide to undergo dissipation and/or degradation to 50% the initial concentration applied (Wauchope et al., 1992). Each bar in Figure 12-17 represents the average half-life in soil for the given herbicide. These half-life values are very general and assume normal application rates and were determined under average temperate climate conditions. Of course, many factors may cause persistence of a specific herbicide to be quite different from that shown. The four factors previously discussed for their effect on success—adsorption, leaching, volatilization, and degradation—are primary mechanisms for removal. Uptake by plants and mechanical removal in runoff and erosion are secondary mechanisms. The several factors that may influence rate of loss under field conditions were reviewed by Walker (1987). Some of the more important factors are briefly described.

Rate of application.   There are herbicides and circumstances for which the rate of disappearance is influenced by initial amount applied. However, persistence only increases disproportionately if unusually high rates of application are used.

Soil type.   Because all primary mechanisms for removal may be influenced by soil type with more than one type involved at the same time, it is difficult to identify specific effects. Organic matter, through its effect on microbial populations and on adsorption, has been shown to be an aspect of soil type that influences persistence. However, the relationship between degradation (including microbial activity) and adsorption as influenced by organic matter may be antagonistic or complementary, depending upon the herbicide.

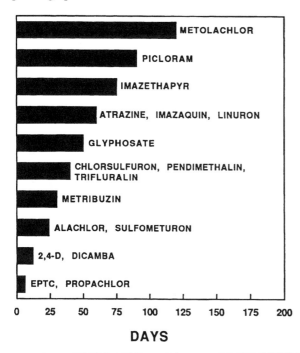

FIGURE 12-17. Relative persistence of selected herbicides in soil.

Source: Data adapted from Wauchope et al., 1992; WSSA, 1994.

Adsorption. There is no clear-cut, consistent effect of adsorption on persistence. It does not always protect against degradation nor always result in faster loss. For example, some herbicides can be degraded at the surface of clays by "adsorption-catalyzed hydrolysis" (Walker, 1987).

Soil pH. Soil pH affects persistence. The effect varies with the herbicide, the pH range involved, and other treatment of the soil. The variable results expected from interactions of herbicide type at different soil pH levels is apparent where degradation of triazines via hydrolysis occurs at a low soil pH yet a high soil pH causes degradation of alachlor (Walker, 1987).

Soil amendments. There are instances of an increase in degradation from additions of soil amendments, such as plant residues. Conversely, instances have been recorded where such additions slowed degradation. Effects of additions of mineral nutrients (nitrogen, phosphorus, and potassium) were also variable.

Temperature and moisture. The effect of temperature approaches a constant because of its effect on the rate of chemical reactions. Similarly, degradation increases with moisture content up to field capacity. The significance of these factors in the field is that herbicide persistence is greater in the cooler northern and dryer regions than in semitropical regions.

Distribution in soil. Incorporation of herbicides subject to loss through volatilization, photodecomposition, or both—trifluralin and EPTC, for example—markedly increases persistence. Rate of degradation is less in subsoil than in surface soil, and the relationship with depth tends to be linear for some herbicides studied.

Repeat treatment. As mentioned under factors affecting success, some herbicides are degraded more rapidly in repeat applications. The mechanism is discussed in the next section. At the other extreme, no evidence exists that repeat applications at recommended rates leads to accumulation of toxic levels.

Cropping. Cropping practices can have various effects on herbicide persistence in soil. As mentioned earlier, the number of weed and crop seeds available to take up an applied herbicide may affect the degree of weed control. It follows that the amount of weed and crop biomass present may influence the removal of an applied herbicide and, thus, its persistence. Indeed, this effect has been shown in some instances, although the amount so removed is probably not very large. Further, the extent of plant cover may well influence both soil temperature and soil moisture. A dense crop cover might reduce both, thus decreasing the rate of decomposition. Thus, the influences of cropping on herbicide persistence may well be counteracting.

Herbicide formulation. Granular formulations of volatile herbicides are more persistent than emulsions, miscible liquids, or wettable powders. Persistence can also be greater for nonvolatile herbicides formulated as granules. Encapsulation formulations that enclose the herbicide within a matrix of starch, alginate, or other polymerizing materials have been developed to allow a slow release of the herbicide into the soil. Thus, these formulations promote persistence of the herbicide well into the crop-growing season and are suggested to be more efficient in weed control compared to other formulations. Weather conditions largely determine the importance of formulation effect. Increased persistence of granules is likely to be significant mainly for herbicides such as atrazine that already pose a potential carryover hazard for certain crops such as soybeans.

Pesticide combinations.   There is evidence that herbicide persistence can be extended by applications of fungicides, insecticides, and other herbicides. For example, chlorpropham degradation is inhibited when combined with methyl carbamate insecticides (Walker, 1987). There is also evidence that degradation of some herbicides can be increased by other pesticides. Where herbicide loss is influenced by other pesticides, it apparently is due to combined effects on microbial activity.

Microbial Degradation

Irrespective of the mechanism by which a herbicide is removed from a given site, degradation, or decomposition, ultimately determines its fate in the environment at large. Further, microorganisms are able to degrade literally any compound introduced into the soil, but the time required and, as we just saw, the rate of breakdown may vary. Thus, it is useful to have a general understanding of microbial degradation of herbicides. Microorganisms mainly involved in herbicide degradation are bacteria, fungi, and actinomycetes capable of metabolizing naturally-occurring carbonaceous substrates (Bollag and Liu, 1990). The ability to utilize carbon of an applied herbicide largely determines the shape and slope of the decomposition curve.

Direct metabolism of herbicide.   If microorganisms are able to develop the enzymes necessary to directly metabolize a given herbicide, then availability of the herbicide results in increases in that microbial population. Under these circumstances, an initial lag period is followed by rapid disappearance of the herbicide. When a herbicide is utilized as an energy source by microorganisms resulting in complete degradation to inorganic compounds of CO_2, H_2O, NH_3, and so on, the process is known as *mineralization*. Also, such herbicides commonly break down more rapidly in subsequent applications. Among the herbicides that serve as direct sources of energy are 2,4-D; dalapon; EPTC; butylate; and cycloate.

Incidental metabolism of herbicide.   If production of the necessary enzymes is not induced, microbial degradation that does occur is dependent upon the size and activity of the existing microbial population and concentration of the herbicide. In this situation, degradation is coincidental to the growth of microorganisms on some other carbon source and is known as *cometabolism*. The rate of loss tends to be constant and un-affected by previous application. Some of the herbicides degraded in this manner are simazine, linuron, monuron, triallate, and diuron. Additions of utilizable carbon sources might be expected to increase the rate of disappearance of such herbicides by increasing populations of the microorganisms responsible for cometabolism. Indeed, adding microbial nutrient broth, sucrose, plant residues, and manures to soil increased the rate of disappearance for some herbicides (Hurle and Walker, 1980). For a detailed discussion of microbial processes in soil as factors in herbicide persistence, see Bollag and Liu (1990).

REFERENCES

Aldrich, R.J. 1950. Factors affecting the practicability of the preemergence use of 2,4-D on corn, Ph.D dissertation. Ohio State University, Columbus.

Aldrich, R.J., and C.J. Willard. 1951. Factors affecting the preemergence use of 2,4-D in corn. Weeds 1:338-345.

Ashton, F.M., and A.S. Crafts. 1981. Mode of action of herbicides, 2nd ed. New York: John Wiley & Sons.

Ashton, F.M., and K. Dunster. 1961. The herbicidal effect of EPTC, CDEC, and CDAA on *Echinochloa crusgalli* with various depths of soil incorporation. Weeds 9:312-317.

Ashton, F.M., and T.J. Sheets. 1959. The relationship of soil adsorption of EPTC to oats injury in various soil types. Weeds 7:88-90.

Bollag, J.-M., and S.-Y. Liu. 1990. Biological transformation processes of pesticides. In H.H. Cheng, ed., Pesticides in the soil environment: Processes, impacts, and modeling. Soil Sci. Soc. Am., pp. 169-211, Madison, WI.

Burnside, O.C., G.A. Wicks, and C.R. Fenster. 1971. Protecting corn from herbicide injury by seed treatment. Weed Sci. 19:565-568.

Burrill, L.C., and A.P. Appleby. 1978. Influence of Italian ryegrass on efficacy of diuron herbicide. Agron. J. 70:505-507.

Chang, F.Y., G.R. Stephenson, and J.D. Bandsen. 1973. Comparative effects of three EPTC antidotes. Weed Sci. 21:292-295.

Cheng, H.H. 1990. Pesticides in the soil environment: Processes, impacts, and modeling. Soil Sci. Soc. Am., Madison, WI.

Crosby, D.G., and C.S. Tang. 1969. Photodecomposition of 3-(p-chlorophenyl)-1,1-dimethylurea (monuron). J. Agric. Food Chem. 17:1041-1044.

Devine, M.D., S.O. Duke, and C. Fedtke. 1993. Physiology of herbicide action. Prentice Hall, Englewood Cliffs, NJ.

Fang, S.C., P. Theisen, and V.H. Freed. 1961. Effects of water evaporation, temperature, and rates of application on the retention of the ethyl-N, N-di-n-propylthiolcarbamate in various soils. Weeds 9:569-574.

Hance, R.J., ed. 1980. Interactions between herbicides and the soil. New York: Academic Press.

Harris, C.J., and G.F. Warren. 1964. Movement of dicamba and diphenamid in soils. Weeds 12:112-115.

Hartley, G.S. 1960. Physicochemical aspects of the availability of herbicides in soils. In E.K.

Woodford and C.R. Sagar, eds., Herbicides and the soil, pp. 63-78. Oxford, UK: Blackwell Scientific.

Hatzios, K.K., and R.E. Hoagland. 1988. Crop safeners for herbicides: Development, uses, and mechanisms of action. Orlando, FL: Academic Press.

Hatzios, K.K., and D. Penner. 1982. Metabolism of herbicides in higher plants. Minneapolis: Burgess Publishing.

Hoffman, D.W., and T.L. Lavy. 1978. Plant competition for atrazine. Weed Sci. 26:94-99.

Hoffman, O.L. 1962. Chemical seed treatments as herbicidal antidotes. Weeds 10:322-323.

Holly, K. 1976. Selectivity in relation to formulation and application methods. In L.J. Audus, ed., Herbicides: physiology, biochemistry, ecology, pp. 249-277. New York: Academic Press.

Holstun, J.T., and O.B. Wooten. 1964. A promising new concept: Triband application of herbicides. Agric. Chem. 19:123-124.

Hurle, K., and A. Walker. 1980. Persistence and its prediction. In R.J. Hance, ed., Interactions between herbicides and the soil, pp. 83-122. New York: Academic Press.

Jordan, L.S., B.E. Day, and W.A. Clerx. 1963. Photodecomposition of triazine. Weeds. 12:5-6.

Kearney, P.C., T.J. Sheets, and J.W. Smith. 1964. Volatility of seven s-triazines. Weeds 11:300-307.

Kemper, H.M., J.H. Miller, and L.M. Carter. 1963. Preemergence herbicides incorporated in moist soils for control of annual grass in irrigated cotton. Weeds 11:300-307.

Messersmith, C.G., O.C. Burnside, and T.L. Lavy. 1971. Biological and nonbiological dissipation of trifluralin from soil. Weed Sci. 19:285-290.

Moyer, J.R. 1987. Effect of soil moisture on the efficacy and selectivity of soil-applied herbicides. Rev. Weed Sci. 3:19-34.

Nalewaja, J.D., E. Kolota, and S.D. Miller. 1987. Flax response to trifluralin. Weed Technol. 1:286-289.

Obrigawitch, T., R.G. Wilson, A.R. Martin, and F.W. Roeth. 1982. Influence of temperature, moisture, and prior EPTC application on the degradation of EPTC in soils. Weed Sci. 30:175-181.

Racke, K.D. 1990. Pesticides in the soil microbial ecosystem. In K.D. Racke and J.R. Coats, eds., Enhanced biodegradation of pesticides in the environment. Am. Chem. Soc. pp. 1-12, Washington, D.C.

Sheets, T.J. 1963. Photochemical alteration and inactivation of amiben. Weeds 11:186-190.

Walker, A. 1987. Herbicide persistence in soil. Rev. Weed Sci. 3:1-17.

Wauchope, R.D., T.M. Buttler, A.G. Hornsby, P.W.M. Augustijn-Bechers, and J.P. Burt. 1992. The SCS/ARS/CES pesticide properties database for environmental decision-making. Rev. Environ. Contam. Toxicol. 123:1-156.

Wiese, A.F., and R.G. Davis. 1964. Herbicide movement in soil with various amounts of water. Weeds 12:101-103.

Winkle, M.E., J.R.C. Leavitt, and O.C. Burnside. 1981. Effects of weed density on herbicide absorption and bioactivity. Weed Sci. 29:405-409.

WSSA. 1994. Herbicide handbook, 7th ed. Weed Science Society of America, Champaign, IL.

13

Minimizing Competition from Emerged Weeds with Herbicides

CONCEPTS TO BE UNDERSTOOD

1. For herbicides applied to weeds growing with desired plants, selectivity is determined by four linked phenomena acting in sequence: retention, penetration, translocation, and biochemical reactions. The first three are of primary concern to us in weed management.
2. Selectivity may be achieved by both physical and biological factors. Selectivity based on physical factors is accomplished by a separation in space between the weed and the desired plant. Selectivity based on biological factors is accomplished by gross morphological differences, differences in leaf surface microstructure, and differences in internal physiology and metabolism between the weed and the desired plant.
3. Factors that affect success of herbicides applied postemergence include: stage of development, growth form, and growth rate of weeds and of desired plants; weed density; and temperature and humidity.
4. Postemergence application methods have been modified in a number of ways that expand the usefulness of herbicides in minimizing competition from growing weeds. These include directed applications, sequential applications, herbicide combinations, reduced herbicide rates, safeners, and growing herbicide-resistant cultivars.
5. Some hard-to-kill perennial weeds in selected crops can often best be dealt with at other times or in other crops.
6. Modern herbicides have made possible the rapid move to conservation tillage in crop production.
7. Available herbicides can be used to prevent seed production by weeds.

As with soil application, applying herbicides to weeds growing with desired plants has some clear-cut advantages. One obvious advantage is that the weed problem is known both in terms of species and their numbers. There is a growing list of herbicides that can be used to selectively remove either grass or broadleaf weeds from most desired plants; some can selectively remove both. Therefore, the choice of herbicide and, in fact, the decision to apply it can be governed by the specific weed problem present. By contrast, soil applications must be based on the assumption that specific weeds that can be controlled with a given herbicide will be present and in sufficient numbers to warrant control. Climatic conditions, time of planting, and other factors may indeed affect both the numbers and kinds of weeds present, as we have already learned.

A second advantage with treatment of the growing weeds is that smaller dosages are usually needed than with soil application. This result is largely due to the diluting effect of soil on soil-applied herbicides.

A third advantage is less direct but highly important nevertheless, especially to a long-term weed management effort: Herbicide treatment of the growing weeds can provide a more feasible way to reduce weed seed production. Rates even lower than those needed for postemergence control may be adequate for preventing weed seed production. Further, with proper timing, it may be possible to combine treatment to minimize competition and treatment to prevent seed production. With soil treatment, in addition to the need for higher rates, unless weed emergence is completely prevented, a second treatment will be needed if seed production by the weed escapes is to be prevented.

STEPS IN HERBICIDE ACTION

In minimizing competition from growing weeds, the effect of the herbicide is the result of four steps or phases: (1) retention, (2) penetration, (3) translocation, and (4) biochemical reaction. The first three steps are of primary concern to us in weed management since they largely determine the effectiveness and extent of expression of inherent selectivity (biochemical reactions) of a given herbicide. That is, variations in retention, penetration, and translocation largely explain variations from what is expected based on the known mode and mechanism of action of a particular herbicide. Finally, retention, penetration, and translocation are the steps over which we have some control. As we turn to an examination of these three as factors affecting selectivity, we should keep in mind that these steps are not separate and self-standing but rather are links in a sequence of events. That is to say, they are mutually dependent, not mutually independent, processes.

HERBICIDE SELECTIVITY

In order to clarify the roles of retention, penetration, and translocation in herbicide selectivity, we will assume that the effect of the herbicide on the desired plants and on the weeds is a direct reflection of the amount each absorbs. Nevertheless, in the final analysis, it is what happens after the herbicide enters the symplast system and reaches the site of biochemical reaction that is the final determinant of herbicidal activity. Retention and penetration simply determine how much is taken up.

The quantity of herbicide retained per unit weight of plant tissue is the important differentiating measure in selectivity, rather than the amount per plant or per leaf, since both of the latter may vary a good deal in size. Just as was true for soil applications, selectivity with foliar applications may be based on physical or biological differences or both. However, these differences come into play at a very different point relative to the respective root and leaf structures. As we learned with soil application, uptake through the soil is not a very discriminating process. Thus, differentiation among species in effect begins with translocation. Uptake by the foliage, on the other hand, varies greatly among species due to differences in both retention and penetration of the applied herbicides. Therefore, it is helpful to understand the roles of retention and penetration in both physical and biological selectivity.

Retention

Differential retention of a herbicide by the foliage of weeds and of desired plants influences the quantity of herbicide available for absorption. Such differential retention provides the basis for both physical and biological selectivity.

Physical basis. Physical selectivity based on differences in retention is accomplished by a separation in space between the weed and desired plants—that is, on a difference in size. If the weed is either shorter or taller than the desired plants, methods of applying herbicides have been developed that provide effective contact with the weed without harmful dosages contacting the desired plants. In other words, selectivity is obtained by providing opportunity for weeds—but not desired plants—to retain the herbicide.

If the desired plant is somewhat taller than the weeds, herbicides may be successfully used by directing spray beneath the foliage of desired plants directly onto the weed foliage. This method is called *directed postemergence*. Many situations offer the size differential necessary for this approach to work, allowing selectivity with herbicides for which the desired plant may not possess adequate biological tolerance. These situations range from applications to weeds in fruit tree orchards to those in annual row crops such as cotton. This approach will be expanded upon later in the chapter. Selectivity is predetermined because of the separation in space occupied by the weed and the crop. Nonetheless, other aspects of selectivity enter into success with this approach. The herbicide must be one that at the rate used will not damage the desired plant if it reaches the soil where it can be picked up by the plant's root system. Neither can it be so volatile as to present an opportunity for excessive uptake through the desired plant's foliage.

If the desired plant is somewhat shorter than the weeds, selectivity may also be attained by providing opportunity for weeds—but not the desired plant—to retain the herbicide. Two general types of equipment to accomplish such selective application are shown in Figure 13-1. Figure 13-1A is a recirculating sprayer that directs the spray horizontal to the ground and into a collecting box. The principle involved is that the weeds growing above the desired plant retain sufficient herbicide to be killed, and any unused or unabsorbed herbicide is caught by the box and recirculated. Figure 13-1B shows a selective application by using a so-called rope-wick applicator. The principle here is one of actually wiping herbicide droplets onto the weeds that overtop the desired plant from a fabric (rope) surface wetted by capillary flow of the herbicide from its source in the plastic pipe. Either of these applications gains a measure of selectivity over sprays directed to weeds beneath the desired plant in that presumably in the former case, only as much chemical is actually applied as can be retained by the weed. Even so, if injury to the desired plant is to be avoided, the herbicide must not volatilize in amounts toxic to the desired plant. Nor can it be one that is easily washed with rain from the weed foliage onto the desired plant or into the soil where the roots of the desired plant can pick it up.

Biological basis. As we learned earlier, plants vary greatly in their aboveground growth form (see Figure 3-9) and leaf surface microstructure (see Figure 11-5). These variations influence the amount of herbicide retained and thereby selectivity. It is important to understand that except with very small weeds, only a fraction—that is, commonly less than 25%—of the leaf surface on an area will actually be contacted by herbicide spray droplets. This fact serves to remind us that foliar applications fall far short of providing complete coverage of all leaves on plants we are trying to control. In such instances, the herbicide usually needs to move to the parts not contacted for satisfactory control to be obtained.

A. Recirculating sprayer directing spray horizontal to the ground. Spray not intercepted by weeds is caught in box opposite nozzles and recirculated.

B. Rope-wick applicator wiping herbicide droplets onto weeds. Rope-wick is continually wetted via capillary action by flow of herbicide from plastic pipe.

FIGURE 13-1. Equipment for selectively applying herbicides to weeds overtopping a growing crop; in this case, johnsongrass overtopping soybeans.

Source: Courtesy of C.G. McWhorter, USDA, Agricultural Research Service, Stoneville, Mississippi.

Differences in leaf surface area exposed to incoming spray droplets provides a basis for selectivity. The broadleaf weed depicted diagrammatically in Figure 13-2, for example, retains more herbicide than does the desired plant both because more leaf surface is exposed and because the horizontal leaves retain the spray droplets while the upright leaves of the desired plant do not. This difference, in part, was the basis for selective use of sulfuric acid and other phytotoxic chemicals for mustard control in cereals in the early 1900s.

Leaf arrangement may also affect the relative leaf surface area exposed to incoming spray droplets. This effect can best be explained by comparing one species with opposite leaves to another that has alternate leaves. When viewed directly from above and assuming leaves are the same size, the species with alternate leaves exposes twice the leaf area as the one having opposite leaves.

The surface of leaves of different species differ in a number of ways, including differences in pubescence, waxiness, and surface roughness. Differences in pubescence

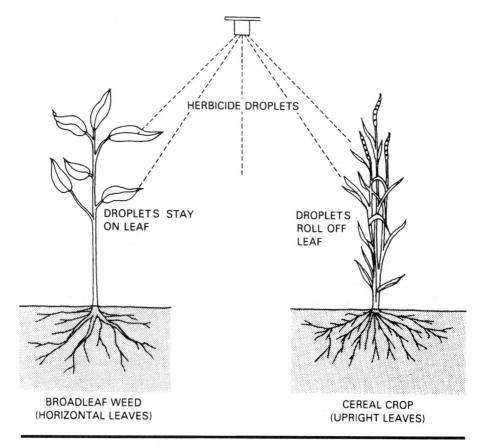

HERBICIDE DROPLETS

DROPLETS STAY
ON LEAF

DROPLETS
ROLL OFF
LEAF

BROADLEAF WEED
(HORIZONTAL LEAVES)

CEREAL CROP
(UPRIGHT LEAVES)

FIGURE 13-2. Influence of plant growth form and leaf attitude on herbicide retention.

and surface roughness were seen in Figure 11-5. Each leaf characteristic may affect retention of a herbicide droplet. Pubescence can be visualized to either increase or decrease actual contact between the spray droplet and the leaf, depending upon the number of such hairs. If hairs are numerous, as with common milkweed in Figure 11-5, they may actually keep a spray droplet from reaching the leaf surface. If sparse, as with velvetleaf in Figure 11-5, they may serve to hold the droplet on the leaf surface against a contact angle that might otherwise allow the droplet to run off.

Differences in leaf size, leaf arrangement, leaf attitude, and the nature of the leaf surface itself, therefore, provide a basis for biological selectivity among species. The usual situation is for more than one of these factors to come into play in such biological selectivity. Table 13-1 shows differences in spray retention on leaves of five plant species. The amount retained, of course, is a consequence of all leaf attributes combined. For example, mustard leaves are numerous, comparatively small, form a fairly dense canopy, and are rough and pubescent. These characteristics combined account for the high retention observed. As can be seen, wild mustard retains approximately 8 times as much spray per gram of dry weight as does barley and 6 times as much as pea. Thus, for a given rate of application, mustard might be selectively controlled by a contact herbicide in these crops.

TABLE 13-1.   Plant species differing in retention of herbicide sprays.

Species	Stage of Growth	Retention in mL g⁻¹ Dry Weight of Shoot
White mustard	2 leaves, 5-7 cm high	2.5
Sunflower	2 leaves, 6 cm high	2.0
Linseed	2 leaves, 5 cm high	1.1
Pea	2 leaves, 5-7 cm high	0.4
Barley	3 leaves, 15-20 cm high	0.3

Source: Data from Blackman et al., 1958.

Characteristics of the spray droplet.   Droplet size and droplet surface tension also influence retention on the leaf. Droplets in the 100 µm size were much more readily retained on pea leaflets than larger droplets (Figure 13-3A). Coverage of the target leaf also decreases as droplet size increases. Although more herbicide is retained with smaller droplets, the potential for drift increases dramatically as size drops below 100 µm (Bode, 1987). The trend for herbicide spraying is to use nozzles that produce droplets larger than 100 µm to reduce drift but smaller than 300 µm to get acceptable coverage. Nozzles are now available that provide droplet size most appropriate for the type of herbicide (systemic or contact) and treatment (broadcast, band, or directed).

Retention increases as droplet surface tension decreases. In simple terms, *surface tension* is a measure of the force with which particles on a liquid surface are held together. This is commonly reported in dynes but may also be reported in other units. A *dyne* is equal to the force required to impart an acceleration of 1 cm s⁻² to a mass of 1 g. A high surface tension indicates that the particles on the surface are held tightly together. We say that a liquid with high surface tension does not *wet* another surface very well. Water has a relatively high surface tension of 71 or 72 dynes cm[-1]. As can be seen in Figure 13-3B, very little of a spray would be retained if it had the surface tension of water. Thus, it is common practice to include an additive to reduce water surface tension if the herbicide formulation itself does not adequately do so.

Surfactant is the term applied to such substances and comes from the fact that the substances have surface activity. Surfactants include such materials as emulsifiers, de-

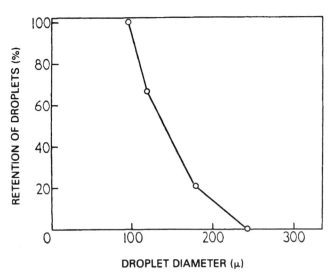

A. Droplet size inversely related to retention

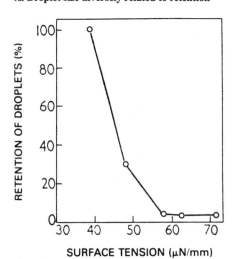

B. Droplet surface tension inversely related to retention on leaf

FIGURE 13-3. Effects of spray droplet size and surface tension on herbicide retention on the leaf; in this case, pea leaflets.

Source: Holly, 1976. Reproduced with permission of Academic Press, Inc.

tergents, stickers, and wetting agents. Surfactants have other properties, but only their effect on surface tension is of interest to us here. A common characteristic is to have both lipophilic and hydrophilic properties. That is, one part of the surfactant is compatible with lipid and lipid-like materials, and the other part is compatible with water. This compatibility tells us that surfactants are active at interfaces. The dual characteristic and its function relative to reducing surface tension are illustrated graphically in Figure 13-4A and Figure 13-4B, respectively. That part compatible with lipid or lipidlike substances—an oil droplet in this case—provides a bond with them, and that part compatible with water ties to water, thus serving to bond these two otherwise opposing liquids.

In effect, then, the surfactant serves to reduce internal cohesiveness of the spray droplet by providing an interfacing capability of that droplet with the leaf surface. Rather than remaining as droplets, the spray spreads as a thin layer on the leaf. Both droplet size and surface tension affect physical selectivity as they interact with the plant morphology and leaf surface microstructure. Postemergence herbicides are commonly marketed containing the appropriate surfactant.

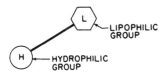

A. Lipophilic and hydrophilic properties of a surfactant

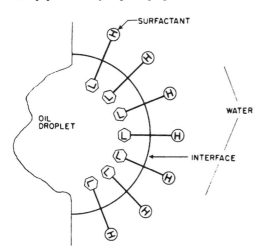

B. Interfacing capability of a surfactant

FIGURE 13-4. Schematic illustrations of the characteristics and function of a surfactant in increasing herbicide spray retention.

Source: Adapted from Behrens, 1964. Reproduced with permission of the Weed Science Society of America.

Penetration

With respect to penetration, the basis for selectivity is the ease with which a species' leaves are penetrated. The extent of contact between the spray droplets and leaf is integrally related to penetration: the greater the contact, the larger the gradient force that can serve to move the herbicide through either the apoplast or symplast system.

Penetration, thus, cannot be neatly separated from retention, even though each is a distinct step in the sequence of events leading to ultimate effect from the herbicide. Anything that increases retention can be expected to increase penetration as well. Here, too, we must be reminded that although penetration is assumed to be synonymous with activity, it is what occurs within the plant that is the final determinant of effect. As we saw when we examined entry and transport in detail in Chapter 11, leaves differ in their cuticle composition, numbers and location of stomata, and internal structure, all of which may influence herbicide penetration. Here we simply need to recognize that these characteristics are altered somewhat by the physical environment and may provide opportunities for selective use.

For herbicides not readily taken up through the leaves of otherwise susceptible weeds in crops tolerant to that herbicide, altering the spray so as to obtain penetration can provide selective control. For example, small amounts of naphthinic or paraffinic oils are added to wettable powder formulations of atrazine for selective postemergence control of weeds in growing corn. Without the addition of the oils, the atrazine does not penetrate the weeds in sufficient quantities to be effective, even though the weeds are susceptible once the atrazine enters. For its part, corn is tolerant of the atrazine because of its ability to detoxify the herbicide. Thus, it is not damaged, even though penetration may also be increased by addition of the oils.

Thus, we see that retention and penetration of herbicides is a function of both the nature of the leaf surface and the physical properties of the liquid applied. New application equipment allows droplet size to be rather precisely controlled and may provide opportunities for using this capability to obtain selectivity not possible with equipment previously available, which delivered droplets whose size varied over a wide range.

Translocation

After the herbicide has penetrated the plant, translocation is the next step in the sequence of events. As discussed in Chapter 11, certain aspects of herbicide usage to minimize competition are influenced by or interact with translocation. Retention is one aspect. Retention plays the role of maintaining a sufficient concentration on the epidermis to assure a gradient force for movement into the leaf until it ultimately reaches the symplast. Once the herbicide is inside the symplast, it moves as a passenger with photosynthate from the leaf to the site of biochemical reaction. Thus, the state of photosynthetic activity influences results. Actively growing weeds may be selectively removed from dormant or relatively inactive desired plants.

FACTORS IN HERBICIDE EFFECTIVENESS

Many things can intervene to reduce control with foliar-applied herbicides, just as was true with soil applications. In seeking an understanding of factors affecting success, the difference between soil and foliar applications is found in the fact that the main modifier of soil applications is the soil itself, whereas the plant is the primary modifier of foliar applications. Variations from the expected can be due to five general factors: (1) stage of weed development, (2) weed growth form, (3) weed growth rate, (4) direct temperature and humidity effects, and (5) weed density.

Stage of Weed Development

As discussed in Chapter 2, weeds cannot grow with crops for very long without causing permanent loss in crop yield. It follows that herbicides must be applied before this loss occurs if they are to fully meet the objective of minimizing competition. It happens that most annual weeds are more easily killed while young. Figure 13-5 shows the pronounced effect stage of development, as measured by leaf number, can have on the amount of herbicide needed for satisfactory control. As can be seen, better than 90% control of cocklebur, a species highly susceptible to the applied herbicide, could be expected with only about 25 g ha^{-1} if applied when the weed has 1 to 2 leaves. Delaying treatment until the weed has 6 leaves would require about 70 g ha^{-1} to obtain comparable control. Results with the moderately susceptible seedling johnsongrass and moderately tolerant red rice (*Oryza sativa*) show how important early treatment is when the weed is not highly sensitive to the herbicide.

Growth Form of Weeds

Chapter 3 discussed the fact that growth form is a very plastic characteristic in plants. The previous section pointed out that growth form very much influences contact and retention of spray droplets. Not surprisingly then, growing conditions prior to spray application may affect overall results. The extent of branching, leaf size, and leaf attitude may all be influenced by density of the plant community. In a very dense stand of weeds with the desired plants, the weeds may be less branched and more upright than in a sparse stand. Low temperature preceding and during leaf development tends to cause the leaves to be small. Moisture stress during leaf development also leads to relatively smaller leaves. Pubescence tends to be greater under high temperature, high light intensity, and low moisture. Influenced by growing conditions, all of these aspects of growth form may modify the effects of the spray application.

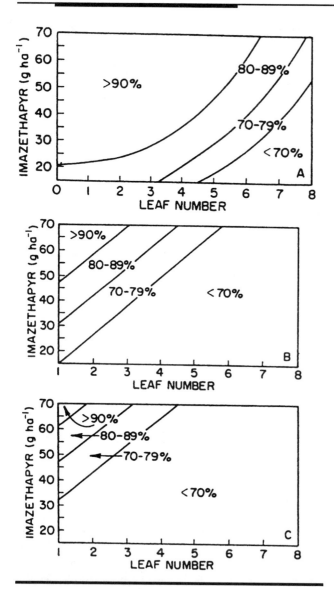

FIGURE 13-5. Predicted percent control of weeds with different sensitivity as affected by weed size. Most susceptible cocklebur (A); moderately susceptible seedling johnsongrass (B); moderately tolerant red rice (C). The contour graphs were drawn from regression curves of control with 17, 35, and 70 g ha^{-1} of imazethapyr applied 3, 6, 12, and 24 days after weed emergence.

Source: From Klingaman et al., 1992.

Growth Rate of Weeds

Growth rate, as affected by growing conditions, can best be visualized as having its effect on penetration, translocation, and ultimate biochemical reaction at the sites of activity. This general statement is predicated on the assumption that the extent of penetration, translocation, and biochemical reaction is quite directly related to movement in the symplast system that in turn is influenced by the degree of photosynthetic activity in the leaf. There is ample evidence that herbicidal activity is related to photosynthetic activity.

Growth rate may also influence results from herbicide applications indirectly through its effect on the leaf cuticle. The slow growth associated with low temperature usually leads to less cuticle. Such plants are more easily penetrated by a herbicide. Slow growth as the result of extended drought, on the other hand, may lead to more cuticle, thus reducing penetration of a herbicide.

Temperature and Humidity

High humidity and high temperature commonly lead to an increase in stomatal aperture. Entry of herbicides through the stoma may be severalfold greater than through the cuticle itself. High humidity is commonly associated with an increase in transfer of the herbicide from the leaf surface to the phloem. Table 13-2 indicates the pronounced effect of relative humidity on translocation of glyphosate. Five to six times as much glyphosate was translocated at 100% than at 40% relative humidity. In addition to its affect on stomatal opening, high relative humidity prevents desiccation of the spray droplet, thereby augmenting penetration.

Temperature influences success through its affects on physicochemical processes such as diffusion and viscosity and on physiological processes such as evapo-transpiration, photosynthesis, phloem translocation, and growth. In a general way, temperatures optimum for growth are also optimum for effectiveness of systemic herbicides. For example, it was shown in very early research that 2,4-D was most effective against buckhorn plantain when temperatures were at 18°C to 24°C, the range most favorable for growth, than at higher or lower temperatures (Marth and Davis, 1945).

TABLE 13-2.   **Translocation of 0.15 μCi ^{14}C-glyphosate in bermudagrass after 48 hours.**

Temperature (°C)	Relative Humidity (%)	Amount Translocated (%)
22°	40	9
	100	58
32°	40	14
	100	62

Source: Data from Jordan, 1977.

Weed Density

Weed control is frequently poorer with herbicides applied postemergence in reduced tillage than in conventional tillage systems. This may be due in part to poorer coverage of more dense populations of weeds under reduced tillage. As can be seen in Table 13-3, weed densities may be much greater under reduced than conventional tillage. Smallest weeds in extremely dense weed stands may be protected from the herbicide spray.

TABLE 13-3.   Weed density as affected by tillage.

	Conventional Tillage	No Tillage
	(plants m^{-2})	
Giant foxtail	148	174
Common lambsquarters	21	64
Pigweed	23	306

Source: Data from Buhler, 1992.

Of course, sensitivity of desired plants may also be influenced by these factors that affect herbicide effectiveness. In view of the many factors that can influence effects of herbicides applied postemergence, it is not surprising that results have sometimes varied throughout the modern era.

POSTEMERGENCE HERBICIDE APPLICATION METHODS

It is beyond the scope of this text to cover the individual herbicides used for postemergence weed control. Here we simply want to examine the general ways herbicides may be used to control growing weeds and consider modifications of postemergence applications that expand usefulness of such applications in a total weed management context. Available options are numerous. For example, more than 25 herbicides are available for postemergence weed control in soybeans alone. Which herbicide(s) to use depends upon weeds to be controlled, growing conditions, stage of development of weeds and of the desired plants, cost, and possible other factors. Herbicides are now available to control most problem weeds encountered in production of desired plants. It must be remembered from what was learned in Chapter 3 that new problem weeds may emerge following successful management of the current weed problem.

Three general ways herbicides may be used to control growing weeds are shown in Figure 13-6. Application before planting the crop (Figure 13-6A) and application after planting the crop but before crop emergence (Figure 13-6B) are rather special uses and are based on a separation in time between the weed and the crop. Since the crop plant is not present, a spectrum of herbicides otherwise toxic to the crop may be safely used.

The first way (Figure 13-6A) is commonly used in zero-tillage systems. In this usage, the crop is planted directly into the killed weeds with no seedbed preparation to bring new weed seeds to the surface. This usage will be explored in detail later, where we consider postemergence herbicide use in conservation tillage.

The second way (Figure 13-6B) may be used where the crop is planted so deeply that the shallow-germinating weeds emerge first. This method was the basis for selective control of weeds in potatoes with dinoseb (Aldrich et al., 1954). For both methods, crop damage is avoided by using herbicides not readily absorbed from the soil.

The third, and most widespread usage by far, is application to a mixture of weeds growing with desired plants, as shown in Figure 13-6C. Control may be equally as striking as with soil applications. For example, we see in Figure 13-7 excellent control of a mixture of annual grass and broadleaf weeds in soybeans. Control was accomplished with a combination of 0.22 kg ha⁻¹ of the selective grass killer sethoxydim and 1.1 kg ha⁻¹ of bentazon.

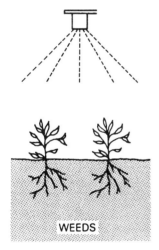 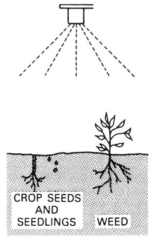 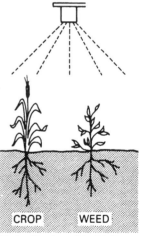

A. Application of herbicide to growing weed before crop has been planted

B. Application of herbicide to growing weeds after the crop has been planted

C. Application of herbicide to growing crop and weeds

FIGURE 13-6. Ways herbicides may be used to minimize competition from growing weeds.

A. Untreated soybeans

B. Soybeans treated with herbicides applied postemergence

FIGURE 13-7. Mixture of broadleaf and annual grass weeds selectively controlled with a combined application of bentazon and sethoxydim.

APPLICATION TO WEEDS GROWING WITH A CROP

As just mentioned, choice of herbicide(s) is influenced by several factors. These factors are evident in differences in recommendations for postemergent weed control in some crops in selected states shown in Table 13-4. Recommendations for specific weed problems can best be obtained from such local sources as State University Extension publications, commercial applicators, and herbicide suppliers.

TABLE 13-4. Recommended rates of selected herbicides for postemergence weed control in major crops vary between states. Numbers in parenthesis are pounds per acre.

Crop	Weed Problem	Herbicide	Rate Recommended (g ha^{-1}) State A	State B
Corn	Annual grasses and broadleaf	Cyanazine	1344-2240 (1.2-2.0)	1120-2688 (1.0-2.4)
Soybeans	Annual broadleaf	Acifluorfen	280-426 (0.25-0.38)	280-560 (0.25-0.5)
Cotton	Annual grasses	Sethoxydim	105-140 (0.094-0.125)	213-426 (0.19-0.38)
Wheat	Broadleaf	2,4-D	280-739 (0.25-0.66)	280-1120 (0.25-1.0)
Sorghum	Annual grasses and broadleaf	Atrazine	1120-2240 (1.0-2.0)	2240-3360 (2.0-3.0)

Source: State recommendations.

Directed Postemergence Applications

Directing the herbicide spray onto weeds beneath the crop canopy provides a way of selectively controlling weeds in crops not adequately tolerant of over-the-top broadcast applications. The concept is illustrated in Figure 13-8. Several types of equipment will deliver the spray in this way (Kleppe and Harvey, 1991a). One type that has been successful in controlling wild-proso millet with sethoxydim in corn is shown in Figure 13-9. Table 13-5 shows better than 90% control with sethoxydim applied to wild-proso millet with this sprayer following butylate and cyanazine applied preplant and incorporated. The preplant herbicide combination was needed to suppress early weed growth and provide a height differential between the wild-proso millet and the sweet corn. Postemergence-directed herbicides have been successfully used to selectively control a variety of weeds in several additional crops, including cotton (Hamilton and Arle, 1970), soybeans (Wilson and Burnside, 1973), and sugarcane (Dill and Martin, 1978).

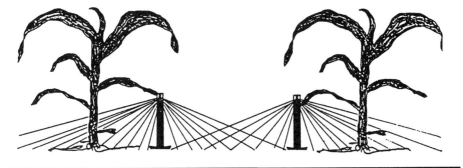

FIGURE 13-8. Postemergence spray directed beneath crop leaves by using fan spray pattern nozzles at appropriate heights above the ground.
Source: Kleppe and Harvey, 1991.

FIGURE 13-9. A postemergence directed sprayer. Above = rear view. Below = side view. The sprayer shown is mounted on the tractors three-point hitch and sprays four 76 cm (30 in.) corn rows. The sprayer has 200 cm-long (5 ft), pivoted, spring-loaded arms spaced 38 cm (15 in.) apart that move independently and extend back from the toolbar to skids. A spring between the toolbar and each arm provides tension to stabilize the skid to the ground.

Source: Courtesy R. Gordon Harvey, University of Wisconsin, Madison.

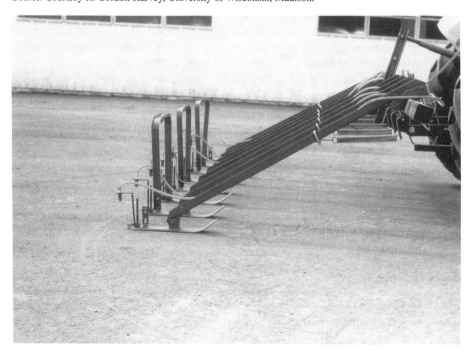

TABLE 13-5. Improvement in wild-proso millet control in corn with a postemergence directed application of sethoxydim.

Herbicides	Rate of Application (g ha^{-1})	Crop Injury (%)	Wild-Proso Millet Control (%)
Butylate + cyanazine preplant incorporated	4450 + 2270	0	64
As above preplant incorporated plus sethoxydim postemergence directed	110	0	93

Source: Data from Kleppe and Harvey, 1991b.

Sequential Applications

Applying part of the herbicide at one time and part several days to a few weeks later may improve weed control and reduce weed loss compared to the same total rate as a single application (Table 13-6). In this example, the herbicide spray was directed onto the weeds beneath the pepper plants. Better weed control with the split application resulted in about one-fourth more peppers than either single treatment. Splitting the herbicide application increases the chances of controlling earliest emerged weeds by treating them while they are small and also provides treatment of weeds that may emerge after the initial application.

TABLE 13-6. Improved weed control in chile peppers with sequential applications of a herbicide.

Herbicide Rate (kg ha^{-1})	Time of Application after Thinning (wk)	Weed Control		Pepper Yield Loss (%)
		Wright Groundcherry (%)	Prostrate Pigweed (%)	
0.28	2	71	18	19
0.28	4	92	45	24
0.14, 0.14	2,4	99	91	-3

Source: Data from Schroeder, 1992.

Mixture of Herbicides

Weed problems commonly include one or more grass and broadleaf species. It follows that a combination including a herbicide especially effective against grasses and one especially effective against broadleaf weeds may provide better control with less risk to desired plants than a single application of a broad-spectrum herbicide. Table 13-7 shows how striking the effect may be. Either herbicide alone provided control of the species it was especially effective against, resulting in nearly a 60% increase in safflower yield over no control. The combination, with chlorsulfuron at only one-half the single application rate, controlled both weeds resulting in a 225% increase in safflower yield. The results point up an additional advantage for herbicide combinations: Herbi-

TABLE 13-7. Better control of a grass and broadleaf weed in safflower with a mixture of herbicides.

Herbicide	Rate (kg ha⁻¹)	Wild Oats (g m⁻²)	Common Lambsquarters (g m⁻²)	Safflower Yield (g m⁻²)
None		2223	1676	76
Sethoxydim	0.25	143	1623	121
Chlorsulfuron	0.011	1965	0	118
Sethoxydim + chlorsulfuron	0.25 0.0055	96	0	171

Source: Data from Blackshaw et al., 1990.

cides may frequently be effective at lower rates in combination than if they are applied alone. Many herbicides are now packaged as mixtures.

Antagonism among herbicides. Most herbicides have no effect on phytotoxicity of other herbicides with which they may be mixed. However, there is a growing number of herbicides that may antagonize activity of other herbicides. The antagonism may be pronounced, as can be seen in Figure 13-10 for bromoxynil against control of large crabgrass by several grass herbicides. Bromoxynil may also antagonize phytotoxicity of glyphosate (O'Donovan and O'Sullivan, 1982). It may be possible to avoid antagonism by applying the herbicides a few days apart rather than as a tank mixture.

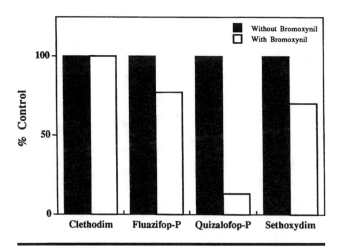

FIGURE 13-10. Antagonism of bromoxynil toward the activity of some selective grass herbicides in controlling large crabgrass.

Source: Data from Jordan et al., 1993.

Reduced Herbicide Rates

Herbicide rates recommended by the manufacturer are the result of extensive and careful testing and are correct for most uses. However, the recommendation must usually cover a variety of weed species and a range in stage of development of weed and crop plants, climatic and weather conditions, soil types, tillage practices, and possible other production practices that may influence the rate needed for effective control. Under optimum growing conditions, rates below the general recommendation may be adequate to control very small weeds and especially sensitive species. Successful control is possible with one-half or even one-fourth of the full rate (Baldwin and Oliver, 1985; Devlin et al., 1991; DeFelice and Kendig, 1994; Griffin et al., 1992; O'Sullivan and Bouw, 1993). Early treatment is essential for success. Figure 13-11 shows the relationship between weed size and herbicide rate that may provide adequate control. Early application of one-fourth the full rate may require a second application if additional weeds emerge. Nevertheless, the total amount of herbicide could be expected to be less than the full rate in a single application. Planning to use reduced rates provides flexibility in a postemergence weed control program; if weather or other events interfere with a planned reduced application, there is still time for a full-rate application.

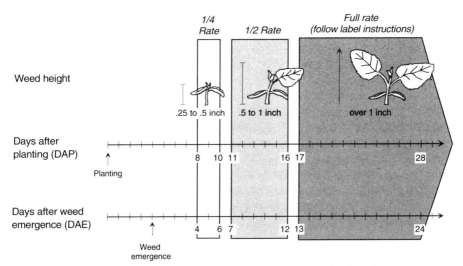

*The degree to which the three timing methods overlap varies somewhat with tillage, weed species and weather.

FIGURE 13-11. Relationship between weed size and herbicide rate.

Source: DeFelice and Kendig, 1994. From University Extension Publication MP 686, University of Missouri, Columbia. Reproduced with permission of the University of Missouri.

Safeners

Chemicals have been developed that protect some desired plants from otherwise toxic herbicides. It appears the safening chemicals induce enhanced herbicide detoxification in protected plants (Hatzios, 1983) or enhanced glutathione conjugation with the herbicide (Fuerst and Gronwald, 1986). Safeners may protect some crops from selected herbicides applied postemergence, although it appears they may have their greatest use in connection with preemergence application. They may be applied to crop seed prior to planting, to transplants, to soil, or with the herbicide.

Herbicide-resistant Crops

It was pointed out in Chapter 1 that some weeds have become resistant to certain herbicides. It follows that desired plants could be developed through appropriate breeding programs to tolerate otherwise toxic herbicides. Indeed several private companies have or are about to introduce herbicide-tolerant cultivars of several important crops (Niebling, 1995). Included are soybeans, canola, cotton, corn, and sugarbeets tolerant of glyphosate; corn tolerant of imidazoline herbicides; and cotton tolerant of bromoxynil. Development of herbicide-tolerant crops expands the usefulness of existing herbicides and may facilitate control of particularly troublesome weeds in certain crops. However, it should be remembered that growing herbicide-tolerant crops could encourage repeated exclusive use of the herbicide, thereby hastening the possible development of resistance on the part of the weed. Steps that can be taken to minimize or slow development of resistance on the part of weeds are discussed in the next chapter.

CONTROLLING HARD-TO-KILL PERENNIAL WEEDS

It is often difficult to prevent competition from perennial weeds growing with desired plants. As we learned in Chapter 5, perennials regenerate from a variety of vegetative (perennating) parts. If competition is to be avoided, the herbicide must prevent regrowth from the perennating part. Merely destroying aboveground growth, adequate for control of annual weeds, is generally inadequate for perennial weeds. Additionally, unlike many annuals that have a "flush" of germination, perennials commonly resume growth over an extended time making control with a single herbicide application unsatisfactory. Furthermore, herbicides effective against perennials often may not be selective for the crop with which the weed is growing. Information on control of specific perennial weeds with available selective herbicides can best be obtained from local sources. Here, we want to recognize some general approaches for dealing with hard-to-kill perennial species.

Crop and Herbicide Rotations

Changing the crops grown to include one in which an effective herbicide can be safely used provides a way of dealing satisfactorily with some perennial weeds. For example, johnsongrass, a serious problem in corn in the southern United States, was prevented from becoming a problem by use of trifluralin in the cotton in a corn–cotton–cotton–corn rotation (Table 13-8). Herbicides usable in corn allowed johnsongrass to increase to competing densities in continuous corn.

TABLE 13-8. **Johnsongrass seeds and seedlings after 4 years of a corn–cotton–cotton–corn and continuous corn production system. Cotton in the rotation allowed use of trifluralin, a herbicide effective against johnsongrass.**

Production System	Seed in 0-15 cm Depth (no. kg^{-1})	Seedlings (no. m^{-2})
Corn[a]–cotton[b]–cotton[b]–corn[a]	0	0
Continuous corn[a]	57	20

[a]2.24 kg ha^{-2} of atrazine was applied preemergence and 0.84 kg ha^{-2} of linuron was applied postemergence directed in corn.
[b]0.56 kg ha^{-2} of trifluralin was applied preplant incorporated in cotton.
Source: Data from Dale and Chandler, 1979.

Herbicide Application After the Crop Is Mature

Adopting a preventive approach can be an effective way of using herbicides in dealing with problem perennial weeds in some crops. For example, Canada thistle and quackgrass, perennial weeds widely distributed throughout temperate regions of the world, have been effectively prevented from regrowth for at least the following year by glyphosate applied after wheat and other crops were ripe but before harvest (Davis and Orson, 1987; O'Keefe, 1980; Kirkland, 1990). Glyphosate is a nonselective postemergence-applied herbicide. Applying glyphosate and other herbicides to perennial weed regrowth in the fall following crop harvest offers another way of using herbicides to prevent interference with desired plants in subsequent years. However, this approach may require at least 3 years of treatment to reduce Canada thistle density to an acceptable level (Donald, 1993).

HERBICIDE USE IN CONSERVATION TILLAGE

Weed control has historically been the main purpose of tillage. Availability of modern herbicides in the mid-1900s offered another means to control weeds. This allowed tillage practices to be designed for other desired objectives, especially soil and water conservation. Moldboard plowing was the initial tillage operation in crop production from the early 1800s until the mid-1900s. Plowing followed by implements such as harrows to prepare a seedbed was the common conventional tillage system. Conventional tillage is still the most widely used practice in crop production. However, it is rapidly being replaced by so-called conservation tillage. *Conservation tillage* is any practice that reduces soil or water loss compared to moldboard plowing (Resource Conservation Glossary, 1976). There are many different conservation tillage systems. The single factor that distinguishes conventional from conservation tillage is how plant residues are handled. In conventional tillage, plant residues are plowed under. In conservation tillage, some or most of the plant residues are left on the soil surface. Plant residues dissipate the impact of raindrops, increase water infiltration, and decrease water runoff. In the Corn Belt, 75% reduction in soil loss is commonly realized with conservation tillage using corn residue (Griffith et al., 1977; Moldenhauer et al., 1983). In drier regions, crop residues can dissipate effects of wind, thereby reducing soil loss. Here, too, the reduction in soil loss may be great. Soil loss due to wind has been reduced from 32 metric tons ha^{-1} with conventional tillage to 2 metric tons ha^{-1} with conservation tillage for wheat production (Moldenhauer et al., 1983).

Thus, there are two aspects of conservation tillage that pose additional challenges for weed management. One is the loss of weed control provided by conventional tillage. In effect, weed control with herbicides must replace control otherwise provided by conventional tillage. The other is the effect of the plant residues. Residues may intercept some of the preemergence-applied herbicide, thereby increasing the amount needed for satisfactory control and increasing the chances for "escapes" from the herbicide treatment. Escapes plus delayed weed seed germination due to altered soil temperature and moisture conditions may necessitate a follow-up postemergence herbicide application.

Herbicides are needed to "burn down" any existing vegetation at planting time under conservation tillage. Two herbicides commonly used to kill existing vegetation are paraquat, a contact herbicide, and glyphosate, a translocated herbicide. Both are essentially nontoxic in soil so crops can be planted immediately before or after treatment. Nevertheless, burn down of existing vegetation with herbicides usually is not enough by itself to prevent development of a competing weed population arising from seeds that germinated after the burn down. Three alternatives for dealing with potential new weeds include: (1) application prior to planting of a herbicide effective in the soil (preemergence or so-called "residual" herbicide); (2) application of a residual herbicide at planting; and (3) application of a postemergence herbicide to weed seedlings after emergence. Some herbicides are effective when applied either pre- or postemergence. Providing the crop is tolerant, a herbicide with such dual activity could replace the burn down herbicide plus prevent emergence of new weeds. Thus, many herbicide combinations have been developed to fit the different crop, weed, climatic, and soil situations. Often, however, more herbicide may be needed for satisfactory weed control in conservation than in conventional tillage.

PREVENTING ADDITION OF VIABLE PROPAGULES

A distinction is made here between the prevention of propagule additions as a result of herbicides that destroy the weed to avoid competition, and prevention as a result of chemicals applied for the express purpose of reducing production of propagules. Destruction of the weed seedling or the growing plant prior to reproduction, of course, serves to prevent the weeds so treated from adding propagules to the reservoir in the soil. This result may be an important side benefit from herbicides used to prevent weed emergence with crops and those used to minimize competition from weeds growing with crops. Also, certain situations may justify the cost of applying chemicals to destroy weed plants for the sole purpose of preventing production of propagules.

Irrespective of the method employed, some way of preventing periodic additions to the propagule bank in soil must be found for a preventive approach to be fully successful. Eliminating only those propagules already present in the soil has limited value in agriculture unless addition of seeds or perennating parts can be prevented. On the other hand, preventing additions to the propagule bank in the soil by itself in time reduces the population of even the most persistent weed to a level that does not interfere with normal crop production (Figure 4-8 and related discussion). Chemicals may be an especially valuable tool for this purpose. Mechanical methods frequently may be impractical because of the advanced stage of development of the crop at the time when action needs to be taken against the weed to prevent production of viable seed or perennating parts.

Concern for possible herbicide damage to crop seed prompted examination of direct effects of herbicides on seed viability soon after the growth regulator–type herbicides were introduced in the 1940s (Aamisepp, 1966). The finding that herbicides indeed could affect viability and germination of crop seed encouraged research to determine if herbicides could be effectively used in this way against specific weeds. In the early 1950s, maleic hydrazide was found to be effective in preventing seed production of annual bluegrass (Engel and Aldrich, 1960) and wild oat (Carder, 1954; Friesen and Walker, 1956). Maleic hydrazide applied at 1.1 kg ha^{-1} (1 lb per acre) was enough to greatly reduce production of annual bluegrass seed heads in bentgrass turf. As little as 0.56 kg ha^{-1} (0.5 lb per acre) was enough to reduce the production of viable wild oat seeds to less than 1%. The fluoro-substituted phenoxyacetic acids were found to be effective in preventing production of annual bluegrass and crabgrass seed without appreciable damage to the turf grasses (Anderson and McLane, 1958).

In more recent research, 2,4-D was found to be effective in preventing seed production in leafy spurge if applied at the appropriate time (Table 13-9). It was most effective if applied at the start of flower bud development. The importance of application early in flower development agrees with findings of 2,4-D on curly dock (Maun and Cavers, 1969), maleic hydrazide on wild oat (Friesen and Walker, 1956), amitrole and maleic hydrazide on medusahead (Evans et al., 1963), fluoro-substituted phenoxyacetic acid on *Poa annua* and crabgrass (Anderson and McLane, 1958), and chlorimuron and imazaquin on sicklepod (Isaacs et al., 1989).

TABLE 13-9. Effect of time of 2,4-D application on production of viable seed by leafy spurge.

Time of 2,4-D Application (days after bud initiation)	Viable Seed per Plant (no.)
0	<1
7	4
14	7
21	31
28	53
35	62
untreated	173

Source: Data from Al-Henaid et al., 1993.

The effectiveness of herbicides to prevent weed seed production has been well established. However, use of herbicides for this purpose in crop production is limited. There are two main reasons: (1) we are not yet geared to weed prevention; and (2) such usage represents an additional operation and cost with little if any yield benefit for the current crop. With respect to the second point, it follows that such usage is most apt to occur as an adjunct to other weed management efforts. One such opportunity is in connection with herbicides applied to aid harvest where weeds escaped early-season weed control in such crops as cotton, soybeans, and peanuts. It has been shown that herbicides can be effectively used for the dual purpose of aiding harvest and preventing sicklepod seed production in soybeans (Ratnayake and Shaw, 1992). Chapter 15 examines the place for such usage in a weed prevention approach.

Preventing Dormancy

Preventing dormancy is a special aspect of the prevention of viable propagules. As already mentioned, dormancy of seed and of perennating parts is common in most weeds. Dormancy assures survival of weeds from the stress imposed by both tillage of the land and by the environment. The fact that a large part of the population of a given crop of weed seeds may, in fact, be dormant is a major reason means of preventing or controlling weeds are necessary every year crops are grown. If dormancy could be prevented, the problem of dealing with the particular weed would be very much simplified. There is evidence that treatment of plants can alter the development of dormancy in both buds and in seeds.

A number of growth regulator herbicides applied to wild garlic when the offset bulbs (see Figure 5-2) were forming in the spring resulted in the bulbs having little dormancy, whereas unsprayed plants had bulblets that were dormant for at least 6 months (Parker, 1976). Additional research with growth regulators has shown them to be effective against other species in reducing seed dormancy as well.

Growth regulator–type herbicides have also been shown to affect dormancy in weed seeds. However, results have been inconsistent. In early work (Rojas-Garciduenas and Kommedahl, 1960), treatment with 2,4-D reduced dormancy in redroot pigweed seed. In later work (Fawcett and Slife, 1978a), 2,4-D and dalapon were found to affect dormancy in redroot pigweed, lambsquarters, and giant foxtail. But effects on initial dormancy were sometimes different than effects on dormancy after overwintering. Further, in this instance, initial dormancy of redroot pigweed seed was increased by 2,4-D.

Nitrogen fertilization of common lambsquarters plants resulted in seed with reduced dormancy (Fawcett and Slife, 1978b). Whereas germination of seed from unfertilized plants was 3%, that from plants fertilized with nitrate was 7 to 12 times greater. In this same study, velvetleaf was unaffected by nitrogen fertilization.

The magnitude of the reduction in dormancy of the rather limited work that has been done falls short of making this a practical way of completely preventing the carryover of viable propagules in the soil. Nevertheless, these studies have shown that chemicals can alter dormancy of seed and of perennating parts. As more is learned about the processes of dormancy, development, and the initiation of growth, and as more of the relatively large numbers of chemical moieties are examined for this effect, we can reasonably expect that this approach may also find a place in a total weed management program.

REFERENCES

Aamisepp, A. 1966. Herbicide effects on plants from seeds from treated plants. Vaxtodling, 22:1-147.

Al-Henaid, J.S., M.A. Ferrell, and S.D. Miller. 1993. Effect of 2,4-D on leafy spurge (*Euphorbia esula*) viable seed production. Weed Technol. 7:76-78.

Aldrich, R.J., G.R. Blake, and J.C. Campbell. 1954. Cultivation and chemical weed control in potatoes. Circular 557. New Jersey Agricultural Experiment Station, New Brunswick, NJ.

Anderson, B.R., and S.R. McLane. 1958. Control of annual bluegrass and crabgrass in turf with fluorophenoxyacetic acids. Weeds 6:52-58.

Baldwin, F.L., and L.R. Oliver. 1985. A reduced rate intensive management soybean weed control program. Proc. South. Weed Sci. Soc. 38:487.

Behrens, R.W. 1964. The physical and chemical properties of surfactants and their effects on formulated herbicides. Weeds 12:255-258.

Blackman, G.E., R.S. Bruce, and K. Holly. 1958. Studies in the principles of phytotoxicity, V. Interrelationships between specific differences in spray retention and selective toxicity. J. Exp. Bot. 9:175-205.

Blackshaw, R.E., D.O. Derksen, and H.-Henning Muendel. 1990. Herbicide combinations for postemergent weed control in safflower (*Carthamus tinctorius*). Weed Technol. 4:97-104.

Bode, L.E. 1987. Spray application technology. In C.G. McWhorter and M.R. Gebhardt, eds., Methods of applying herbicides, pp. 85-110. Monograph Series No.4. Weed Sci. Soc. Am.

Buhler, D.D. 1992. Population dynamics and control of annual weeds in corn (*Zea mays*) as influenced by tillage systems. Weed Sci. 40:241-248.

Carder, A.C. 1954. The selective control of wild oats in cereal crops by use of maleic hydrazide. Research Report, vol. 11, p. 50. North Central Weed Control Conference, Fargo, ND.

Dale, J.E., and J.M. Chandler. 1979. Herbicide-crop rotation for johnsongrass (*Sorghum halepense*) control. Weed Sci. 27:479-485.

Davis, C.J., and J. H. Orson. 1987. The control of *Cirsium arvense* (creeping thistle) by sulfonylurea herbicides and a comparison of methods of assessing efficacy. 1987 Br. Crop Prot. Conf.-Weeds, pp. 453-460.

DeFelice, M., and A. Kendig. 1994. Using reduced herbicide rates for weed control in soybeans. University Extension MP686, University of Missouri, Columbia.

Devlin, D.L., J.H. Long, and L.D. Maddux. 1991. Using reduced rates of postemergence herbicides in soybeans (*Glycine max*). Weed Technol. 5:834-840.

Dill, G.M., and F.A. Martin. 1978. Postemergence-directed applications of paraquat in Louisiana sugarcane. Proc. South. Weed Sci. Soc. 31:70.

Donald, W.W. 1993. Retreatment with fall-applied herbicides for Canada thistle (*Cirsium arvense*) control. Weed Sci. 41:434-440.

Engel, R.E., and R.J. Aldrich. 1960. Reduction of annual bluegrass, *Poa annua,* in bentgrass turf by the use of chemicals. Weeds 8:26-28.

Evans, R.A., B.L. Kay, and C.M. McKell. 1963. Herbicides to prevent seed set or germination of medusahead. Weeds 11:273-276.

Fawcett, R.S., and F.W. Slife. 1978a. Effects of 2,4-D and dalapon on weed seed production and dormancy. Weed Sci. 26:543-547.

_____. 1978b. Effects of field applications of nitrate on weed seed germination and dormancy. Weed Sci. 26:594-596.

Friesen, H.G., and D.R. Walker. 1956. Selective control of wild oats in Olli barley with MH. Research Report, vol. 13, p. 54. North Central Weed Control Conference, Chicago, IL.

Fuerst, E.P., and J.W. Gronwald. 1986. Induction of rapid metabolism of metolachlor in sorghum (*Sorghum bicolor*) shoots by CGA-92194 and other antidotes. Weed Sci. 34:354-361.

Griffin, J.L., D.B. Reynolds, P.R. Vidrine, and A.M. Saxton. 1992. Common cocklebur (*Xanthium strumarium*) control with reduced rates of soil- and foliar-applied imazaquin. Weed Technol. 6:847-851.

Griffith, D.R., J.V. Mannering, and W.C. Moldenhauer. 1977. Conservation tillage in the eastern Corn Belt. J. Soil Water Conserv. 32:20-28.

Hamilton, K.C., and F.A. Arle. 1970. Directed applications of herbicides in irrigated cotton. Weed Sci. 18:85-88.

Hatzios, K.K. 1983. Herbicide antidotes: development, chemistry, and mode of action. Adv. Agron. 36:165-316.

Holly, K. 1976. Selectivity in relation to formulation and application methods. In L.J. Audus, ed., Herbicides: physiology, biochemistry, ecology, 2nd ed., vol. 2, pp. 249-277. New York: Academic Press.

Isaacs, M.A., E.C. Murdock, J.E. Toler, and S.U. Wallace. 1989. Effects of late-season herbicide applications on sicklepod seed production and viability. Weed Sci. 37:761-765.

Jordan, D.L., M.C. Smith, M.R. McClelland, and R.E. Frans. 1993. Weed control with bromoxynil applied alone and with graminicides. Weed Technol. 7:835-839.

Jordan, T.N. 1977. Effects of temperature and relative humidity on the toxicity of glyphosate to bermudagrass (*Cynodon dactylon*). Weed Sci. 25:448-451.

Kirkland, K.J. 1990. Preharvest quackgrass (*Agropyron repens* (L.) Beauv.) control. In Proc. Quackgrass Symp. Oct. 24-25, pp. 127-134. London: Ontario.

Kleppe, C.D., and R.G. Harvey. 1991a. Postemergence-directed sprayers for wild-proso millet (*Panicum miliaceum*) control. Weed Technol. 5:185-193.

_____. 1991b. Precision postemergence-directed sprayer equipment for herbicide application in field and sweet corn. University of Wisconsin, Extension Publication no. A3528.

Klingaman, T.E., C.A. King, and L.R. Oliver. 1992. Effect of application rate, weed species, and weed stage of growth on imazethapyr activity. Weed Sci. 40:227-232.

Marth, P.C., and F.F. Davis. 1945. Relation of temperature to the selective herbicidal effects of 2,4-dichlorophenoxy acetic acid. Bot. Gaz. 106:463-472.

Maun, M.A., and P.B. Cavers. 1969. Effects of 2,4-D on seed production and embryo development of curly dock. Weed Sci. 17:533-536.

Moldenhauer, W.C., G.W. Langdale, W. Frye, D.K. McCool, R.J. Papendick, D.E. Smika, and D.W. Fryrear. 1983. Conservation tillage for erosion control. J. Soil Water Conserv. 38:144-151.

Niebling, K. 1995. Agricultural biotechnology companies set their sights on multi-billion $$ markets. In Genetic engineering news, pp. 1, 20, and 21. July, 1995, vol. 15 (13).

O'Donovan, J.T., and P.A. O'Sullivan. 1982. Haloxyfop-methyl toxicity to johnsongrass reduced when sprayed in combination with 2,4-D. Weed Sci. 38: 103-107.

O'Keefe, M.G. 1980. The control of *Agropyron repens* and broad-leaved weeds pre-harvest of wheat and barley with the isopropylamine salt of glyphosate. Proc. 1980 Br. Crop. Prot. Conf.-Weeds, pp. 53-61.

O'Sullivan, J., and W.J. Bouw. 1993. Reduced rates of postemergence herbicides for weed control in sweet corn (*Zea mays*). Weed Technol. 7:995-1000.

Parker, C. 1976. Effects on the dormancy of plant organs. In L.J. Audus, ed., Herbicides: Physiology, biochemistry, ecology. 2nd ed., vol. 1, pp. 169-190. New York: Academic Press.

Ratnayake, S., and D.R. Shaw. 1992. Effects of harvest-aid herbicides on sicklepod (*Cassia obtusifolia*) seed yield and quality. Weed Technol. 6:985-989.

Resource Conservation Glossary. 1976. Soil Conservation Society of America, Ankeny, IA.

Rojas-Garciduenas, M., and L. Kommedahl. 1960. The effect of 2,4-D on germination of pigweed seed. Weeds 8:1-5.

Schroeder, J. 1992. Oxyfluorfen for directed postemergence weed control in chile peppers (*Capsicum annuum*). Weed Technol. 6:1010-1014.

Wilson, R.G., and O.C. Burnside. 1973. Weed control in soybeans with postemergence-directed herbicides. Weed Sci. 21:83-85.

14

Production Practices Affect Weeds

CONCEPTS TO BE UNDERSTOOD

1. Weeds on a given farm or field change as a result of introduction from outside, as a result of genetic and physiologic alteration within a species, and as a result of production practices followed.
2. In terms of numbers and the time in which marked changes may occur, changes brought about by production practices are the most important for weed management.
3. A combination of herbicides is frequently more effective than a single herbicide against a mixture of weed species.
4. Some weeds have developed resistance to certain herbicides. Practices that lessen selection pressure—herbicide rotation, reduced herbicide rates, and herbicide combinations—will slow this phenomenon.
5. Each new crop production or managerial practice will ultimately have its own complement of weeds.
6. Cropping, tillage, and herbicide practices interact to determine weed composition on an enterprise unit, but the effects of herbicides commonly mask the other effects.
7. Conservation tillage often results in an increase in perennial and small-seeded, shallow-germinating annual weeds.

In Chapter 1, reference was made to the changes in weed composition that have occurred in response to herbicides. In the intervening chapters, we have examined competition, weed reproduction, germination and dormancy, allelopathy, biological relationships, and herbicides. One or more of these factors may be involved in shifts in the makeup of weeds important in crop production. With this background, we are ready to examine the interrelationships between these factors and crop production practices. This is a necessary first step in predicting future weed problems. Today's computer technology and systems modeling provide weed science the necessary tools to predict weed problems if the effects on weed composition are known and well documented. Thus, the student of weed science should approach the science with the expectation that prediction of changes in weeds and the design of programs for dealing with such changes will be an integral part of weed science in the future.

First, to develop an understanding of changes that production practices cause in weeds, it is desirable to review the three broad ways in which weeds change. In terms of weeds as a problem with which the farmer must deal, changes can occur as a result of: (1) introduction; (2) genetic and physiological modification of existing species; and (3) shifts within the community of weeds in response to changes in the environment imposed by production practices. Humans are an integral force with respect to each of the ways weeds may change.

WEED INTRODUCTION AND SPREAD

Examination of the ways in which weeds are introduced and spread clearly shows the extent to which humans are involved. Possibly two-thirds of the problem weeds in the United States are species introduced from other countries. Europe has been a major contributor, coincidental with being the major source of early settlers. Table 14-1 shows the source by plant family of weeds introduced into North America from Europe. These families are recognized as being of relatively recent evolutionary origin. Is this not what we would expect based upon what we know about response to change— that is, for natural selection to result in species and populations adjusted to the new environment? A second speculation may be in order. Does the preponderance of weeds among families of relatively recent evolutionary origin suggest that the long-term trend is toward increasingly troublesome weeds?

Asia has also contributed some weeds to the United States. Japanese honeysuckle (*Lonicera japonica*) and Japanese knotweed (*Polygonum cuspidatum*) are two examples. They were introduced as ornamentals and later escaped to become pests.

Introduction has not been a one-way street, however. Even though the United

TABLE 14-1. **Seven families whose species comprise 60% of the 700 species of weeds introduced into eastern North America from Europe, with the number of such species and a representative species in each.**

Family	Number of Species	Representative Species
Compositae	112	Common ragweed *(Ambrosia artemisiifolia)*
Gramineae	65	Quackgrass *(Elytrigia repens)*
Cruciferae	62	Wild mustard *(Brassica kaber)*
Labiatae	60	Henbit *(Lamium amplexicaule)*
Leguminosae	54	Sicklepod *(Cassia obtusifolia)*
Curyophyllaceae	37	Common chickweed *(Stellaria media)*
Scrophulariaceae	30	Common mullein *(Verbascum thapsus)*

Source: Data from Fogg, 1966.

States has contributed far fewer species to other nations than it has received, horseweed, tumbling pigweed (*Amaranthus albus*), and sunflower are three species apparently native to the United States that have been introduced to become problem weeds in Europe.

Weeds were introduced in a number of ways. The most important were as contaminants in crop seeds, with feed for livestock, in nursery stock, and in the days of colonization with sailing vessels as contaminants of soil used for ballast. The ballast was dumped at any convenient port to make room for the return cargo.

Aliens

Why are so many U.S. weeds aliens? The answer is provided by ecology. First, clearing of the land on the new continent provided environments ecologically comparable to those of the Old World where weeds had been provided an opportunity to become established and to evolve. Second, the new continent provided an environment free of some of the constraints imposed by that of the Old World. In effect, this fact recognizes a basic concept discussed in connection with biological control. The search for biological control agents has been concentrated in the country from which the weed species originated. This is so because the weed in its new environment can be expected to be relatively free of natural enemies, as was learned in Chapter 9. Third, some aliens were successful weeds in a new environment simply because there were unfilled niches that the alien was able to fill.

This review of introduction leads to the conclusion that it will not be a major source of change in weeds in the future, partly because ample time and enough opportunities have been provided for most species to have been introduced, and partly because none of the vehicles for introduction are significant factors today. This does not mean that introduction can be ignored, only that the flush of introduction as a major source of change is past. Now, the appearance of each new species is noteworthy. For example, witchweed (*Striga lutea*), a native of tropical and subtropical regions in the Eastern Hemisphere that was discovered in isolated areas in North Carolina and South Carolina in 1956, has been the subject of concerted efforts at eradication. Itchgrass, a native of India found in Louisiana and Florida in the early 1970s, has been the subject of study to determine its potential spread. By 1981, common crupina (*Crupina vulgaris*), a native of the Mediterranean region, had a known infestation of 23,000 acres in Idaho and is now the subject of special containment efforts. However, these additions as a source of change are relatively much less important than changes among species already present.

Local Spread

The following is a list of mechanisms for weed seed dispersal over relatively short distances:

Crop seed
Livestock feed
Birds and other animals
Machinery
Wind
Surface water
Crop and livestock waste

Before briefly examining each of these mechanisms, we need to remind ourselves that the movement of weeds was mainly from east to west with the pioneers, so much so, that the Indian name for plantain was White Man's Foot. We should also keep in mind that movement within the United States is important largely as a source of reinfestation rather than as a way of establishing a species previously not a problem. Our present spread, however, is extremely significant for any long-term weed management effort. Finally, we need also recognize that spread by birds and animals and by wind—relatively unimportant in total numbers—are the only mechanisms of the seven listed over which we have little or no control.

Crop seed. Infested seed is by far the most important way in which weeds were spread in the past and is still a major source today. The potential is shown in results reported by Dunham (1972) of a survey of weed seeds in farmer's drills in Manitoba, Canada. In the survey, 28% of the farmers were planting seed too contaminated with weeds to pass inspection. Contamination of crop seed is the only mechanism of movement, with the possible exception of movement in irrigation water, that can introduce enough weed seed to cause appreciable economic loss in the crop in the planting year. With the availability of certified seed under the state seed improvement programs, there is little reason now for crop seed to be a major source of weed infestation.

Livestock feed. Since feed is not subjected to the same cleanup and regulation as is seed, it can be an important source of weed spread. Feed, even hay, commonly may be transported many miles from the point of production to where it is fed to livestock. It has been well established that seeds of many weeds can survive passage through the animal gut. Thus, weed seeds in feed can be expected to add to the seedbank when and wherever the manure is used.

Birds and other animals. Birds and small animals may be a source of spread as a result both of seed ingestion and subsequent release in droppings and of seed attachment to feathers or hair. The spread of St. Johnswort seed in the western United States in the early part of this century was associated with the movement of cattle and sheep in the area. Similarly, the spread of johnsongrass was apparently associated with the movement of horses and their feed during the Civil War.

Machinery. Machinery can be a particularly important source of spread over fairly short distances. As machinery moves from field to field, and in some cases even from farm to farm, weed seed, rhizomes, and so forth, may be moved with it in mud on the wheels or in trash attached to cultivator teeth, left in the combine, and in other such ways. Although no data are available to identify the effect, it seems likely that many weeds are moved over relatively great distances with the advent of custom combining that may involve movement of machines with grain harvest from the Gulf of Mexico into the Canadian provinces in North America.

Wind.   Wind is among the more obvious vehicles for weed seed dissemination. Dandelion seed with its parachute-like transport system is commonly observed to be moving with the wind. Many seeds like dandelion have special structures to assure dissemination of their seed (Figure 4-2).

Surface water.   Weed seeds may also move in significant quantities in surface water. Table 14-2 shows the number of seeds of selected weeds found in water from the Columbia River and in an irrigation lateral during the irrigation season. A total of 137 different species were represented in weed seed in the irrigation lateral and 77 plant species in water from the Columbia River. The fact that more seeds are found in the irrigation lateral than in the Columbia River, which is the water source, is a reflection of the irrigation bank as a source of weed seed.

Crop and livestock waste.   Crop screenings and livestock waste may be the vehicle for spreading large numbers of weed seeds. In the handling of sugarbeets, for example, dirt attached to beets delivered to a processing plant is normally returned to the farm. However, the dirt returned is not necessarily that associated with beets from that par-

TABLE 14-2.   **Kind and number of weed seeds in water from the Columbia River and an irrigation lateral.**

Weed	Seeds per 254 Kiloliters Water[1]		Percent Germination[2]
	Columbia River	Irrigation Lateral	
Barnyardgrass	3.53	72.87	19
Cattail	0.12	8.87	34
Cutgrass, rice	0.69	8.99	9
Dandelion	0.12	1.24	65
Dock, curly	2.40	6.08	37
Dropseed, sand	0.79	6.89	66
Flixweed	1.01	22.66	52
Foxtail, green	0.69	7.56	2
Foxtail, yellow	0.12	4.97	52
Horseweed	3.76	15.20	80
Lambsquarters	21.72	307.24	5
Lettuce, prickly	8.18	16.83	57
Mustard, tumble	36.40	65.46	70
Pigweed, redroot	6.82	66.10	4
Quackgrass	0.00	10.97	80

[1]254 kiloliters is equivalent to an irrigation of 2.47 inches per acre.
[2]Average of all sources.
Source: Adapted from Kelley and Bruns, 1975.

ticular farm. Centralized feed lots with local distribution of the associated manure also represent a potential source of weed spread.

Thus, we see that introductions from long distances likely will occur only infrequently, but local introductions occasionally may be important sources of change in weed composition today. The principles involved in dispersal are beyond the scope of our interest here. Refer to van der Pijl (1982) for a thorough treatment of such principles.

GENETIC AND PHYSIOLOGICAL MODIFICATION OF WEEDS

In our agricultural endeavors, we are continually providing forces necessary for speciation to occur. Resulting genetic and physiological alteration of existing species, although important, is usually a comparatively long-term process, as discussed in Chapter 3. An exception is the development of herbicide-resistant weed populations mentioned in Chapter 1. Strictly speaking, this is the buildup of resistant biotypes through natural selection. Herbicide-resistant populations have now been identified in about 75 weed species (LeBaron and McFarland, 1990). Time necessary for development of a resistant population is influenced by a number of aspects of herbicide use. Practices that reduce amount of herbicide applied and selection pressure slow development of resistance. Resistant populations are usually developed over a period of years but can develop quickly. Studies showing wind-blown pollen may travel more than 150 m indicate that a herbicide-resistant ecotype could spread quickly (Mulugeta et al., 1994). A population of rigid ryegrass (*Lolium rigidum*) in Australia resistant to the herbicide diclofop developed after about 4 years (Heap and Knight, 1982). Clearly, the potential for development of herbicide resistance needs to be considered in formulating systems of weed management.

EFFECTS OF CROP PRODUCTION PRACTICES ON WEED CHANGES

In terms of its implication for weed management, the most important source of change is the shifts in composition in response to production practices, including weed control practices. Like speciation, these ecologically caused changes are also occurring continuously. However, the time frame within which an effect can be manifested ordinarily will be much shorter—as soon as 3 to 5 years—than for speciation. Even so, the effects are subtle and inconstant from year to year because of the interactions with environmental factors that in themselves are not constant. In this regard, remember that *each new production or managerial practice ultimately has its own complement of weeds*. Unless this fact is recognized and steps taken to deal with it, new production practices may fail to attain the potential envisioned for them. To a degree, this experience has occurred with minimum tillage. The merits of this production practice have been documented for a variety of crops, soils, and geographic locations. However, evidence now suggests that the weed problems are different and may be more difficult to deal with than under conventional tillage. The weed problems with minimum tillage serve to remind us that it is essential to have an understanding of the shifts that may be expected and why they occur with specific changes in crop production practices.

Production practices may be divided into three broad types: (1) cropping practices, (2) tillage practices, and (3) herbicide practices. Each is examined separately, although a change in one is usually associated with a change in one or both of the other practices. For example, the move to monoculture corn in the Corn Belt occurred in part because herbicides were available to effectively control weeds. The availability of effective herbicides was also largely responsible for a shift to reduced or minimum tillage in the production of several crops. Examination of each of these categories develops an understanding of the underlying forces involved.

Cropping Practices

As discussed in Chapter 2, crop species and varieties differ in their competitiveness toward weeds. Thus, we would expect weed populations to change with changes in crops and varieties produced. Some of these specific ramifications are explored in conjunction with an examination of crop sequence and monoculture versus rotation.

Crop variety.   Most studies of the competitiveness of varieties have been concerned with crop yield rather than with effects on weeds and have been relatively short-term. Such studies provide only circumstantial evidence that variety influences weed composition over time. More importantly, since varieties are continually changing, it is difficult and somewhat academic to trace such effects. To the extent that changes in varieties represent a general change in growth form, the effect should be traceable. Over the years, plant breeders' objectives have changed, leading to the development of new varieties quite different in growth form from their predecessors. For example, a general shift toward shorter- and stiffer-strawed varieties of wheat leads us to expect a change in the makeup of the weed community associated with wheat production. Indeed, short-strawed cultivars had about 4 times as many annual grass weeds just before wheat harvest as did taller cultivars (Table 14-3). Not only do shorter cultivars provide less competition for such weeds but they also produce less residue than taller ones (Ramsel and Wicks, 1988). Under conservation tillage, less residue could result in more weeds in the following crop.

TABLE 14-3. Density of annual grass weeds as affected by wheat cultivar height, North Platte, Nebraska.

Cultivar Height	Summer Annual Grass Density[a] (no. m^{-2})
Medium tall and medium	37
Medium short and short	164

[a]Barnyardgrass, green foxtail, stinkgrass (*Eragrostis cilianensis*), longspine sandbur (*Cenchrus longispinus*), and witchgrass (*Panicum capillare*).

Source: Data from Wicks et al., 1994b.

The effect on weediness of a change in crop growth form has been demonstrated in potato production on Long Island. A switch from the Green Mountain variety to the Katahdin variety was followed by a substantial increase in yellow nutsedge (Sweet, 1976). As discussed in Chapter 5, yellow nutsedge is relatively sensitive to shading. In mid-July, the Katahdin variety intercepted only 25% of the incident light compared to 65% interception by Green Mountain. By the end of August, the interception percentage was 20% and 58%, respectively. Thus, the Katahdin variety provides less competition for light than the Green Mountain variety. Since this area of Long Island has been in monoculture potato production for many years, all other factors could be ruled out as explanations for the insurgence of yellow nutsedge.

Previous crop. The sequence in which crops are grown in a rotation influences the makeup of the weed community over time. The crop that preceded sugarbeets in studies in Colorado had a marked effect on weed numbers and community composition in the beets (Dotzenko et al., 1969). The number of all weeds except lambsquarters was highest where corn was the preceding crop and least where beans was the preceding crop, with barley in between these two (Figure 14-1). The numbers are those that germinated in a 400 g sample of soil after 3 years of each sequence. Lambsquarters was highest following barley. The explanation is found in the opportunity each preceding crop offers weeds to become established. Barley is the earliest planted in the spring and is seeded before soil temperatures are ideal for germination and growth of all species except lambsquarters. Thus, the total number of weeds is somewhat restricted, but lambsquarters is maximal. Corn is planted in mid-April when temperatures are more optimum for weed germination and growth; as a result, weed numbers are high. Beans are planted in early June after the flush of germination of many summer annual weeds. Seedbed preparation thus destroys many of these weeds. In fact, the effect of the preceding crop often is likely that of time of seedbed preparation rather than of the crop itself. Whatever the explanation, information on weed associations with specific crops allows choice of crop sequence to encourage presence of weed species most easily managed in succeeding crops.

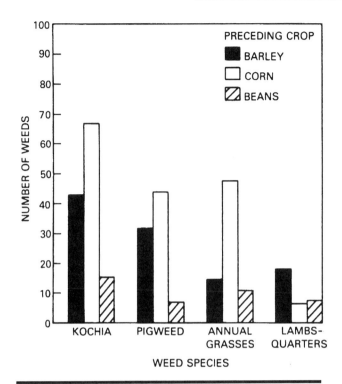

FIGURE 14-1. Effect of preceding crop on weed numbers and community composition. Weed numbers are those that germinated in a 400 g sample of soil following 3 years of each sequence.

Source: Data from Dotzenko et al., 1969.

Monoculture and crop rotation. The effects of a preceding crop are accentuated if continued for several years or for several complete rotations. This fact is shown in Figure 14-2 for corn, wheat, and soybeans after 6 years grown in monoculture and in rotation in Illinois. Five of the more than 30 species identified are included in the figure. The area was in alfalfa for several years prior to establishing this study. Standard cultivation—but no herbicides—was used throughout the 6-year period. Thus, the results represent the accumulative effect of crops grown. Six years provided for two complete cycles of the multicrop sequences. In just 6 years, clear-cut and substantial differences in weed composition were apparent among the crop sequences. Velvetleaf and giant foxtail, which overtop soybeans, were favored by the continuous soybean rotation. Wild buckwheat (*Polygonum convolvulus*), which germinates in the fall, was favored by continuous wheat, and even 1 year in 3 planted to wheat was sufficient for the weed to increase, compared with crop sequences that did not include wheat. Crabgrass was clearly discouraged by tillage since it decreased in all crop sequences that involved tillage every year. This conclusion is also supported by the fact that it, along with fox-

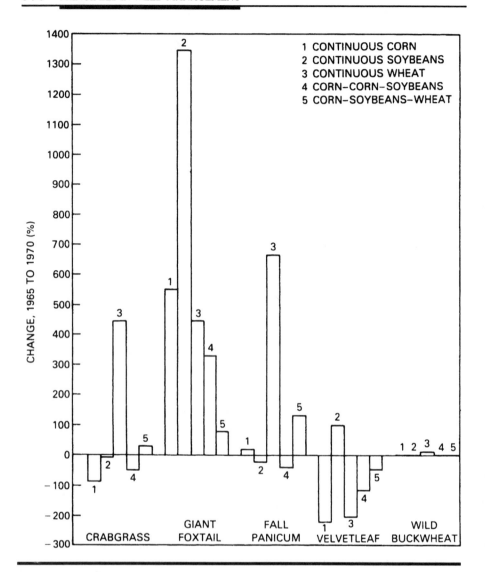

FIGURE 14-2. Effect of monoculture and crop rotation on weed seed composition of soil.
Source: Data from MacHoughton, 1973.

tail and panicum, increased substantially in continuous wheat. Crop rotation may also influence weed density in dryland crop production (Table 14-4). In just 5 years, downy brome numbers were more than 10 times greater in continuous wheat than in a wheat–canola rotation in Colorado.

Note also in Figure 14-2 that the instances in which the number of weed seeds increased is twice that for instances when the number of weed seeds decreased. In other

TABLE 14-4. Rotation affects downy brome in winter wheat in Colorado.

| Year | Downy Brome Plants | |
	Continuous Wheat	Wheat–Canola
	(no. m^{-2})	
1988	23	23
1993	800	60

Source: Data from Blackshaw, 1994.

words, the shift from alfalfa to annual crop production was followed by a general increase in weed seed numbers. This result is one more example of the fact that disturbed conditions increase the opportunities for weeds to increase.

The degree of change indicates the relative ecological stability of the individual cropping sequences. If we ignore the minus and plus signs and simply add the numbers for percent change in seven species that underwent major shifts (crabgrass, foxtail, panicum, pigweed, smartweed, velvetleaf, and wild buckwheat) in MacHoughton's (1973) study, the sum provides a measure of ecological stability. The resulting sums are 773, 1365, 869, 443, and 179 for continuous corn, continuous soybeans, continuous wheat, corn–corn–soybeans, and corn–soybeans–wheat, respectively. In effect, each sum provides for the cropping system an index of the degree of change in weed numbers and rate. Greater changes occurred under monoculture than with a rotation. The explanation lies in the density relationships associated with disturbed environments, covered in Chapter 3. As discussed there, simple (annual row-crop) agroecosystems tend to have violent fluctuations in weed numbers and to have fewer species over time than undisturbed environments. In effect, the higher degree of change with monoculture than with the rotation simply reflects the fact that monoculture provided the maximum opportunity for the best-suited species to increase.

Intensity of crop production. Intensity of the production enterprise may also affect weed composition. Figure 14-3 shows the effect on composition of seeds of selected weeds following a change from agronomic crop agriculture to vegetable crop agriculture in England. Prior to the initiation of the vegetable cropping study in 1953, the land had been in agronomic crops predominated by cereals for many years. The more intensive management associated with vegetable crops was reflected in an 87% reduction in weed seeds after 9 years. As can be seen, species varied greatly in their contribution to the overall reduction. Some actually increased.

Cropping history. The shifts in weed populations identified in these examples are largely explained by the combined effect of tillage and competitiveness of the crops toward specific weeds. These examples serve as a reminder that at any point in time, the weed problem is a consequence of the cropping history of the land. The corollary is that reversion to the former agriculture and crop sequence can be expected to see a return of those weeds associated with the former crops. In other words, if the fields in vegetable crops in England were reverted to cereal crops in 1963, the weeds associated

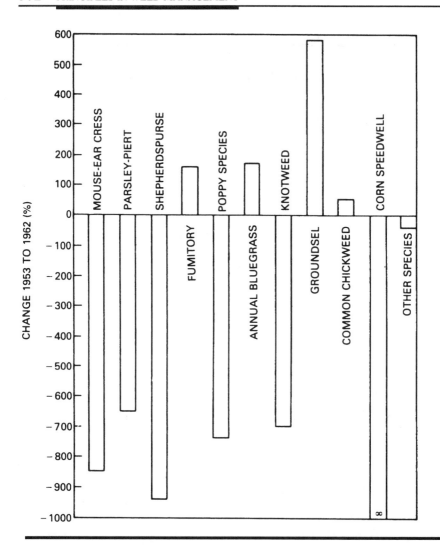

FIGURE 14-3. Changes in weed seed composition in soil under vegetable cropping begun in 1953.

Source: Data from Roberts and Stokes, 1965.

with cereals in 1953 would soon become the predominant ones again. Data such as these, when accumulated for sufficient locations and crop sequences, should make possible reasonably accurate predications of weed problems for specific crop sequences in response to changes in them.

Crop population. It is well known that density of the crop, as determined by seeding rate, is a factor in competitiveness toward weeds. For example, research with soybeans (Staniforth and Weber, 1956), cereals (Godel, 1935), and flax (Gruenhagen and Nalewaja, 1969), showed the depressing effects of increasing crop density on weeds. However, the research did not examine composition of the weed community.

Moss and Hartwig (1980) did show lambsquarters to be reduced more than other weeds by competition from corn and soybeans interseeded in the row. In their study, as the combined stand of corn and soybeans increased from 54,340 plants to 113,620 plants per hectare, the kilograms of dry matter of lambsquarters decreased from 2,211 to 760. Dry matter of other weeds increased from 376 kg to 553 kg, respectively. Marx and Hagedorn (1961), studying the effects on weeds of pea spacing in the row, observed a differential effect among the weed species. As spacing was increased from 3.6 cm to 8.7 cm, green foxtail increased 4.4 times, while the broadleaf species increased only 3 times. Thus, some evidence exists that crop seeding rate influences weed composition, but the data are insufficient for predictive purposes. The overall suppressing effect of increased crop density on weeds is quite clear-cut, however.

Crop pattern and spacing. Plant spacing, of course, is affected by seeding rate. However, the effect of the *pattern* of crop spacing is an issue distinct from seeding rate. That is, the same seeding rate per land area can be obtained by either an equidistant planting pattern or by planting in rows. For example, a desired soybean stand of 500,000 plants per hectare can be obtained by spacing plants 20 cm apart in all directions or by spacing plants 4 cm apart in rows 100 cm apart. Theoretical analysis has shown that crop plants in an equidistant planting pattern are about twice as competitive toward weeds as in a rectangular planting pattern in which distance between rows is 3.5 times distance within the row; assuming simultaneous emergence and equal numbers of weed and crop plants (Fischer and Miles, 1973). The greater competitiveness for light of the solid-planted (narrow-row) soybeans over those planted in wide rows was pointed out in Chapter 7. Because of the effect of row width on competitiveness, an effect on weed composition is to be expected. Spacing of corn plants has changed greatly over the years. Corn populations roughly doubled following introduction of hybrid corn (Aldrich et al., 1986). Additionally, row width has narrowed; in Illinois 40% of the corn was grown in 76 cm (30 in.) rows in 1984 compared to only 17% in 1970. These changes in spacing have no doubt contributed to changes in weed composition known to have occurred over the past 40 years, but the effects can not be separated from those of tillage and herbicides that have also changed during this period.

Although studies of differential effects on weeds are limited, those that have been done indicate that weed species are affected. Broadleaf weeds tended to be decreased and grassy weeds increased as row spacing of soybeans increased from 25.4 cm to 101.6 cm, as can be seen in Figure 14-4. No herbicide was used but the plots were cultivated. The weeds involved were giant foxtail, smooth pigweed, crabgrass, and prickly sida, plus a light infestation of velvetleaf. Although there is some hazard in extrapolating from dry matter production to seed production because of compensation by individual plants, it is nevertheless realistic to predict higher seed production by the taller-growing broadleaf annual weeds, thus leading to their increase where soybeans are planted in narrow rows over a period of years. If the growth form of the weed, especially the height attained, is known for each weed in the community of weeds associ-

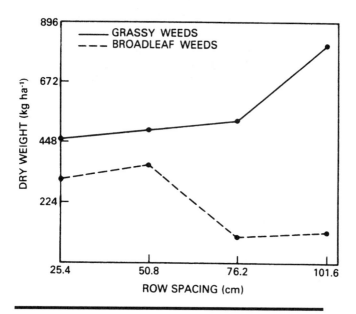

FIGURE 14-4. Different effects of crop row spacing on weeds.

Source: Data from Wax and Pendleton, 1968.

ated with a particular crop, it would seem that fairly accurate predictions of the shift in community makeup in response to plant spacing can be made for other crops. This supposition assumes that light is the growth factor most usually competed for, particularly during the early part of the growing season. This is not to say that the weed problem is expected to be the same for a given row spacing in a given crop wherever that crop is grown since many other factors also affect shifts in composition of the weed community. What this statement does say, however, is that when the makeup of a weed community on a given area (field) is known at the time a shift in row spacing is planned, the shift in weed composition should be predictable.

Soil fertility. Fertility level is expected to affect composition of the weed community because of differences in competitive ability among weeds. As stated in Chapter 3, early colonizers in ecological succession are those species whose survival strategy depends upon large numbers of seed. Later stages are represented by species of increasing competitive ability. Thus, we would expect that the most competitive species would be favored over the less competitive species by high fertility.

Research in pasture crops provide good examples that this expectation is the case. Studies of perennial ryegrass longevity in England (Smith and Allcock, 1978) showed that the relatively less aggressive grasses [bentgrass (*Agrostis tenuis*) and *Poa trivialis*] invaded over time under low nitrogen (188 kg ha[-1]), while the more aggressive quackgrass was the most prevalent under high nitrogen (376 and 752 kg ha⁻¹). Peters and Lowance (1974) found that broomsedge (*Andropogon virginicus*), a frequent species in poor, rundown pastures, could be eliminated from permanent pastures in Missouri in 4 years by drilling in fescue, fertilizing with nitrogen, phosphorus, and potassium, and mowing each winter. Just fertilizing the bluegrass sod gradually reduced the broomsedge after 5 years.

Weeds often accumulate more mineral nutrients than crops, thereby gaining a competitive advantage from applied fertilizer. It follows that fertilizer practices could influence the competitive relationship for nutrients between crops and weeds. Several fertilization practices have been shown to affect this relationship. It can only be speculated that practices that favor mineral uptake by the crop will alter weed composition over time. Clearly, however, these are practices that can be utilized as part of an effective total weed management program. This will be considered in the next chapter where we examine systems of weed management. Here we need only identify the crop-favoring practices.

Banding. In a review of fertilization practices (DiTomaso, 1995), it was pointed out that banding within the crop row lowered weed populations compared to broadcast applications and increased crop yield of beans, soybeans, peanuts, wheat, alfalfa, bromegrass, and littleseed canarygrass. Placing fertilizer phosphorus in a band 7 cm below bean seeds in the row resulted in about 40% less weed biomass than a surface band or broadcast application (Otabbong et al., 1991). Deep placement of fertilizer in rice had a similar effect on the crop–weed relationship (Moody, 1981). Placing the fertilizer below the crop seed presumably allows the crop seedling readier access to the nutrients than to the shallow-germinating weed seedlings. Recall from Chapter 6 that a majority of annual weeds originate from seeds in the surface 2–5 cm of soil. Although deep banding may favor the crop over shallow-germinating annual weeds, bear in mind that over time such a practice may lead to an increase in deeper germinating, harder to kill weeds.

Nitrogen formulation and nitrification inhibitors. There is evidence that plants differ in response to nitrogen form (Teyker et al., 1991). Corn responded equally to nitrogen applied as nitrate or as ammonium plus an experimental nitrification inhibitor. However, pigweed shoot dry weight was 75% less with the ammonium plus inhibitor treatment than with nitrate. Here, too, it can be speculated that choosing a nitrogen formulation to favor the crop over specific weeds in time could change the weed community.

Timing fertilizer application. Applying nitrogen during the fallow period in a winter wheat–fallow system lessened downy brome interference with wheat compared with application at planting (Anderson, 1991). Application during fallow allowed the N to move deep enough into the soil profile to be relatively inaccessible to the shallow-rooted downy brome.

Duration of crop cover.   Intercropping and relay cropping, which are forms of *multiple cropping*—that is, the growing of two or more crops on the same field in one year—have been practiced for centuries, mainly in tropical regions with high rainfall. Pressures for more food and to reduce capital investment in food production have pushed science to develop technology to extend multiple cropping to other areas. In the United States, growing a soybean crop in the year wheat is harvested has become a common practice on many farms south of the southern Corn Belt, and the practice is slowly moving northward. This practice has obvious implications for the makeup of the weed community since weeds are provided relatively less time to grow in the absence of competition from crops. In effect, the second crop is filling the niche that would otherwise be filled with one or more weed species. Those species that commonly germinate after wheat harvest can be expected to be discouraged in favor of winter annuals or perennial species that commonly germinate or make considerable growth in the fall following soybean harvest. Data are not yet available to identify the precise effects.

Planting cover crops to suppress weeds by filling the void in crop cover is a type of multiple cropping with potential for long-term weed management. Although this idea is expanded upon in Chapter 15, we can appreciate here that such a cropping practice would be expected to lead to a change in the weed community.

Tillage Practices

Shifts in weed composition in response to tillage practices are a result of differences in one or more of three factors: (1) the severing and segmenting of perennating parts of perennial weeds; (2) the distributing of weed seeds and perennating parts in the soil profile; and (3) the mixing of plant residues in the soil. All of these effects can serve to differentiate among weed species because of different requirements and opportunities offered for germination of seeds and seedling establishment and for regrowth of perennating parts. Changes in tillage practices embodied in the move to conservation tillage could be expected to differentially affect the three factors listed above. Of course, specific species' response to these effects could be expected to vary depending upon a number of other factors (climate, cropping practices, and herbicide practices). However, some general responses have been observed.

Perennial weeds.   Some perennial weeds increase under reduced tillage. Conventional tillage severs and breaks up perennating parts and, depending upon the secondary tillage used for seedbed preparation, may bring some of the parts to the surface where they are subject to desiccation and low temperature. We learned in Chapter 5 that viability of perennating parts is reduced by such conditions. We also learned that there is a time after separation from the parent plant when the perennating part is destroyed by removal of its topgrowth. Furthermore, segmenting rhizomes and stolons of some perennials may lessen correlative inhibition thereby allowing regrowth of more perennating parts that can be removed by further tillage. Follow-up tillage to moldboard plowing could be expected to destroy some perennating parts in this way. There is much evidence, in fact, that periodic removal of new growth for 1 or 2 years will eradicate some perennial weeds from a given area. All or some of those suppressing effects of conventional tillage are lost under reduced tillage. Some perennial weeds observed to increase under reduced tillage are listed in Table 14-5.

TABLE 14-5.   Some perennial weeds that increase following several years of reduced tillage in different regions in the United States.

Regions	Weed Species
North Dakota (no-till)	Canada thistle Perennial sowthistle *(Sonchus arvensis)*
Illinois, Indiana (no-till)	Johnsongrass in southern portions Quackgrass in northern portions Milkweeds Hemp dogbane *(Apocynum cannabinum)* Bindweed *(Convolvulus arvensis)*
Oklahoma (conservation tillage)	Horsenettle *(Solanum carolinense)* Bindweed Hemp Dogbane Milkweeds
Tennessee, North Carolina, Mississippi, Georgia (conservation tillage)	Trumpet creeper *(Campsis radicans)* Milkweeds Horsenettle Nutsedges Bermudagrass *(Cynodon dactylon)* Johnsongrass

Source: Data from Koskinen and McWhorter, 1986.

Annual Weeds.   Shallow and early-germinating annual broadleaf and grass species may increase under reduced tillage. This is due in part to concentration of seeds at or near the soil surface. Plowing tends to distribute the seeds more or less uniformly throughout the soil profile (Figure 14-5). Species such as giant foxtail, crabgrass, and redroot pigweed that germinate best at or very near the soil surface have relatively more of their seeds in a position favorable to germination than is true if tillage distributes them throughout the soil profile. Thus, reduced tillage provides them an opportunity to gain in numbers over species that germinate from greater depths, such as wild oat, cocklebur, the annual morning-glories, and many others. Early-germinating species are favored by no-tillage because some of them will already have germinated by the time the crop is planted, thus giving them a competitive edge over the later germinators.

It was pointed out in Chapter 3 that temperatures are lower and soil moisture higher under residues than under bare soil. Soils can be 3°C to 4°C colder at the seed level under conservation than under conventional tillage (Gebhardt et al., 1985). Tillage type affects the amount of residue on the soil surface. Moldboard plowing leaves essentially no residues on the surface but chisel-plowing leaves 75% to 80% (Aldrich et al., 1986). Because of different seed germination optima among weed species, these environmental effects of reduced tillage also contribute to observed shifts in species.

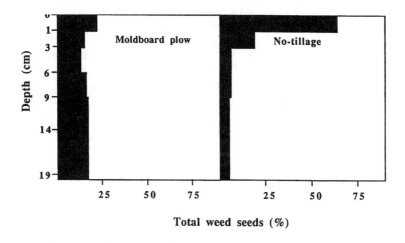

FIGURE 14-5. Effect of tillage on distribution of weed seeds in soil.
Source: Data from Yenish et al., 1992.

The differential effects on germination and establishment of downy brome and tumble mustard (*Sisymbrium altissimum*) in Nevada rangeland, shown in Table 14-6, indicate that trash can indeed have a pronounced effect upon weed species. The downy brome seed is large but light with large awns. The better moisture environment provided by litter offers more safe sites for its germination than does bare soil. The tumbling mustard has a small, smooth, dense seed with a seed coat that surrounds itself with mucilage when wetted. These seeds find proportionately more safe sites on a smooth, bare surface than on one even partially covered with litter.

Time of seedbed preparation.   In view of what we learned in Chapter 6 about periodicity of weed seed germination, when the seedbed is prepared could be expected to affect composition of the weed community. A study of the effect of tillage time on wild oat numbers, shown in Figure 14-6, indicates that indeed it does. Wild oat numbers increased about 400 times under 6 years of early tillage and decreased at about the same rate under later tillage. Of course, the effect of a preceding crop discussed earlier is often a reflection of tillage time.

TABLE 14-6.   Frequency of downy brome and tumble mustard the second year after planting as affected by surface litter.

Species	Characteristics of Soil Surface	Observed Frequency
Downy brome	Litter-covered	48
	Smooth	6
Tumble mustard	Litter-covered	0
	Smooth	90

Source: Adapted from Evans and Young, 1970. Frequency was determined at 100 random points in 5 transects of each plot.

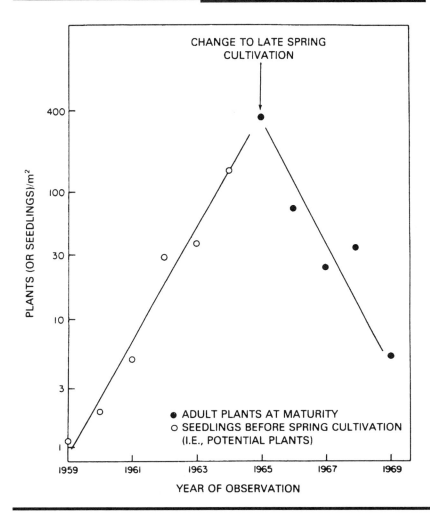

FIGURE 14-6. Effect of tillage time on weed composition. Results shown are for wild oat plants in continuous spring barley. During the period 1959-1964, cultivation was done in early spring; during the period 1965-1969, cultivation and sowing were done late.

Source: Harper, 1977. Reproduced courtesy J.L. Harper. © J.L. Harper.

Herbicide Practices

Our interest here is on changes in the weed community brought about by specific herbicide practices. As we examine these effects, we should keep in mind that the objective in herbicide use should be to minimize crop losses due to weeds and to reduce costs of weed management in the plant production system over time. For this goal to be achieved, changes in the weed community must be minimized or slowed. Any major change in the weed community likely will be accompanied by losses in production and increased costs of weed management. Changes in the weed community, of course, will necessitate a change in production or in weed management practices or both. There will be a period of time before a change in management when the new weed community takes a toll. Additionally, the new weed community commonly may be harder to control, thereby increasing costs. Too often herbicides have been used with the more restricted objective of controlling a specific weed or weeds in the current year. Exclusive reliance on herbicides to control weeds can create quite volatile conditions for weed composition. This is because environmental conditions that encourage a given weed community may be kept from exerting their influence as long as the herbicide is used and is effective. However, the conditions remain and can again exert their effects if there is a herbicide failure or a change in herbicides with a change in the production system.

Single herbicide. Early in the modern herbicide era, it was common practice to use a single herbicide year after year to control mostly susceptible weeds in a given plant production enterprise. Ideally, a herbicide should control all weeds equally well. Practically, however, most selective herbicides were somewhat more effective against some weeds than against others. Thus, continued use of a given herbicide could be expected to result in a shift in weed composition. Many such changes have been observed. For the most part, shifts occurring as a result of use of a single herbicide can be viewed as a response to competition within the community of weeds. It has been observed many times in many different situations that selective control of only some weeds in a community of species leads to a community dominated by the uncontrolled species. As already discussed, this shift occurred in corn in the United States when effective control of broadleaf species by 2,4-D led to a weed community dominated by grasses, and later, following use of the s-triazine herbicides, to a community dominated by late-germinating weeds, such as panicum. In asparagus, lambsquarters was observed to build up under the use of chloramben and diphenamid, and redroot pigweed to build up under DCPA (Welker and Brogdon, 1972). In orchards (Schubert, 1972), weeds of a few species increased following continued use of certain herbicides. In soybeans, trifluralin, widely used as a preemergence herbicide, was more effective against annual grasses than against annual broadleaf species (Wax and Pendelton, 1968). Where it was used for a period of years, there was an increase in broadleaf weeds. By the mid-1980s, broadleaf signalgrass (*Brachiaria platyphylla*) was a predominant annual grass problem in peanuts in North Carolina because general use of chloroacetamide herbicides effectively controlled fall panicum and large crabgrass but was marginally effective on the signalgrass (Johnson and Coble, 1986).

Herbicide combinations. Shifts in weed composition as a consequence of differential effectiveness of herbicides have led to the now common practice of combining two or more herbicides. MacHoughton (1973) examined the effects of a single herbicide and a rotation of herbicides on weed seed composition of soil in a number of different crop sequence systems common to the Corn Belt in the United States. Figure 14-7, which shows some of the results in the continuous corn production system, indicates

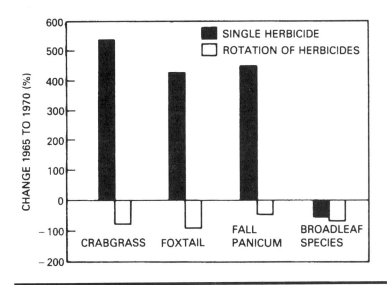

FIGURE 14-7. Weed seed composition in soil after 6-year treatment with a single herbicide or a rotation of herbicides in corn.

Source: Data from MacHoughton, 1973.

the extreme differences that can occur. In this case, the use of atrazine as the single herbicide for 6 years resulted in more than a fourfold increase in seeds of the three problem annual grasses. The rotation of herbicides resulted in a 48% to 95% reduction in seeds of these species in the soil. Broadleaf weeds collectively were reduced by both herbicide treatments but somewhat more by the rotation. Of course, the opposite can occur. If the single herbicide used is the most effective available chemical toward certain species, using other less-effective herbicides in some of the years would lead to an increase. The important point here is that rotation of herbicides will slow a shift in weed composition compared to use of a single herbicide.

Rotating crops of course may allow, or may necessitate, use of different herbicides in each crop. Data in Table 14-7 show the striking effect on buildup of selected peren-

TABLE 14-7.   Selected perennial weed populations after 14 years of cropping systems using applicable herbicides.

Rotation	Hemp Dogbane	American Germander	Field Bindweed
		(*no. 0.04 ha⁻¹*)	
Continuous corn[a]	110	0	0
Corn–soybeans[b]	4	1573	105

[a] Treated with alachlor plus atrazine each year.
[b] Corn treated with alachlor plus cyanazine; soybeans treated with alachlor plus metribuzine plus ropewick application of glyphosate to overtopping perennial weeds.
Source: Buhler et al., 1994.

nial weeds of using different herbicides in the two rotations. The results make two points relative to the effect of herbicide combinations on changes in the weed community: (1) Crop monoculture provides the best opportunity to reduce infestations of weeds susceptible to an effective herbicide or herbicide combination, and (2) crop rotation may provide the best opportunity to prevent buildup of a species not adequately controllable with available herbicides in one of the crops.

Development of herbicide resistance. A combination or a rotation of herbicides may slow development of resistant weed populations compared to repeated use of a single herbicide. Development of resistance will be slowed with the combination because the rate of application of each herbicide in the mixture commonly will be less than if only one herbicide is used. Lowering the rate reduces selection pressure thereby slowing the selection process. With a combination, herbicides are chosen for their effectiveness against certain weeds in the weed community. Commonly, one herbicide in a two-herbicide combination may be especially effective against grass weeds and the second especially effective against broadleaf weeds. In combination, each can be used at only the rate necessary to control its target specie(s), which will be a lower rate than would be needed to control all species with one herbicide or the other. A rotation of herbicides may slow development of resistance by reducing the years that selection pressure is applied by the particular herbicide. Reduced herbicide rates may also slow development of resistance compared to full rates by reducing selection pressure.

Crop residues. In considering effects of herbicide usage on weed composition over time, the effect of crop residues associated with conservation tillage must be borne in mind. Crop residues may intercept an appreciable part of the herbicide applied for weed control. For example, about 50% soil cover with wheat straw intercepted up to 50% of metolachlor applied for preemergence weed control in no-till corn (Table 14-8). Herbicides not lost by volatilization or degradation when intercepted by residue might be more effective against surface-germinating weed seeds by virtue of the fact that the residues retain them in the vicinity of the germinating seeds. In any case, trash normally associated with reduced tillage offers an additional source for shifts in weed populations as the result of its potential effect upon herbicide effectiveness.

TABLE 14-8. Interception of metolachlor by wheat straw mulch in no-till corn.

Soil covered by mulch (%)	Metolachlor Intercepted	
	Site #1 (%)	Site #2 (%)
0	0	0
58	14	50
77	55	67
92	83	88

Source: Data from Wicks et al., 1994a.

COMBINED EFFECTS OF CROPPING, TILLAGE, AND HERBICIDE PRACTICES

The three production practices—that is, cropping, tillage, and herbicides—that influence weed composition are commonly exerting their effect simultaneously, not independently. This fact is summarized schematically in Figure 14-8. Furthermore, aspects of the three production practices do not represent mutually independent decisions for the enterprise operator. That is, the specific levels of the several aspects listed under tillage are influenced by one or more of those aspects listed under cropping practices. Similarly, the choice of herbicide is influenced at least in part by the crop sequence and possibly by the variety chosen under cropping practices, and quite likely by timing and residues chosen under tillage practices.

The dominating effect of weed removal by herbicides is emphasized in Figure 14-8. In this connection, recall that removal of a dominant (the objective in most herbicide usage) is in itself a disruptive influence. That is, removal provides a new environment (niche) for other species.

Differential response of weeds to the herbicides used is viewed as an override on the effects of cropping and tillage practices. Although their effects may be masked by herbicides, the influence of cropping and tillage practices remain as latent forces that may express themselves with a change in the herbicide. Thus, although herbicides may be the dominant influence, weed composition over time is the result of the combined influences of all aspects of the production system.

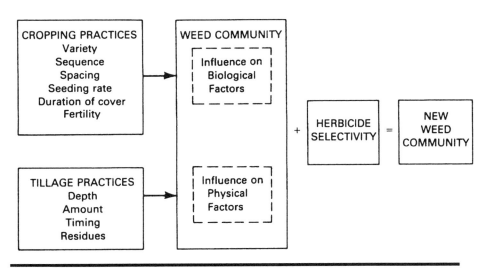

FIGURE 14-8. Schematic representation of simultaneous effects of the three types of production practices—cropping, tillage, and herbicides—on weed ecology.

REFERENCES

Aldrich, S.R., W.O. Scott, and R.G. Hoeft. 1986. Modern Corn Production. A and L Publications, Inc. Champaign, IL.

Anderson, R.L. 1991. Timing of nitrogen application affects downy brome (*Bromus tectorum*) growth in winter wheat. Weed Technol. 5:582-585.

Blackshaw, R.E. 1994. Rotation affects downy brome (*Bromus tectorum*) in winter wheat (*Triticum aestivum*). Weed Technol. 8:728-732.

Buhler, D.D., D.E. Stoltenburg, R.L. Becker, and J.L. Gunsolus. 1994. Perennial weed populations after 14 years of variable tillage and cropping practices. Weed Sci. 42:205-209.

DiTomaso, J.M. 1995. Approaches for improving crop competitiveness through manipulation of fertilization strategies. Weed Sci. 43:491-497.

Dotzenko, A.D., M. Ozkan, and K.R. Storer. 1969. Influence of crop sequence, nitrogen fertilizer, and herbicides on weed seed populations in sugar beet fields. Agron. J. 61:34-37.

Dunham, R.S. 1972. The weed story. St. Paul: Agricultural Extension Service, University of Minnesota.

Evans, R.A., and J.A. Young. 1970. Plant litter and establishment of alien annual weed species in rangeland communities. Weed Sci. 18:697-703.

Fischer, R.A., and R.E. Miles. 1973. The role of spatial pattern in the competition between crop plants and weeds. A theoretical analysis. Math. Biosci. 18:335-350.

Fogg, J.M., Jr. 1966. The silent travelers. Plants and Gardens 22:4-7.

Gebhardt, M.R., T.C. Daniel, E.E. Schweizer, and R.R. Allmaras. 1985. Conservation tillage. Science 230:625-630.

Godel, G.L. 1935. Relation between rate of seeding and yield of cereal crops in competition with weeds. Sci. Agr. 16:165-168.

Gruenhagen, R.D., and J.D. Nalewaja. 1969. Competition between flax and wild buckwheat. Weed Sci. 17:380-384.

Harper, J.L. 1977. Population biology of plants. New York: Academic Press.

Heap, I., and R. Knight. 1982. A population of ryegrass tolerant to the herbicide diclofop-methyl. J. Aust. Inst. Agric. Sci. 48:156-157.

Johnson, W.C. III, and H.D. Coble. 1986. Crop rotation and herbicide effects on the population dynamics of two annual grasses. Weed Sci. 34:452-456.

Kelley, A.D., and V.F. Bruns. 1975. Dissemination of weed seeds by irrigation water. Weed Sci. 23:486-493.

Koskinen, W.C., and C.G. McWhorter. 1986. Weed control in conservation tillage. J. Soil and Water Conserv. November-December, pp. 365-370.

LeBaron, H.M., and J. McFarland. 1990. Herbicide resistance in weeds and crops: overview and prognosis. In M.B. Green, W.K. Moberg and H.M. LeBaron, eds., Managing resistance to agrochemicals: From fundamental research to practical strategies, pp. 336-352. Am. Chem. Soc. Symp. Ser. #421.

MacHoughton, J. 1973. Ecological changes in weed populations as a result of crop rotations and herbicides, Ph.D. dissertation. University of Illinois, Urbana.

Marx, G.A., and D.J. Hagedorn. 1961. Plant population and weed growth relations in canning peas. Weeds 9:494-496.

Moody, K. 1981. Weed-fertilizer interactions in rice. International Rice Research Institute (IRRI), Int. Rice Paper Ser. No. 68, 35 pp.

Moss, P.A., and N.L. Hartwig. 1980. Competitive control of common lambsquarters in a corn–

soybean intercrop. In Proc. of NEWSS, vol. 34, pp. 21-28. Grossinger, NY.

Mulugeta, D., B.D. Maxwell, P.K. Fay, and W.E. Dyer. 1994. Kochia (*Kochia scoparia*) pollen dispersion, viability, and germination. Weed Sci. 42:548-552.

Otabbong, E., M.M.L. Izquierdo, S.F.T. Tolavera, U.H. Geber, and L.J.R. Ohlander. 1991. Response to P fertilizer of *Phaseolus vulgaris* L. growing with or without weeds in a highly P-fixing mollic andosol. Trop. Agric. 68:339-343.

Peters, E.J., and S.A. Lowance. 1974. Fertility and management treatments to control broomsedge in pastures. Weed Sci. 22:201-205.

Ramsel, R.E., and G.A. Wicks. 1988. Use of winter wheat (*Triticum aestivum*) cultivars and herbicides in aiding weed control in an ecofallow-corn (*Zea mays*) rotation. Weed Sci. 36:394-398.

Roberts, H.A., and F.G. Stokes. 1965. Studies on the weeds of vegetable crops, V. Final observations on an experiment with different primary cultivations. J. Appl. Ecol. 2:307-315.

Schubert, O.E. 1972. Plant cover changes following herbicide applications in orchards. Weed Sci. 20:124-127.

Smith, A., and P.J. Allcock. 1978. Hurley, England: Grassland Research Institute Annual Report, 1977.

Staniforth, D.W., and C.R. Weber. 1956. Effects of annual weeds on the growth and yield of soybeans. Agron. J. 48:467-471.

Sweet, R.D. 1976. When it comes to competing with weeds, some are more equal than others. Crops and Soils 28:7-9.

Teyker, R.H., H.D. Hoelzer, and R.A. Liebl. 1991. Maize and pigweed response to nitrogen supply and form. Plant Soil 135:287-292.

van der Pijl, L. 1982. Principles of dispersal in higher plants, 3rd ed. New York: Springer-Verlag.

Wax, L.M., and J.W. Pendleton. 1968. Effect of row spacing on weed control in soybeans. Weeds 16:462-265.

Welker, W.V., Jr., and J.L. Brogdon. 1972. Effects of continued use of herbicides in asparagus plantings. Weed Sci. 20:428-432.

Wicks, G.A., D.A. Crutchfield, and O.C. Burnside. 1994a. Influence of wheat (*Triticum aestivum*) straw mulch and metolachlor on corn (*Zea mays*) growth and yield. Weed Sci. 42:141-147.

Wicks, G.A., P.T Nordquist, G.E. Hanson, and J.W. Schmidt. 1994b. Influence of winter wheat cultivars on weed control in sorghum (*Sorghum bicolor*). Weed Sci. 42:27-34.

Yenish, J.P., J.D. Doll, and D.D. Buhler. 1992. Effects of tillage on vertical distribution and viability of weed seed in soil. Weed Sci. 40:429-433.

15

A Total Weed Management System

CONCEPTS TO BE UNDERSTOOD

1. Preventing production of propagules is fundamental to a maximally effective long-term weed management system.
2. Weed prevention and weed control have complementary parts in a total weed management system.
3. Machinery is available to facilitate plowing, planting, and cultivating for a wide range of weed management systems.
4. Mathematical models are available to help the enterprise operator determine cost effectiveness of weed management practices.
5. Utilizing what is known about ecological relationships, a weed management program can be planned for an enterprise unit that will minimize weed numbers and associated costs over time.

The underlying theme for our approach to weeds in the future must be the development of a long-term plan. Such a plan must have as its focus the production unit or possibly even the individual field. The latter might be particularly applicable for a monoculture cropping system. In either case, this approach to weeds must recognize the dynamic relationships that exist between weeds and desired plants at the individual unit level. These relationships can best be observed and interpreted by the owner/operator on the land. Contingencies that are bound to arise can best be dealt with by this individual.

Viewed in this way, our program for the future can be most effective if it is founded on the principles of weed prevention, including a reduction in weed propagules, a reduction in emergence of weeds with our desired plants, and a lessening of competition from those weeds that are present. Such a plan for dealing with weeds can then embrace all tools available, including what is known about competition, about the weeds themselves, allelopathy, biological agents, and, of course, herbicides. In such a context, weed control becomes a capstone for an effective total weed management effort. Control will be relied upon to provide the final measure of protection rather than the only measure, as is now frequently the case.

PREVENTION AS THE FOUNDATION

A weed management system founded on reduction of seed and perennating parts must be the ultimate goal to be maximally effective over time. This can be seen by considering the three times during a weed's life history when action may be taken to reduce the effect on desired plants: (1) while it is a seed in the soil; (2) when it is growing and competing with desired plants; and (3) when it is producing seed and perennating parts. There is a building relationship among these times, beginning with production of propagules, that can be represented schematically by a pyramid (Figure 15-1). At the bottom of the pyramid, success in preventing production of propagules with the associated reduction of their soil reservoir will relieve some of the pressure on the next two levels. In fact, it has been shown in Minnesota that weed seedbanks less than 1,000 m^{-2} in the top 10 cm will produce populations that can be controlled with tillage; seedbanks larger than this will require herbicides (Forcella et al., 1993). Furthermore, effective weed management for a succession of years can reduce seedbanks to levels below 1,000 m^{-2} in this depth (Schweizer and Zimdahl, 1984). Modest success in reducing propagules improves chances for success from action taken to prevent weed emergence. Success in reducing propagules and in preventing weed emergence lessens the need for, and improves the chances for success from, control by reducing both weed numbers and weed spectrum. For example, seeds of five of the most prevalent weed species present at the outset of the study cited above (Schweizer and Zimdahl, 1984) were not found after 6 years of effective weed control. Failure to reduce propagules still leaves preventing weed emergence and control of growing weeds as ways of avoiding weed interference. Failure to reduce propagules and to prevent weed emergence leaves only weed control as a way to avoid interference. Fewer opportunities are available to avoid interference the further up the pyramid one goes. Viewed in this way, prevention can be seen as the necessary foundation for a successful total weed management effort in order that all potential tools for dealing with weeds may be used.

ROLE OF WEED CONTROL

Weed control will continue to be an important part of a total weed management effort. However, it too needs to be seen as having an important place during each of the three stages of a weed's life cycle. That is, control of the weed plant can be accomplished to reduce propagules, weed emergence, and competition with the crop. Control will make

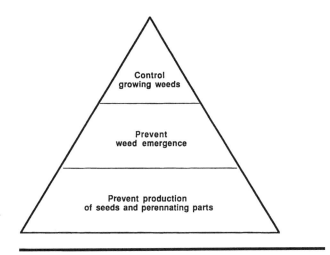

FIGURE 15-1. A three part total weed management system.

its greatest contribution to a total management effort when viewed in this expanded context.

COMPLEMENTARY RELATIONSHIP OF PREVENTION AND CONTROL

Prevention and control can best be seen as complementary parts of a total program, not as alternatives. Success with one does not eliminate the need for the other but can be expected to improve results from the other. This view of prevention and control also helps to show why weed management necessitates a commitment to an on-going effort. That is to say, we must accept that weeds will be a factor to be reckoned with, each year and indefinitely on into the future. We may reduce the numbers through effective management, but it is unrealistic to expect we will ever eliminate a problem weed. Acceptance of this fact provides the necessary setting for long-term planning of steps to be taken at each stage.

Of course, we must always remember that the objective of an effective weed management program is maximum, sustainable crop production. This objective simply recognizes that weed prevention and weed control by themselves have value only as they contribute to the purpose of the particular production enterprise. They are not ends in themselves. Thus, for any given contemplated weed prevention or weed control effort, a decision of what to be done needs to be made in terms of this broader perspective.

Similarly, the relationship between prevention and control indicates why the focus of a weed management approach must be the production unit. That is, since this approach embraces an on-going effort with interactions among approaches to weeds in the entire plant production system or crop rotation, it necessarily involves the individual fields and desired plants on an individual production unit (farm, plant nursery, golf course, and so on).

WEED MANAGEMENT DECISION MAKING

The operator of a production unit often makes three weed management decisions: what tillage to use, if any; whether or not to apply preplant or preemergence herbicides; and whether or not to apply postemergence herbicides. Accumulated knowledge about weeds and their relationship to desired plants has led to technological advances that can help the operator make the most productive decisions in each of these areas. Technological advances in two general areas have been especially important: (1) tillage and planting equipment and (2) bioeconomic weed management modeling.

New Equipment

Drills and planters. The development of machinery to facilitate reduced tillage demonstrates the adaptability and ingenuity of the farm machinery industry. Drills and planters have been especially designed for planting directly into sod and untilled ground. Some examples are shown in Figure 15-2. Attachments and special features like those shown in Figure 15-3 make precision planting possible even in heavy residues. Precision planting is accomplished by building into one machine (Figure 15-3A) the essential elements of the several machines commonly involved in preparing a complete seedbed. A single pass over the field results in a miniseedbed (Figure 15-3B). From what we have learned about weed seed survival and longevity and about reproduction in perennial weeds, we know that these practices have an influence on weed composition and numbers.

Plows. The chisel plow shown in Figure 15-4 is now widely used throughout agriculture to disrupt hardpans without turning under crop residues that serve to hold the soil against erosion. The traditional moldboard plow, as can be seen in Figure 15-5, is available in different forms to facilitate desired depth of plowing, among other things.
The NU (New Universal) 36 cm (14 in.) bottoms have a long, slow-turning moldboard. They are primarily for sod, alfalfa, and grassland at speeds upward from 4 kph (2.5 mph) and depths of 13 cm to 20 cm (5 in. to 8 in.). Gentle inversion of the furrow slice reduces buckling and leaves a ribbon-like look. The NU 41 cm (16 in.) bottoms are designed for speeds ranging up to 8 kph (5 mph) and depths to 30 cm (12 in.). Features include unusually light draft, easy scouring, and effective tillage in stalk, bean, and stubble fields. The NU 46 cm (18 in.) bottoms for general-purpose plowing offer

A. Tye stubble drill

B. Buffalo till planter

C. Marliss no-till drill

FIGURE 15-2. Drills available for planting directly into sod and untilled ground.

Source: Part A reproduced with permission of the Tye Company; Part B reproduced with permission of Fleischer Manufacturing, Inc.; Part C reproduced with permission of Marliss Industries.

A. Hiniker Econ-O-Till planter

B. Miniseedbed

FIGURE 15-3. Precision planting in heavy residues by preparation of a miniseedbed.
Source: Reproduced courtesy of Hiniker Company.

light draft, good scouring, and moldboard capacity to handle big furrow slices at speeds to 9.6 kph (6 mph) depending on conditions. They plow to 36 cm (14 in.) deep and leave a wide furrow floor for large tractor tires. Proper plow selection, setting, speed, and soil moisture should make it possible to place nearly all of the weed seeds deposited on the soil surface at the bottom of the plow furrow. Thus, plows are available to utilize what is known about the effects of burial depth on survival and emergence of seeds and of perennating parts to minimize or reduce weed interference.

FIGURE 15-4. Chisel plow for soil penetration without mixing.

Source: Reproduced with permission of John Deere and Company.

A. NU 14 in. New Universal bottoms

B. NU 16 in. New Universal bottoms

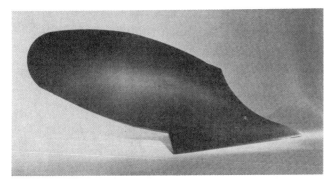

C. NU 18 in. New Universal bottoms

FIGURE 15-5. Moldboard plow bottoms designed for different soils, plant cover, plowing depths, and tractor speeds.

Source: Reproduced with permission of John Deere and Company.

Cultivators.   Progress has also been made in designing cultivators for narrow rows and for heavy residues. Figure 15-6 shows a cultivator designed for use in heavy residues. A depth-control disk between each row cuts through the residue and maintains uniform penetration. Precise control on the depth of cultivation will be needed in many instances if the objectives of a long-term weed management plan are to be fully met. Thus, this feature needs to be taken into account in the development of narrow-row cultivators and other tillage equipment.

Cultivators have been made for removing weeds within the crop row (Figure 15-7). Spyders (Figure 15-7A) can be set to move soil away from crop plants in the first cultivations and then move soil into the row to bury weed seedlings in follow-up cultivations. Torsion weeders (Figure 15-7B), spinners (Figure 15-7C), and spring hoe weeders (Figure 15-7D) uproot weeds in the row. The torsion and spring hoe weeders flex vertically and horizontally and can be set at a depth sufficient to uproot weed seedlings but not deep enough to uproot crop seedlings. Weed densities in corn with in-row cultivation were 43% and 99% less than with standard between-row cultivation in 2 years of a Colorado study (Schweizer et al., 1992).

Ridge tillage.   Knowledge that most annual weeds emerge from the top few centimeters in soil and that tillage encourages germination has been used to develop a

FIGURE 15-6. Cultivators designed to handle heavy residues found in conservation tillage systems.

Source: Reproduced with permission of Fleischer Manufacturing, Inc.

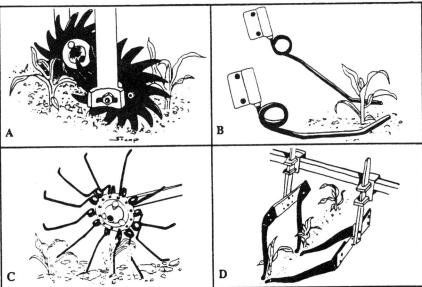

FIGURE 15-7. An in-row cultivator with spyders, torsion weeders, half sweeps, and spinners mounted to separate tool bars. Enlargements of a pair of spyders (A), a pair of torsion weeders (B), and a spinner (C). A pair of spring hoe weeders (D) replaced the pair of spinners after the first cultivation.

Source: Schweizer et al., 1992. Reproduced with permission of the Weed Science Society of America.

tillage system and associated equipment to reduce weed numbers within the crop row (Wicks and Somerhalder, 1971). The system, called *ridge-tillage,* involves movement of soil to and from the crop row during the growing season (Figure 15-8). Immediately before sowing corn, in this case, the top about 5 cm of soil from the previous years corn row is moved to the furrow between rows. At the last cultivation, soil in the furrow is moved back to the ridge crest. Such displacement moves weed seed away from the crop row for the early part of the growing season and may encourage their germination in the furrows between rows where they can be controlled with cultivation or herbicides. Up to 76% of the buried weed seed in the ridge was moved to the furrow in a continuous corn study (Wicks and Somerhalder, 1971).

A phenomenon of modern agriculture in the United States with implications for weed science is that many farmers modify commercially available equipment to meet

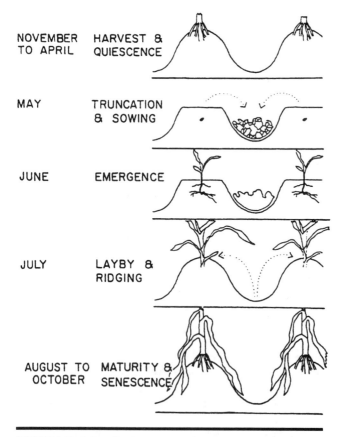

NOVEMBER TO APRIL	HARVEST & QUIESCENCE
MAY	TRUNCATION & SOWING
JUNE	EMERGENCE
JULY	LAYBY & RIDGING
AUGUST TO OCTOBER	MATURITY & SENESCENCE

FIGURE 15-8. Idealized schedule of events in a ridge-till crop production system in the Northwestern Corn Belt of the United States.

Source: Forcella and Lindstrom, 1988. Reproduced with permission of the Weed Science Society of America.

their specific situation. Such entrepreneurship is likely to continue to be a part of the American agriculture scene. Thus, weed scientists must be prepared for a future in which there are relatively more variations, both in how a given crop is produced and in the weed management approaches needed, than has been true in the past. In particular, weed scientists will need to help farmers understand the weed–crop ecological relationships so that farmers' weed management programs will fit their circumstances.

Bioeconomic Weed Management

Identification early in the modern weed control era of the mathematical relationships between weed density and crop yield provided the foundation and stimulus for development of bioeconomic models and programs of weed management. A bioeconomic weed management system integrates weed threshold information and economic information to help the production unit operator decide what action to take relative to weeds. The objective is to prevent losses from weeds that would exceed the cost of their control. Ideally, the system should incorporate actions to be taken at each level of the total weed management pyramid. Progress in development of models and associated programs has actually been from the top down in the pyramid. Those developed first were based upon early efforts that identified the relationship between weed density and crop yield (see Figure 2.2 and related discussion). They were designed to help the operator decide what number of weeds would justify the cost of their postemergence control. Since then models and programs have been developed based upon anticipated weed emergence, thereby providing help to the operator in deciding what action to take prior to crop emergence. Some limited progress has been made in formulating models that consider the weed seedbank along with emerging and growing weeds in deciding actions to be taken. It is beyond the scope of this text to consider the mathematics involved in development of the models. Kinds of models are described to indicate the technology available to assist enterprise operators formulate total weed management systems and to relate such systems to what has been learned in the preceding chapters.

The generalized relationship for a full range of weed densities is that shown in Figure 2-2. As discussed in Chapter 2, many factors, especially time of weed emergence, can alter the effect of weed density on yield. Models developed to integrate the effects of time of weed emergence and density on crop yield (O'Donovan et al., 1985; Cousens et al., 1987) correct for this effect. It is difficult to apply such integrated models because weeds commonly emerge in flushes over time rather than only at one time. This problem is overcome in models that relate crop yield to relative portion of the plant canopy occupied by the weed and the crop (Kropff and Spitters, 1991; Wiles and Wilkerson, 1991). Canopy occupancy takes into account density and flushes of emergence as well as, to some degree, aspects of the environment that influence relative growth rate of the weed and the crop (temperature, response to fertility, soil moisture, and so on). It has been suggested that relative canopy occupancy may in effect integrate the combined effects of competition for light, water, and nutrients (Aldrich, 1987).

Economic thresholds for postemergence herbicide applications. Knowledge of the mathematical relationship between weeds growing with a crop and yield of that crop has provided the foundation and stimulus for development of postemergence herbicide application recommendations based on expected economic returns. Common sense tells us there is some density on the low end of the weed density–crop yield curve where cost of control would equal the value of crop yield saved. This is termed the *economic threshold* (Cousens, 1987).

A method for estimating the economic threshold for multiple weed species in corn and soybeans was developed in North Carolina (Coble, 1985). A mixture of weed species, rather than a single species, is the more usual situation encountered, especially in row crops. This multiple species method deals with the low density end of the weed density–crop yield curve where the relationship is essentially linear. The method is based on the relatively simple regression analysis of the competitiveness of the individual weed species toward the crop. A computer-based economic decision model based on the method made the correct decision 90% of the time. Several decision models for other crops and crop production locations have been developed.

Time of weed emergence models. Knowing when weeds emerge could help the production unit operator decide tillage practices to use and plan pre-planting and pre-emergence weed control. Models have been developed that predict weed emergence fairly accurately based on soil temperature (Bridges et al., 1989). Of course, other factors may affect germination. As we learned in Chapter 4, there must be sufficient soil moisture for imbibition by seed and to sustain the germination process thereafter. A model has been developed that predicts when large crabgrass will emerge based on soil temperature and soil moisture (King and Oliver, 1994). One has been developed that includes soil temperature, temperature fluctuations, soil moisture, and seed depth on weed emergence (Alm et al., 1993), thereby including more of the factors known to affect germination. All of these models were designed to predict time and rate of germination. There are several ways such information could be used to prevent or minimize weed emergence with a crop or other desired plants. Crops could be planted ahead of the predicted flush of weed emergence thereby giving the crop a competitive advantage. Or, crop planting could be delayed until after the flush of weed seed germination, which would allow the operator to destroy the weeds with seedbed preparation tillage or with herbicides. Knowing when annual turf weeds such as large crabgrass or common chickweed will begin to germinate could assure application at the proper time. Applying a preemergence herbicide too early can result in poor weed control due to its loss by leaching or decomposition. If applied too late, herbicides only effective against germinating weed seeds would be ineffective.

Predicting seedling numbers. Attempts to predict seedling numbers based on known weed seedbanks have been relatively unsuccessful (Ball and Miller, 1989; Wilson et al., 1985). The operator needs to know how many weed seedlings will emerge, not just when they will emerge to make the most cost-effective decision about all weed management practices made prior to emergence of the crop. A major obstacle to adequate prediction of number of weeds to emerge is the effect of seed dormancy. As was learned in Chapter 6, dormancy may be inherited or may be brought about by environmental conditions. A model has been developed for johnsongrass that attempts to deal with dormancy (Benech Arnold et al., 1990). The model proposes that johnsongrass seeds are in one of three states relative to response to temperature: (1) highly dormant and will therefore not respond; (2) those that will respond once their fluctuating temperature requirement is met; and (3) those that have had their fluctuating temperature requirement met and will respond by changes in germination rate. Thus, we see that progress is being made in technology to predict weed emergence, but available information is inadequate for fully satisfactory cost-effective pre-crop emergence weed management decisions.

Production system weed management decisions. Even though numbers of weeds to emerge can not be accurately predicted, potential numbers can be estimated based on seedbanks. By using computers and simulation technology, such estimates coupled with economic threshold information have been used to devise bioeconomic weed management strategies for several cropping systems, including corn (Lybecker et al., 1991), sugarbeets (Shribbs et al., 1990), and a barley-wheat-pinto bean-sugarbeet rotation (Schweizer et al., 1988). The strategies are designed to evaluate weed management practices for controlling weeds that emerge before, with, or after crop planting. A soil-applied treatment submodel is included that selects the most cost-effective weed management practice based on predicted number of weeds to emerge, by species, from known seedbanks. After weeds and crop emerge, a postemergence treatment submodel selects the most cost-effective weed management practice based on the number of weed seedlings present, by species, within the row. Both tillage and herbicides are considered. It should be noted that weed seedbanks tended to increase under the most cost-effective systems.

Including seedbanks in the decision-making system. If the cost factor associated with an increase in the seedbank is included, the economic threshold is lowered (Cousens et al., 1986; Doyle et al., 1986). For example, the optimal economic threshold density of wild oat in winter wheat, on a 10-year basis, was 2 to 3 seedlings m^{-2} compared to 8 to 10 seedlings m^{-2} if only the economic threshold the year of herbicide application was considered (Cousens et al., 1986). These studies evaluated the effect of different levels of postemergence herbicide control that allowed different numbers of the selected weeds to produce seed. Results would no doubt be very different if practices specifically designed to reduce seedbanks were considered. For example, simulated-roller applications of glyphosate to velvetleaf and giant foxtail overtopping soybeans prevented 99% and 96% production of seed, respectively, and reduced germination of seed produced by 58% and 95%, respectively (Biniak and Aldrich, 1986). Using a roller applicator can result in savings of 80% or more compared with a broadcast spray application (Schepers and Burnside, 1979).

Risk in decision making.   Risk associated with a particular weed management decision is important to the enterprise unit operator. For many such operators, minimizing the risk may be at least as important as maximizing economic returns. That is, an operator may prefer to err on the side of better weed control than to risk the effects of not obtaining the expected control from a lower level of management. Acceptance of the economic threshold approach in itself involves risk since it relies exclusively on postemergence control. As pointed out at the beginning of this chapter, failure at this level means economic loss.

Thus, an impressive body of knowledge has accumulated to help the operator decide what weed management action to take, although there are significant gaps in that knowledge. In the final analysis, the operator must decide based not only on economic but also on such other factors as the degree of risk acceptable (per hectare value of the crop), importance of aesthetics, and species of weeds.

APPLYING WEED PREVENTION KNOWLEDGE

This section links what has been learned about ecological relationships, weed control, machinery, and modeling to practices that incorporate such knowledge. Concepts developed in each chapter are used as a background and are listed in Table 15-1 for each of the ecological relationships studied.

The ecological relationships include reproductive capacity, longevity, and germination and emergence of seed and perennating parts, competition, allelopathy, biological agents, and enterprise operators. Herbicides are included under operators since their use is entirely dependent upon them. The other phenomena, although influenced by operators, have effects that can be expressed independent of operator efforts.

The concepts listed in the table are generalizations. Specific weed and crop situations may express different relationships from these generalizations. It is necessary to draw upon generalizations to develop an appreciation of how knowledge about ecological relationships can be used in the prevention component of weed management. With this knowledge as background, each of the following sections concludes with a summary showing how control and prevention might fit together in weed management.

Because the reproductive unit (seed and perennating part) is basic to a management approach, the discussion begins with ways to reduce propagules. The conceptual approaches to prevention are developed by considering each ecological relationship in turn. However, the desired degree of prevention cannot be attained from application of a single concept, but rather from the use of several concepts.

Reducing Weed Propagules

Reproduction capacity.   The concepts in Table 15-1 collectively point to the need for efforts aimed specifically at preventing reproduction. Approaches designed only to

TABLE 15-1.   Ecological relationships and concepts that may be utilized in weed prevention.

Reproductive Capacity—Seed
a. Many weeds are capable of producing a normal seed crop even when their numbers are reduced.
b. Seed production commonly starts at an early age in weeds and may occur over an extended period.
c. Some very immature weed seeds may be viable.
d. Seed production of many weeds is curtailed by shade.
e. Influx from outside does not add significant numbers to the weed seedbank.
f. Weed seeds face many hazards before becoming a part of the seedbank.

Reproductive Capacity—Perennating Parts
a. Production commonly starts within a few weeks after emergence from either seed or an old perennating part.
b. Factors that encourage vegetative growth commonly discourage production of perennating parts.
  (1) High nitrogen levels
  (2) Shade
  (3) Ample to excessive water supply
  (4) Cool temperatures
c. Production is likely to be influenced by growth regulators.
d. Shading may reduce the size of the perennating part.

Longevity—Seed
a. Dormancy is common for some seeds of many weed species; such seeds are highly tolerant of temperature and moisture extremes.
b. A majority of weed seeds either germinate or deteriorate in 2 to 3 years after entering the soil, but a few may live much longer.
c. Longevity is commonly increased by burial in the soil.
d. Longevity is commonly reduced by tillage.

Longevity—Perennating Parts
a. Some dormancy is not uncommon, but it is less common and less pronounced than in seeds.
b. A majority of perennating parts either resume growth or die the year following their production, but some may live longer.
c. Perennating parts are vulnerable to desiccation and some to temperature extremes encountered in nature.
d. Longevity is commonly increased by burial in soil.
e. Tillage reduces longevity.

Germination and Emergence—Seed
a. Emergence is inversely related to depth of burial in soil: For many weed seeds, emergence falls sharply below 4 cm (1 in.). Germination will occur at any depth providing requirements have been met.
b. Many weed seeds have dormancy induced by burial in soil and some by leaf shade.
c. Alternating temperatures and several days of freezing or near-freezing temperatures frequently are needed for best germination.
d. Many weed seeds have chemical inhibitors of germination.
e. Many weed seeds require light for best germination.
f. Many weed seeds have the capacity to anticipate the best season in which to germinate and complete the life cycle.
g. The effects of factors are largely a matter of degree rather than being absolute.

Germination and Emergence—Perennating Parts
a. Emergence is inversely related to depth of burial in soil; successful emergence commonly occurs from greater depths than for seeds.
b. Emergence is directly related to the size of the perennating part.
c. Sprouting is directly related to N content.
d. Some form of correlative inhibition is common among buds on perennating parts; dominance by the apical meristem is most common.
  (1) Separating the part from the parent often increases such dominance.
  (2) Fragmenting the rhizome reduces but does not eliminate dominance.
  (3) High N levels reduce dominance.

Competition
a. Weeds must be controlled early.
b. Weeds that emerge after about one-third of the crop growing period usually do not cut crop yields.
c. The effect of weed numbers on crop yield tends to be curvilinear and that of weed weight tends to be linear.

TABLE 15-1. (*continued*)

d. Most annual weeds are relatively intolerant of competition.
e. Crop species and varieties differ in competitiveness.
f. Annual weeds tend to cause more losses in annual crops and perennial weeds more in perennial crops.
g. The leaf is the site of aboveground competition; factors that influence quantity and quality of light absorbed determine competition.
h. The root is the site of belowground competition; extent of competition is largely determined by root volume occupied.
i. The life cycle of many annual weeds is shorter than that of the crop with which it may be competing.
j. Competition for one growth factor frequently leads to competition for others.
k. Competition may influence both the production and activity of growth regulators by crops and weeds.

Allelopathy
a. Chemicals are produced by some plants that prevent or inhibit germination and growth of others.
b. Chemicals may be produced in all plant parts.
   (1) Those in crop roots are likely to be the most significant in suppressing weeds in a crop.
   (2) Location in the crop plant is immaterial if the residue is used to suppress weeds.
c. Quantities produced seem to be directly related to stress.
d. Effects may be on any plant process or function and during any stage of development.
e. Allelochemicals enter the environment via volatilization, leaching, exudation, and decomposition of plant parts.
f. Allelochemicals embrace a wide variety of chemical compounds likely traceable to all basic processes of plant metabolism.
g. Crop species and genotypes vary in presence and in quantity.

Biological Agents
a. Natural enemies of weeds are present among plant-feeding insects, plant pathogens, and nematodes.
b. Some biotic agents selectively attack the flowers and seeds of weeds.
c. Greatest success with the inoculation approach is usually achieved with introduced pests of alien weeds.
d. Under the inoculation approach, 3 to 5 years or more may commonly be needed for a biological agent to build to an effective control level.
e. Results with a plant pathogen have shown that immediate control of a weed by a single application is practical.

Enterprise Operators
a. Dependence on a single weed control approach over time leads to a community of weeds adapted to that approach.
b. Monoculture tends to reduce number of weed species but increase weed numbers.
c. Herbicides can be applied to soil to selectively prevent emergence of weeds with desired plants.
d. Reduced tillage favors perennial weeds.
e. Among annual weeds, reduced tillage favors the surface germinators (such as crabgrass) over those that germinate at greater depths (such as morning-glories).
f. Herbicides can be safely applied to weeds growing with desired plants to prevent weed competition.
g. Depth and type of tillage determines distribution of weed seeds in the soil profile; standard moldboard plowing distributes them more or less uniformly throughout the plow layer.
h. Plant residues on the soil surface can affect weed seed survival, germination and species composition, and effectiveness of soil-applied herbicides.
i. Crops and crop cultivars differ in competitiveness and can affect composition of the weed community depending upon the number of years grown.
j. Increasing crop density through higher seeding rates and narrower rows favors taller-growing weed species.
k. The sequence in which crops are grown influences weed number and species.
l. Herbicides can be used to prevent seed production by weeds.
m. The economic threshold can be predicted fairly accurately for some weeds in some crops.
n. Time of weed emergence and numbers can be estimated based on the weed seedbank and environmental conditions.
o. Some hard-to-kill perennial weeds in some crops can best be dealt with before or after the crop growing season or in other crops.
p. A weed management program can be planned that when implemented will minimize weed numbers and associated costs on an enterprise unit over time.

control the weed plant itself are likely to be inadequate since any escapes from such practices can be expected to produce abundant seed or perennating parts. Several approaches might be used in response to the concept that seed and perennating parts of many weeds are curtailed by shade (concepts d and b, respectively). Among the possible approaches are: selection of crop cultivars, and possibly species, for earliness and completeness of canopy; use of row and plant spacings that will develop a complete canopy as early as possible following planting; and filling as many of the growing season niches as possible with desired plants.

Although information is incomplete for most effective use of each of these approaches, still, enough is known about each for reasoned decisions to be made that may reduce the production of propagules. For example, crop varieties and selections in breeding programs vary greatly in rate and form of growth. This information is usually recorded and, in fact, may be included in recommendations to growers to help them select a variety to plant. By choosing varieties and selections for their ability to compete for light, the production of weed propagules can be reduced.

Much information has accumulated on the appropriate seeding rate for both agronomic and horticultural crops. By planning the weed program for a several-year time frame, realistic projections of anticipated weed problems for any one year can be used to choose a seeding rate and row spacing that will maximize crop competition for light. For the most part, selection of varieties and plant spacings to maximize light competition may be viewed as no-cost approaches. That is, it should be possible to select varieties especially competitive for light from those already available for which no sacrifice in yield would be made to other varieties. Similarly, the use of narrow rows with soybeans, for example, as a means of increasing the competition of the crop for light does not represent an additional cost nor a potential loss in production capacity. Quite the contrary, there is reason to believe that solid-planted soybeans have a higher yielding potential than those in wide rows. Of course, for these approaches to be maximally effective, it may be necessary to select for the attributes in appropriate plant breeding programs and incorporate them into overall production programs. Nevertheless, utilization of these approaches may be viewed as relatively low cost in comparative terms.

Using cover crops to fill the niches not occupied by the desired plant may reduce production of propagules. Many annual crops actually do not utilize much more than one-half of the growing season. For example, soybeans in mid-Missouri commonly occupy from 100 days to 120 days of a 188-day growing season. Simply put, there is ample time prior to planting and after soybean harvest for weeds to produce seed or perennating parts. The long growing period available to weeds following wheat harvest provides a particularly good opportunity for weeds to multiply. In view of the sensitivity of annual weeds to competition, as well as the sensitivity of perennials to competition in the production of their perennating parts, judicious use of cover crops clearly offers the potential for reducing propagules during this period.

Use of cover crops to fill the open niches in a cropping system offers the additional advantage of holding soil against erosion where that is a problem. Further, depending upon the cover crop used, this approach may also be a way of providing some of the nitrogen needed by the crop. Prior to the widespread availability of cheap fertilizer nitrogen, legumes were commonly used in this way as a source of nitrogen for such crops as corn, wheat, and other cereals. The escalation in fertilizer nitrogen cost may well provide a financial advantage to cover crops as an approach for providing some of the nitrogen needed. The so-called no-till, or zero-till, approach to crop production partially utilizes this concept, but without full consideration of the weed management aspects. By combining reduced tillage and cover cropping with full consideration of their impact on weeds, the advantage of this production system could be even greater.

Of course, the use of cover crops to fill the niches in the growing season not filled by the crop represents an additional cost. Further, although there is information on the use of cover crops in crop production, much of it may not be directly applicable to their use in preventing production of weed propagules. Nevertheless, this use of cover crops may well offer a relatively effective and inexpensive way of preventing propagule production.

The fact that high nitrogen levels may encourage vegetative rather than reproductive growth in perennials—concept b(1) under perennating parts—may be utilized to time nitrogen applications on crops in order to minimize reproductive parts. Although there are clear limits in the extent to which the time of nitrogen application can be changed to minimize production of perennating parts, there are certain crop and weed situations where this approach could be used. The most obvious possibility is with those crops for which it is common practice to top-dress with nitrogen. In those situations, it is simply a matter of considering the stage of development of both the weed and the crop in order to decide when to apply the nitrogen. Conceivably, there are instances when changing the time of nitrogen application in order to minimize production of perennating parts by the weed would result in little or no sacrifice in crop yield.

Longevity. Table 15-1 shows that longevity of both seed (concept c) and perennating parts (concepts c and d) is commonly extended by burial in the soil. This information may be utilized to reduce propagules. The approach, of course, would be quite different with seeds than with perennating parts since seeds are on the soil surface to start with, whereas the perennating parts of perennials are commonly distributed through several centimeters of the soil, with some of them located at considerable depth in certain species. The specific aspect of burial depth we want to use is the sensitivity to exposure to temperature and moisture extremes at the soil surface. Thus, if a large weed seed crop of annuals is produced, delaying tillage until the following spring and then using only shallow tillage should reduce the number that become a part of the seedbank.

With perennial weeds, appropriate tillage implements can be used to bring reproductive parts to the surface to be destroyed by desiccation and lethal temperatures. Because such an approach means special or extra tillage, however, it is doubtful that it could be justified on large areas under extensive farming. Nevertheless, it might be a feasible approach in intensively grown crops. Further, since perennials frequently occur in patches rather than in uniform distribution across a field, such an approach might be usable by concentrating on such patches.

There may well be situations where tillage by itself (concept d for seeds and e for perennating parts) should be used to reduce propagules in the seedbank. For example, some shallow tillage for 2 to 3 years after production of a large weed seed crop should reduce the seedbank, compared to no tillage. Persistent and timely tillage for 2 to 3 years will destroy some perennial weeds.

Germination and emergence.   The concepts in Table 15-1 are not directly applicable to reducing weed propagules. Indirectly, of course, concepts and associated management that reduce numbers of plants on a given area can be expected to reduce propagules.

Competition.   The concepts associated with competition in Table 15-1 provide little in the way of direct application for preventing the production of seeds and of vegetative reproductive parts. Indirectly, of course, reduced weed growth from the use of competitive crop species and varieties (concept e) could reduce production of propagules.

Allelopathy.   Table 15-1 shows that allelochemicals are effective against a wide variety of plant processes or functions (concept d). This fact suggests that crop selections may be available, or could be developed (according to concept g), that are specific in their inhibition of seed production or the production of vegetative perennating parts. For such crop selection to be effective in this regard, the allelochemicals produced need to be exuded through the crop root and readily translocated upward in the weed. Allelochemicals produced by crops may need only stress neighboring weeds thereby indirectly suppressing seed production by weed cohorts. This statement simply recognizes that leaching from leaves and stems and release as a result of decomposition of residues in the soil, in all likelihood, cannot be depended upon to match up with the time of reproduction in the weeds targeted. This use of allelopathy may not have enough potential to justify research and development exclusively for this purpose. However, it may be worthwhile to look for this special attribute in research already planned or underway to identify allelopathic effects on overall growth characteristics.

Biological agents.   Of the concepts listed in Table 15-1, the use of insects and plant pathogens (concept b) specifically to reduce the production of seeds appears to be especially promising. Such selective activity has already been identified with both types of biological agents. For annual weeds, potential usage may be best as a cleanup of weeds that escape the other weed management approaches. Often, one or only a few species escape to overtop the crop and thus offer the potential of producing a seed crop. As we learned, biological agents are likely to be most useful against single species rather than against a mixture of weeds. Effectiveness of biological control of weed seed production can be improved with the use of multiple biotic agents, as shown for the specific seed-feeding insect and seed-attacking fungus on velvetleaf. This approach is feasible since methods for efficiently rearing insects for mass release into fields infested with weed escapes are available and application of microbial agents is similar to that for chemicals. Fungi that are potential biotic agents for attacking seeds of grass weeds, including large crabgrass, should be available for practical use as constraints to mass culture and specific requirements for infection are overcome.

For perennial weeds, this approach may be most usable against weeds that have not yet reached the economic threshold level (concept d in Table 15-1). Common milkweed in wheat in Missouri, shown in Figure 15-9, is an example. Milkweed has increased in much of the Midwest but likely is not yet a serious competitor. This is the case for a number of perennial weeds where minimum or reduced tillage is being practiced. Such situations provide the necessary time for the introduced pest to build to levels that will keep the perennial in check and prevent it from becoming a problem resolvable only by more costly approaches. Pathogens specific for infection of several problem perennial weeds, including common milkweed, Canada thistle, and johnsongrass, suppress growth and ultimately reduce total seed production. These agents can be applied to the weeds in patches in the field by using methods similar to those for herbicide application.

In systems where weed seeds can remain on the soil surface, predation of the seeds by native insects may be enhanced by maintaining buffer strips of vegetation along field borders to provide cover and to assure high populations of the seed-feeding insects.

Enterprise operators. We have already considered ways to reduce the reservoir of weed propagules in the soil by utilizing concepts relating to their production and longevity. Use of herbicides to prevent seed production (concept 1 under operators) needs to be added to these other concepts. Although herbicide effectiveness has been demonstrated, their use for this express purpose is limited, and there has not been a concerted effort to determine the merits of this approach. Advances in application technology offer new opportunities for use of herbicides for this purpose. Split applications of reduced herbicide rates is an example. As was learned in Chapter 13, there may be fewer escapes with split applications and it may also be possible to use less herbicide or at least no more, by making the first application to very small weeds. Thus, weed seed production may be reduced with no more herbicide than might be used in a single application for weed control.

Modeling technology offers the potential of providing the enterprise operator information on the cost effectiveness. This technology coupled with application of concepts relating to reduction of propagules can be brought together in a long-range plan that will minimize weed propagule reservoirs and associated costs (concept p).

Relationship of prevention to control. Figure 15-10 summarizes the cropping and tillage preventive approaches and adds the control practices that might be followed to fully implement weed management to reduce propagules in the seedbank. Although the technology has been identified for biological control, preventing seed production with herbicides, and hastening germination of seeds with chemicals, their application in a weed management context has yet to be demonstrated. The fifth control practice

FIGURE 15-9. Common milkweed, often the only weed in wheat at harvest time in Missouri.

Source: Reproduced courtesy of L.E. Anderson, Department of Agronomy, University of Missouri, Columbia, Missouri.

listed—that is, the manipulation of soil microorganisms—is only a theoretical possibility at present. Nevertheless, it should be apparent from the numerous approaches suggested that a reasonably effective weed management program could be developed to reduce the numbers of propagules in soil.

Preventing Weed Emergence with Desired Plants

Our interest in this section is in those ecological relationships and concepts in Table 15-1 that might be used to minimize the number of weeds that emerge with the crop.

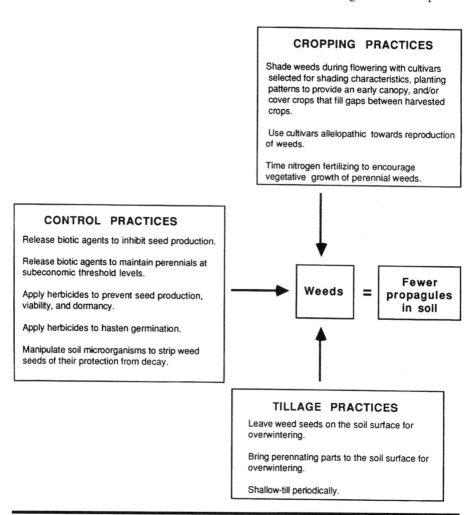

FIGURE 15-10. Summary of conceptual approaches for reducing numbers of propagules in the weed seedbank.

Reproduction capacity. Those concepts developed for reproduction capacity in Table 15-1 are not directly applicable to preventing weed emergence. Indirectly, of course, any success achieved in preventing the production of viable propagules helps reduce the number of weeds that emerge with desired plants.

Longevity. The concepts developed for longevity in Table 15-1 are not directly applicable to preventing weed emergence. However, any steps that shorten the life of propagules in the soil ultimately reduce the number available to emerge with desired plants.

Germination and emergence. The concepts in Table 15-1 relating to depth effect on germination and emergence (concept a) of both seeds and perennating parts offer opportunities for preventing their emergence with the crop. An approach for seeds is depicted in Figure 15-11. This approach buries the weed seeds so deeply in the soil that they cannot successfully emerge even if germination occurs. For many annual weeds common to row crops, placing the seed at the bottom of the plow layer, possible with available plows, would eliminate those seeds as a source of infestation with the crop that year. Only shallow tillage should be practiced in subsequent years to avoid bringing them to the surface where they could germinate and emerge.

There might be special situations where attempts to leach inhibitors (concept d in Table 15-1) from seeds are warranted. With irrigated crops, for example, it might be desirable to irrigate in the spring prior to tillage, both to leach out possible inhibitors as well as to provide the moisture necessary for early germination. Weed seeds depleted of inhibitors may also be more vulnerable to microbial attack in soil. The emerged weeds could then be destroyed before the crop is planted.

The season-anticipating characteristic (concept f) offers possibilities of reducing emergence of weeds with the crop. Emergence can be reduced either by delaying

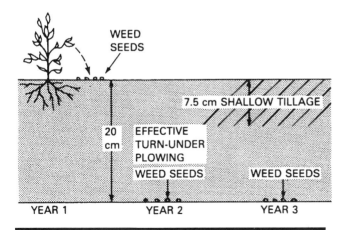

FIGURE 15-11. Use of tillage to minimize weed emergence following production of a large number of weed seeds.

seedbed preparation until the flush of germination has passed or by changing the crops grown to include either an early crop that precedes germination of most weeds or a late crop that is planted after most weed seeds have germinated.

Where perennial weeds are involved, the common occurrence of correlated inhibition (concept d) might be used to limit emergence of new plants. Limitation might be accomplished by deep plowing (30 cm or so) with only shallow tillage thereafter. The objective in this approach is to disrupt as little as possible the dominance effect of both the parent plant and the apical meristem. An alternative approach might be to till several times with a disk or other implement designed to accomplish as much fractionating of the rhizome or tuber chains as possible to minimize correlative inhibition and then, drawing upon concept b, plow them under to a depth from which successful emergence would be minimal.

Competition.   None of the concepts under competition in Table 15-1 would appear to be directly applicable in preventing weed emergence. Indirectly, the reduction of propagule production as a result of use of competitive crop species and varieties (concept e) would reduce the number of seeds available to germinate and emerge with the crop.

Allelopathy.   In Table 15-1, concepts a, b, e, and g for allelopathy may be usable in reducing weed emergence with the crop. In general terms, these concepts may be applicable to both living crop plants and to the residues of crop plants. Where residues are the approach, it is desirable to concentrate them in the soil zone from which the major weeds are expected to originate. If shallow-germinating annual species are the major problems expected, then it is best to concentrate the residue in the surface soil layer (5 cm to 7.5 cm if possible). In this way, the allelochemical produced by the crop is concentrated in the zone from which the weeds originate. This is the same concept involved in the incorporation of herbicides. Of course, if we seek to inhibit a deep-rooted perennial or large-seeded annual, it would be necessary to distribute the residue throughout the plow layer. In either case, it might well be necessary to incorporate the plant material some weeks in advance of planting to allow time for microbial decomposition and release of the toxin.

In no-tillage systems, fall-planted cover crops such as rye can be killed in the spring to provide a layer of undisturbed residue on the soil surface. Allelochemicals released from the residue can suppress germination and emergence of annual weeds. Alternatively, residue from a previous year's crop known to contain allelopathic activity, such as grain sorghum, could be used in a manner similar to cover crop residue. Management systems that include a cereal grain in a crop rotation scheme can reduce emergence of certain annual weeds without increasing herbicide applications. Dramatic decreases in giant foxtail densities have been obtained in no-tillage corn and soybeans when wheat was included in a soybean–wheat–corn rotation compared to either continuous corn or soybean–corn rotation (Schreiber, 1992). In situations where double-cropping is an option, residues of the first crop might be managed for selective suppression of annual weeds in the succeeding crop. Such is the case for soybeans following wheat where wheat straw can provide allelopathic activity.

Use of living crop plants to inhibit emergence may have its best potential against late-germinating weeds, such as fall panicum in corn in the central Corn Belt. By this time in the growing season, an allelopathic corn variety should have had sufficient time to produce enough allelochemicals to have the desired inhibitory effect. Crops such as grain sorghum, oats, and rye actively exude allelochemicals. These crops may be useful in suppressing weeds prior to planting a succeeding crop. However, the possibility for the growing crop to provide season-long weed suppression, as seems to be the case for some rice accessions, indicates that this trait may soon be incorporated into varieties of several crops. An additional incentive for looking to allelopathy to deal with such weeds is the fact that other approaches, such as applying herbicides, pose the hazard of mechanical damage to the crop.

Biological agents.   Several possibilities exist for applying the concepts under biological agents in Table 15-1 to enterprise operations. Several approaches for managing weed emergence have been described (Kremer, 1993). Preparations of selected bacterial or fungal biotic agents can be applied directly to soil, crop residues, or on crop seeds to prevent emergence of weeds in the crop seed–germinating zone. Seeds of cover crops used in some management systems also provide a means of delivering biotic agents into the soil. Certain chemicals known to stimulate seed imbibition or germination can be incorporated into soil combined with selected seed-attacking microorganisms to kill germinating weed seeds.

The soil environment might be manipulated to favor microbial attack on weed seeds. Reduced tillage systems encourage infection of some weed seeds (giant foxtail, wild oat) by soilborne or seedborne microorganisms, resulting in reduced numbers of these weed seeds in the seedbank. Also, in intensively managed enterprises such as vegetable and orchard operations, use of solarization (increasing soil temperatures to $40^{\circ}C$ to $70^{\circ}C$ near the soil surface by covering soil with plastic) can reduce weed emergence and seed viability. Selected microorganisms applied immediately after solarization may enhance reduction of viable seeds even further by attacking those weed seeds that survived solarization but were weakened by the process and thus are more susceptible to microbial attack.

Enterprise operators.   Herbicides (concept c) are already widely used as pre-plant incorporated or preemergence applications to prevent weed emergence with desired plants. Efficacy of such applications might be increased by timing them to predicted weed emergence (concept n). Currently, such applications are made with the expectation that annual species will soon germinate. As we learned in Chapter 6, several things may affect germination. Soil temperature and soil moisture in particular affect the onset and rate of germination. Use of developing technology that considers such environmental variables to predict weed emergence could increase the likelihood of having adequate herbicide in the vicinity of the weed seeds at a time to be most effective.

Weed seeds present in the seedbank might be reduced with soil-applied herbicides that stimulate dormant seeds to germinate. Shallow incorporation of such herbicides enhance emergence of weeds, which can be followed by cultural or chemical methods to eliminate seedlings prior to or at planting. New classes of "seed-killing" chemicals based on attacking seed dormancy mechanisms in soil may be available specifically for management of seedbanks. Application of several such compounds may be necessary to induce a maximum number of seeds of several weed species to germinate at a given time. Another potential approach for reducing seed numbers is the use of chemicals to alter seed properties making them vulnerable to attack by soil- or seedborne microorganisms, essentially causing the seeds to decompose or seedlings to be diseased. Even a modest reduction in weed seedling emergence will not only reduce seedbank size but also make more efficient use of postemergence herbicides applied to lower weed densities resulting from soil treatments. Knowledge that the sequence in which crops are grown influences weed numbers (concept k) might be utilized to reduce weed emergence by switching to later- or earlier-planted crops. Limits to this approach are obvious because climate, available machinery, markets, and the like restrict farmers' crop options. Still, we should be aware that occasional changes of this kind may reduce the numbers of weeds with which we must deal in any given year.

Relationship of prevention to control.   Figure 15-12 summarizes the cropping and tillage preventive approaches and adds the control practices that might be followed to fully implement weed management to minimize weed emergence with crops. Of course, the control practice of applying herbicides to soil is widely used. The technology for utilizing allelopathic plant residues, breaking dormancy and destroying emerged plants, and leaching germination inhibitors from weed seed to destroy weeds that emerge has been at least partially identified. Thus, there is a sizable array of approaches to make a weed management effort to minimize weed emergence with crops even more effective.

Minimizing Weed Competition

For the foreseeable future, herbicides or timely tillage will be needed to avoid serious losses once a weed problem exists. However, ways of managing production to minimize the competitiveness of weeds present with the crop are available. From the discussion about the period of competition and the likelihood that competition for one factor leads to competition for another, we know that effective prevention and control provide a potential compounding effect toward weeds, thus preventing a corresponding compounding of competition on the part of the weed. With this information as background, let us examine possible ways to utilize the concepts developed for each of the factors in Table 15-1 in terms of providing a competitive edge to the crop.

Reproduction, longevity, and germination.   There is no concept in Table 15-1 directly applicable to reducing competition in the areas of reproduction, longevity, and germination. However, success in preventing reproduction, as well as success in reducing the soil bank of propagules, indirectly reduces competition.

Competition.   The key concept in Table 15-1 is that most annual weeds are relatively intolerant of competition (concept d). The fact that crop species and crop cultivars dif-

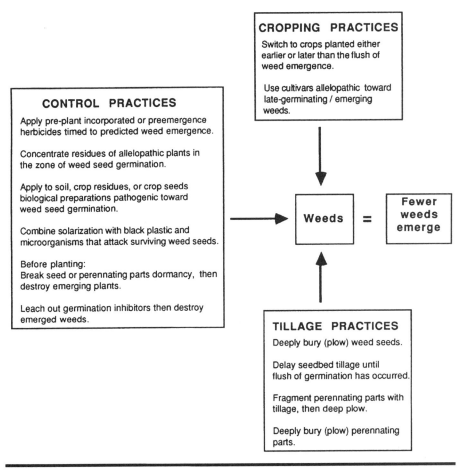

FIGURE 15-12. Summary of conceptual approaches for preventing weed emergence with crops.

fer in competitiveness (concept e) due likely to differences in their aboveground and belowground growth forms (concepts g and h, respectively) suggests that selection for characteristics that contribute to early development of a complete canopy and rapid elaboration of root systems would be worthwhile in plant breeding programs.

As we have already learned, time is important in many ways in the weed–crop competition struggle. An advantage of only a few days in closing a canopy or in occupying the soil with roots can have a pronounced effect on competition. For most agronomic and horticultural crops, available germplasm provides a wide range of growth characteristics from which to choose those that provide the desired type.

Differences in aboveground growth form and size are well known. Differences in root growth form and size may be at least as great. With soybeans, for example, the dry weight of roots (an indication of root volume) of 8 varieties at maturity ranged from

1.52 g to 5.94 g per plant (Mitchell and Russell, 1971). Surface area of roots of two varieties differed by a factor of nearly 2 in other studies (Raper and Barber, 1970).

It should be recognized that those factors that contribute to high yields selected for in a breeding program may also coincidentally contribute to competitiveness toward weeds. However, it seems reasonable to expect that selection for those attributes most effective against the type of weed competition anticipated would be more successful than selection obtained coincidental to selection for yield. This statement simply recognizes that in a general way the differences in root:shoot ratios within the genetic material might be used to develop varieties especially competitive for soil factors or for light.

In situations where competition for water or nutrients is apt to be the basis of competition, selecting for early root elaboration would be advantageous. Conversely, where light is the factor most apt to be competed for, there would be an advantage in selecting for a high shoot:root ratio. For this approach to be most effective, however, the nature of competition between the weeds and the crop involved needs to be well understood. That is, we need to know what it is that the crop and weed are competing for, not just when they are competing.

Utilizing this approach to provide a competitive edge to our crops may not be as much of an additional burden to our present crop breeding programs as might at first be thought. Much of the data currently obtained on genetic material in a breeding program might well be usable in identifying that material most apt to contribute the desired competitiveness toward weeds. This fact can be seen by examining samples of data sheets for recording information on each of three different commodities in a breeding program (Figure 15-13). In any case, selection for competitiveness toward weeds might well provide an important edge to the crop and one that might not be recognized in the selection process unless specifically identified as a goal. Thus, it may be advisable to include this as a goal in more plant breeding programs.

Crop seed treatments.   Under very special situations, such as in production of high-acre value crops, it might be practical to presoak the crop seed or coat it with a hygroscopic material to shorten the time from planting to emergence. Germination 2 or 3 days earlier by such an approach could markedly reduce competition from weeds.

Allelopathy.   Concepts a and g under allelopathy in Table 15-1 can be drawn upon to develop and select cultivars inhibitory toward weeds. Because allelochemicals are somewhat selective in the plants affected, this approach may have its primary place against individual weeds. Further, the approach may be most usable against the dominant perennials, such as those occurring under reduced tillage programs. Where the objective is to suppress such a perennial, there might well be an advantage to releasing the allelochemical by leaching, exudation, and decomposition rather than by exudation only. This practice could serve to simplify the development of allelopathic breeding lines since only total quantities of allelochemicals would be of concern rather than where they are located within the crop plant. Also, this method could serve to extend the duration of effect, which would be especially advantageous with perennials be-

CORN

Plot	Stand	Lodging		Ears		Days to*			Root Pull*	Moisture	Yield
		Root	Stalk	Drop	Usable	Flower	Tassle	Silk			

FORAGE LEGUME

Plant Introduction number	Uniformity*	Habit (Erect/ Prostrate)*	Vigor	Plant*		Leaves*		Maturity*	Winter hardiness*
				Height	Width	No.	Texture		

SOYBEAN

Strain	Date*		Height*	Stem termination*		Branching*	Yield
	Flowered	Mature		Determinate	Indeterminate		

FIGURE 15-13. Examples of data recorded for genetic material in a plant breeding program for selected crops. Starred (*) items identify data potentially useful in identifying breeding material competitive toward weeds.

cause of their common usage of a large part of the growing season.

The fact that quantities of allelochemicals produced are often greater under stress (concept c) suggests it should be possible to time the application of nitrogen to a crop to maximize the production of allelochemicals during the time when such chemicals would be most important as inhibitors of weeds. This approach might be particularly useful in such perennial crops as forages where there may be an option to apply fertilizer in either the spring or the fall of the year.

In those cropping programs where it is desirable to use the crop residue as the source of allelochemicals to suppress weeds, management of the residues needs to be done in such a way as to maximize the effect on the weed. For example, if the residue is intended to suppress a perennial that initiates growth in late spring or early summer, it is probably best to delay incorporation of the crop residue as long as possible prior to planting the crop to maximize the quantities of allelochemicals present when the weed initiates this first flush of growth. Nutsedges are good candidates for this situation. On the other hand, if the perennial is one that makes substantial early spring growth—for example, quackgrass—it might be best to incorporate the residue as early in the spring as possible. Similarly, the depth from which maximum regrowth of the perennial occurs should be taken into account when determining the depth to which the crop residue should be incorporated.

Biological agents.   The most useful role for biological agents may be against developing weed problems. Where a weed problem is developing but has not yet reached the point of seriously competing with the crop, enough time may be available for the biological agent to develop to a control level (concept d in Table 15-1). This approach to the use of biological agents is limited by our inability to predict weed problems. As we learn more about the nature of competition and the response of weeds to changes in production practices, we will have the basis for anticipating those weeds likely to be associated with our crops. With enough advance warning, coupled with additional information about potential biological agents, satisfactory ways of utilizing biological agents to suppress emerging weeds can be developed.

Plant pathogens may offer a way to effectively deal with escapes by utilizing concept e. This approach is important because escapes may well be the most difficult problem to contend with in a total weed management program. Such escapes could contribute significant numbers to the propagule bank but in themselves not pose enough of a threat to the crop in which they occur to warrant control. It follows that the approach used against such weeds must be relatively low cost. In this case, pathogens combined with seed-attacking insects may be effective in reducing seed production, as was noted for velvetleaf. The need for low cost, plus the fact that alternatives are limited, suggests that plant pathogens be carefully considered. Where there is a mixture of such weeds, it may be that a mixture of pathogens could be used. The suggestion that pathogens be considered is not meant to imply that this may be a low-cost approach. Rather, it is felt that all of the limited possibilities need to be explored. As formulations are developed to surmount environmental constraints to effectiveness of current mycoherbicides, biotic agents for a wider range of weeds should become available to the enterprise operator.

Maximum effect of biotic agents in minimizing competition from weeds will depend on skillful integration of biological control with other weed management approaches. For example, successful management of musk thistle in pastures involves applications of herbicides at reduced rates to enhance the effects of insects used in biological control of the weed. Grazing and fertilizer applications are also timed to maximize the effect of competition by desirable forage species on musk thistle. Similar integrative approaches must be devised for maximum efficiency of biotic agents in other enterprises to minimize weed competition.

Enterprise operators.   It is very important to long-term success of a weed management program that the potential effect on weeds of a single weed management approach (concept a) and of crop monoculture (concept b) and sequence (concept k) be considered early in the program. Otherwise, economic losses from the "new" weed community will be unavoidable as pointed out earlier in this chapter. For some, such as a Great Plains wheat farmer, there may be little alternative to the one crop being produced. For many, at least some options are open concerning crops to be grown. Where options are available, occasional changes in crops may prevent, or at least delay, the development of a new and even more difficult weed problem. In another sense, there may be an advantage to growers in having the narrow weed base associated with monoculture. After all, the complications for effective weed management can be expected to increase as the number of species increases.

Thus, farmers interested in growing more than one crop might simplify weed management by raising only crop A on field 1, crop B on field 2, and so on. After a period of years, crops could be shifted among fields and the process repeated. Of course, factors other than weeds must be considered. Crop diseases and insects may increase under such a modified monoculture system, for example. Further, there must be effective ways to deal with the weeds encouraged by such a system. Nevertheless, this example serves to indicate the approaches growers may take if they fully understand the weed–crop ecological relationships.

Herbicides are already widely used to prevent competition from weeds growing with desired plants (concept f). Such usage could be even more effective if steps are taken to slow development of weed resistance, carefully time application to the most sensitive weed stage, utilize split applications and reduced rates, and use herbicides as part of a total weed management plan.

Whenever a major change in the production program is contemplated (tillage, crop, growth form of crop, and so on) operators need to draw upon concepts a, b, d, e, h, i, j, and k to anticipate what change it may cause in weed composition. This approach allows plans to be made for dealing with the new problem before it becomes serious.

It is helpful for users to know the effect of tillage depth and timing on weed seed distribution and plan accordingly (concept g). Any plan, no matter how good, can break down under adverse weather, machinery failures, and other similar factors. Thus, we can expect that there will occasionally be a year when a large number of weed seeds are produced or when a particular perennial weed is able to store large quantities of food reserves for its perennating parts. Understanding the weed and the factors affecting longevity of its parts can provide users the information necessary to till in the right way at the right time and thus avoid tillage that might greatly magnify and extend the problem posed by such escapes.

Similarly, there will be times when users may not be able to manage their crop residues according to plan. Knowledge of the effect of such residues on weed seed longevity and germination (concept h) and on potential weed seed pathogens in soil can help farmers choose the appropriate tillage practice to minimize the potential disruptive effects on the long-term weed management program.

Relationship of prevention to control. Figure 15-14 summarizes the cropping and tillage preventive approaches and adds the control practices that might be followed in a weed management program to minimize competition from weeds growing with crops. The utility of postemergence herbicides and rotation of herbicides have been well established and both are widely used. Even so, weeds still cause annual losses exceeding $6 billion. This fact suggests the need to include the preventive approaches shown if this loss in crop yield is to be reduced significantly.

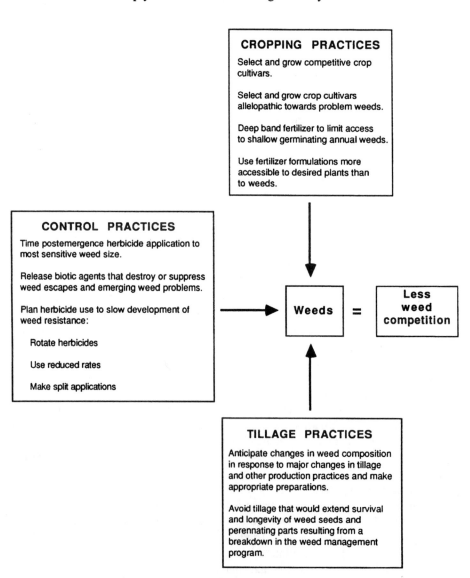

FIGURE 15-14. Summary of conceptual approaches for minimizing competition from weeds with crops.

WEED MANAGEMENT FOR A CROP PRODUCTION SYSTEM

In the previous section, we considered possible approaches to weed prevention in each of three broad stages in weeds' life cycles. In the course of our examination, the range of potential tools for dealing with weeds in a total management effort were identified. Here, these prevention approaches and control approaches are brought together and examined in terms of the entire crop production system. A hypothetical field example is useful for this purpose.

A producer plans to initiate a reduced-tillage production system involving a rotation of soybeans, wheat, and corn. Let us assume that distributed uniformly in the plow layer is a large population of seeds of annual weeds—both grass and broadleaf—common to the southern part of the Corn Belt in the United States. Also, there is a scattered stand of a perennial weed not yet interfering with crop yield. To simplify the example, practices are identified that might be followed each year for 7 years on only one field. This procedure allows us to consider approaches in each crop twice plus consider a change in crops grown. After careful study of this program, it should be possible to use the concepts involved to develop comparable programs for other crops and other localities.

At the outset, the producer needs to make a projection of what the weed problems will be during the first few years. Five years may be a convenient period. Based on what is known about weed seed longevity, the effects of reduced tillage on weeds, and of wheat as a potential source of weed renewal, these general projections are justified: (1) the perennial will increase during the 5 years unless steps are taken to check it; (2) annual weeds will compete during the period even if their seed production is prevented during this time; and (3) annual grasses, especially late germinators, will increase unless steps are taken to check them. The practices identified for each year are based on these projections.

Year 1: Soybeans

Because there is a large population of annual weed seeds uniformly distributed throughout the soil profile, efforts will be needed to limit the expected dense weed stand. One approach might be to prepare a shallow seedbed as early as possible with a disk harrow, delay soybean planting a few days to give weed seeds the maximum time to emerge, then destroy emerged weeds with either a herbicide—for example, paraquat or glyphosate—or shallow tillage just before planting. By direct drilling soybeans into the killed weeds, no new weed seeds would be brought to the soil surface. However, a preemergence herbicide should probably be used because the large initial weed seedbank likely contains some late germinators. The soybean variety chosen should be one that develops a canopy quickly and yields well when planted in narrow rows.

Follow-up weed control needed will be determined by numbers and species of weeds present. A mixture of recommended herbicides should be applied as soon as it is determined that the economic threshold will be exceeded. Application should be made while weeds are still small so a reduced rate of herbicides can be used. The field should be closely monitored to see if a second reduced rate application is needed. If there are some escapes from the first application but not a competing infestation, a rope-wick application of glyphosate or other appropriate herbicide can be made to avoid interference of overtopping weeds with soybean harvest and to prevent their production of seed; this treatment could also serve to check growth of the perennial. Steps should be taken to determine whether there is genetic material known to be allelopathic toward the new perennial weed, if one is anticipated, as well as to determine whether any biological agents have been successfully used against it. If such genetic material or biological agents are available, plans can be made to use them at the appropriate time.

During this first year, all weed species present should be identified in order that a projection can be made of the shift in weed species that may be expected after 5 years under this new production and weed management system. Once the species are known, information can be gathered on their growth form and life cycle, relative response to the herbicides being used, and relative response to the tillage planned. With this information, a first projection of weed composition 5 years hence should be made. The projection should be updated each year thereafter based on a check of weed composition.

Immediately following soybean harvest, the field should be shallowly tilled with an implement such as a field cultivator to minimize storage of carbohydrate reserves in perennating parts of the perennial weed as well as to segment the perennating parts and bring them to the surface where some could be destroyed by low temperatures and desiccation.

Year 2: Wheat

Efforts should focus on checking the potentially troublesome perennial weed and on preventing seed production of annual weeds following wheat harvest. As soon as the wheat is mature, a herbicide such as glyphosate might be applied to the actively growing perennial; this treatment should also destroy the late summer–emerged grass and broadleaf annual weeds. An alternative if the perennial overtops the wheat would be application of an appropriate herbicide with a rope-wick applicator. An alternative approach for dealing with all weeds might be the use of a legume cover crop. It could be broadcast into the wheat in late winter to provide cover for the period from wheat harvest until the field is returned to corn the next year. Following wheat harvest, it may be necessary to mow the stubble to prevent production of seed of the fall panicum and summer-germinating broadleaf species and to prevent carbohydrate storage on the part of the perennial weeds. Whether or not this treatment is needed will be determined by the effectiveness of the cover crop in keeping these weeds in check.

Year 3: Corn

A soil residual herbicide application may not be needed since seed production of annual weeds has been prevented for 2 years, and there has been no tillage to bring new seeds to the surface. The need should be determined by using the latest technology to estimate numbers likely to emerge. Plans should be made to apply postemergence herbicides to prevent competition and seed production of annual weeds and to check growth of the perennial. If the legume cover crop alternative was chosen in the wheat, plowing it under can be expected to bring a large number of seeds of annual weeds to the surface. In this case, plans should be made to apply either a preplant-incorporated or preemergence herbicide(s). Arrangements should be made to use in-row cultivation to destroy annual weeds not prevented by the soil-applied herbicides.

For either approach, consideration should be given to planting the corn in 50 cm (20 in.) rows to provide the earliest possible canopy, as well as rapid occupancy of the soil by corn roots. Fertilizer should be banded below the corn row, using a formulation not as readily used by the weeds as by the corn, to minimize its availability to the shallower-germinating annual weed seeds. If it appears that the seedbank will be increased by surviving annual weeds, the farmer should be prepared to make a postemergence-directed herbicide application to them at the time the final cultivation might otherwise be made.

Year 4: Soybeans

Since seed production has been prevented for 3 years, conceivably the number and species to emerge will not require a soil residual herbicide. This should be verified by estimating emergence from counts of seeds in the surface 5 cm to 7.5 cm. If a residual herbicide is not needed, the farmer could plan to burn down existing vegetation if necessary and plant directly into the residue, thereby minimizing new weed emergence. Plans should be made for an early postemergence application of a reduced rate of herbicides to control the annual weeds if their number will exceed the economic threshold. If numbers do not justify early treatment, plans should be made for a later application to prevent seed production and to further check the perennial. An alternative would be herbicide application with a rope-wick if the weeds overtop the soybeans. The weed composition should be checked to see if the shifts projected in year 1 are taking place. If they are, planning should be started for the management, including the herbicides, to be used against the new weeds a year hence.

Years 5 and 6: Corn–Corn or Corn–Soybeans Rotation

These 2 years might be appropriate ones to change the crop rotation to either corn to be followed by corn, or corn to be followed by soybeans. This change could serve to avoid or at least slow down a shift in weed composition. Zero or only shallow tillage should be used to avoid bringing more weed seeds to the surface. Plans should be made to use reduced rates or sequential applications of herbicides postemergence if the weed economic threshold is exceeded or if there will be enough escapes to replenish the seedbank. Inclusion of winter cover crops that can be used to suppress weed emergence the following spring is an option for consideration.

Year 7: Wheat

Depending upon the success of efforts to find allelopathic crop germplasm and biological agents, this might be the year in which these approaches could be initiated to serve as a check against a new perennial, if one is developing. Also, if efforts to prevent weed seed production have been successful to this point in time, the seedbank will have been drawn down sufficiently to warrant modifying the overall weed management program. The main change would be to eliminate the soil application of herbicides and rely upon postemergence applications to control escapes and prevent seed production.

NEEDED RESEARCH

There are gaps in our knowledge in several areas important to fully effective weed management. Four of the most important are briefly discussed.

1. The cost effectiveness of weed prevention practices needs to be determined for a total weed management program over time. What is it worth to reduce the reservoir of weed seeds and buds a given percentage? What is it worth to have a competitive crop variety developed by the plant breeder? What is it worth to have an allelopathic variety developed? These are important questions in weed management and ones that cannot be adequately answered in terms of any one year. Answers to these and similar questions are needed for enterprise operators to choose the practices that will contribute the most to reducing weed numbers and associated costs. Without such information, weed prevention can not attain its full potential.

2. More precise information is needed about competition. There are nagging, unanswered questions about this subject that put serious constraints on weed management. For example, why are crop yields sometimes reduced when weeds are present for only the first 2 or 3 weeks after crop emergence? Knowing what we do about the plasticity of most of our crops, it is difficult to understand why they would not fully compensate for any competition that might have occurred during this brief, early period. However, the fact remains that such brief exposure to weeds does sometimes cause losses in yield. Can the losses be accounted for by competition for a minor element in scarce supply? Can it be accounted for by the effect of competition during this early time on the production of the crop's inherent growth regulators? At this time, we can only speculate. However, in view of the pervasiveness of growth regulators within a plant, more information clearly needs to be known about their relationship to competition. Of course, there is an urgent need for more precise information on what it is the

weed and the crop are competing for at a particular point in time in their co-development. Until we have such precise information, our approach will continue to be mainly one of dealing with symptoms rather than the cause. Such prophylactic treatment may be unduly costly.

3. More discrete information is needed about shifts in weed composition in response to changes in production practices. This need is exemplified by what has occurred in the move to conservation tillage. There is general knowledge that perennials and some shallow-germinating annuals will increase but precise information is lacking. Only when we are able to predict the weeds likely to be present following a change in a production system will we be able to develop a fully effective plan for dealing with them. The problem is basically one of determining for any given production system the numbers of propagules of individual species already present, the numbers added to the seedbank annually, and their survival—that is, their dormancy, germination, and longevity. Extrapolation of trend lines drawn from data accumulated for a series of years could be used to predict future levels of infestation. Of course, accuracy of the projections would increase with years. Even so, early projections (after 2 to 3 years) could be very useful. They could provide advance warning of the weed problem likely to be faced at some future date. Advance warning would allow plans to be made for dealing with the problem before it becomes severe. We must view precise prediction as an ideal. Because we are dealing with a dynamic system containing many variables, prediction is not likely to ever be a precise technology. However, even as an estimative technology, it has much to offer a weed management approach.

4. More information is needed about dormancy. Dormancy of both seeds and perennating parts is an obvious obstacle to more effective management of weeds. More needs to be learned about the factors causing both the onset and the release from dormancy, especially the effects of production practices.

A PHILOSOPHY FOR WEED MANAGEMENT

Weed management has much to offer in response to societies' concern for sustainable food production and protection of the environment. Weeds will always be a factor in crop production. Tillage and herbicides used to minimize their effect on crop yield make up a large part of crop production costs; tillage also contributes to soil and water erosion and herbicides to soil and water contamination. Effective weed management could reduce weed numbers and associated costs over time, thereby lowering the inputs needed to prevent their interference. For this to happen, weed science must provide the initiative for bringing about the needed changes.

Integration of weed prevention and weed control should be the goal in weed management. Only then will its benefits be fully realized. Attaining this goal must begin with weed science course offerings. This is because implementing the appropriate practices ultimately will entail decisions by the operator on the land. It follows that this individual must have an understanding of basic factors in weed reproduction, survival, and competition and of relationships to production practices. Course offerings must provide this information.

REFERENCES

Aldrich, R.J. 1987. Predicting crop yield reductions from weeds. Weed Technol. 1:199-206.

Alm, D.A., E.E. Stoller, and L.M. Wax. 1993. An index model for predicting seed germination and emergence rates. Weed Technol. 7:560-569.

Ball, D.A., and S.D. Miller. 1989. A comparison of techniques for estimation of arable soil seedbanks and their relationship to weed flora. Weed Res. 29:365-373.

Benech Arnold, R.L., C.M. Ghersa, R.A. Sauchez, and P. Incausti. 1990. A mathematical model to predict *Sorghum halepense* (L.) Pers. seedling emergence in relation to soil temperature. Weed Res. 30:91-99.

Biniak, B.M., and R.J. Aldrich. 1986. Reducing velvetleaf (*Abutilon theophrasti*) and giant foxtail (*Setaria faberi*) seed production with simulated-roller herbicide applications. Weed Sci. 34:256-259.

Bridges, D.C, H. Wu, P.J.G. Sharpe, and J.M. Chandler. 1989. Modeling distributions of crop and weed seed germination time. Weed Sci. 37:724-729.

Coble, H.D. 1985. Development and implementation of economic thresholds for soybean. In R.E. Frisbee and P.L. Adkisson, eds., Integrated pest management of major agricultural systems, pp. 295-306. Texas A&M University, College Station, TX.

Cousens, R. 1987. Theory and reality of weed control thresholds. Plant. Prot. Q. 2:13-20.

Cousens, R., P. Brain, J.J. O'Donovan, and A. O'Sullivan. 1987. The use of biologically realistic equations to describe the effects of weed density and relative time of emergence on crop yield. Weed Sci. 35:720-725.

Cousens, R., C.J. Doyle, B.J. Wilson, and G.W. Cousens. 1986. Modeling the economics of controlling *Avena fatua* in winter wheat. Pestic. Sci. 17:1-12.

Doyle, C.J., R. Cousens, and S.R. Moss. 1986. A model of the economics of controlling *Alopecuris myosuroides* in winter wheat. Crop Prot. 5:143- 150.

Forcella, F., and M.J. Lindstrom. 1988. Movement and germination of weed seeds in ridge-till crop production systems. Weed Sci. 36:56-59.

Forcella, R., K. Eradat-Oskoui, and S.W. Wagner. 1993. Application of weed seedbank ecology to low-input crop management. Ecol. Appl. 3:74-83.

King, C.A., and L.R. Oliver. 1994. A model for predicting large crabgrass (*Digitaria sanguinalis*) emergence as influenced by temperature and water potential. Weed Sci. 42:561-567.

Kremer, R.J. 1993. Management of weed seed banks with microorganisms. Ecol. Appl. 3:42-52.

Kropff, M.J., and C.J.T. Spitters. 1991. A simple model of crop loss by weed competition from early observations on relative leaf area of the weeds. Weed Res. 31:97-105.

Lybecker, D.W., E.E. Schweizer, and R.P. King. 1991. Weed management decisions in corn based on bioeconomic modeling. Weed Sci. 39: 124-129.

Mitchell, R.L., and W.J. Russell. 1971. Root development and rooting patterns of soybean (*Glycine max* L. Merrill) evaluated under field conditions. Agron. J. 63:313-316.

O'Donovan, J.T., E.A. de St. Remy, P.A. O'Sullivan, D.A. Dew, and A.K. Sharma. 1985. Influences of the relative time of emergence of wild oat (*Avena fatua*) on yield loss of barley (*Hordeum vulgare*) and wheat (*Triticum aestivum*). Weed Sci. 33:498-503.

Raper, C.D., and S.A. Barber. 1970. Rooting systems of soybean, I. Differences in root morphology among varieties. Agron. J. 62:581-584.

Schepers, J.S., and O.C. Burnside. 1979. Electronic moisture sensor for monitoring herbicide solution on a roller applicator. Weed Sci. 27:559-561.

Schreiber, M.M. 1992. Influence of tillage, crop rotation, and weed management on giant foxtail (*Setaria faberi*) population dynamics and corn yield. Weed Sci. 40:645-653.

Schweizer, E.E., D.W. Lybecker, and R.L. Zimdahl. 1988. Systems approach to weed management in irrigated crops. Weed Sci. 36:840-845.

Schweizer, E.E., P. Westra, and D.W. Lybecker. 1992. Controlling weeds in corn (*Zea mays*) rows with an in-row cultivator versus decisions made by a computer model. Weed Sci. 42:593-600.

Schweizer, E.E., and R.L. Zimdahl. 1984. Weed seed decline in irrigated soil after six years of continuous corn (*Zea mays*) and herbicides. Weed Sci. 32:76-83.

Shribbs, J.M., D.W. Lybecker, and E.E. Schweizer. 1990. Bioeconomic weed management models for sugarbeet (*Beta vulgaris*) production. Weed Sci. 38:436-444.

Wicks, G.A., and B.R. Somerhalder. 1971. Effect of seedbed preparation for corn on distribution of weed seed. Weed Sci. 19:666-668.

Wiles, L.J., and G.G. Wilkerson. 1991. Modeling competition for light between soybean and broadleaf weeds. Agric. Systems 35:37-51.

Wilson, R.G., E.D. Kerr, and L.A. Nelson. 1985. Potential for using weed seed content in the soil to predict future weed problems. Weed Sci. 33:171-175.

APPENDIXES

1. Common and Scientific Names of Weeds
2. Common and Chemical Names of Herbicides

APPENDIX 1. COMMON AND SCIENTIFIC NAMES OF WEEDS

Common name	Scientific name	Common name	Scientific name
Acacia	*Acacia* spp.	Buttercup, creeping	*Ranunculus repens*
African feathergrass	*Pennisetum macrourum*	Cactus, prickly pear	*Opuntia* spp.
Alder, red	*Alnus rubra*	Canada thistle	*Cirsium arvense*
Amaranth, Palmer	*Amaranthus palmeri*	Carline thistle	*Carlina vulgaris*
Annual bluegrass	*Poa annua*	Carpetweed	*Mollugo verticillata*
Annual morning-glory	*Ipomoea* spp.	Carrot, wild	*Daucus carota*
Anoda, spurred	*Anoda cristata*	Catchfly, nightflowering	*Silene noctiflora*
Barnyardgrass	*Echinochloa crus-galli*	Cattail	*Typha latifolia*
Beggarticks, hairy	*Bidens pilosa*	Chamomile, wild	*Matricaria chamomilla*
Beggarweed, Florida	*Desmoduim tortuosum*	Charlock	*Brassica arvensis*
Bentgrass	*Agrostis tenuis*	Chickweed, common	*Stellaria media*
Bermudagrass	*Cynodon dactylon*	Cinquefoil, common	*Potentilla canadensis*
Bindweed, field	*Convolvulus arvensis*	Cockle, corn	*Agrostemma githoga*
Birdseye speedwell	*Veronica persica*	cow	*Vaccaria segetalis*
Blackgrass	*Alopecurus myosuroides*	Cocklebur, beach	*Xanthium echinatum*
Black knapweed	*Centaurea nigra*	common	*Xanthium strumarium*
Black medic	*Medicago lupulina*	Common burdock	*Arctium minus*
Black mustard	*Brassica nigra*	Common chickweed	*Stellaria media*
Black nightshade	*Solanum nigrum*	Common cinquefoil	*Potentilla canadensis*
Bladder campion	*Silene vulgaris*	Common cocklebur	*Xanthium strumarium*
Blue couch	*Digitaria scalarum*	Common crupina	*Crupina vulgaris*
Bristly foxtail	*Setaria verticillata*	Common milkweed	*Asclepias syriaca*
Broadleaf plantain	*Planatago major*	Common mullein	*Verbascum thapsus*
Broadleaf signalgrass	*Brachiaria platyphylla*	Common purslane	*Portulaca oleracea*
Brome, downy	*Bromus tectorum*	Common ragweed	*Ambrosia artemisiifolia*
Broomsedge	*Andropogon virginicus*	Common reed	*Phragmites communis*
Buckhorn plantain	*Plantago lanceolata*	Common yarrow	*Achillea millefolium*
Buckwheat, wild	*Polygonum convolvulus*	Corn cockle	*Agrostemma githago*
Buffalobur	*Solanum rostratum*	Corn marigold	*Chrysanthemum segetum*
Bull thistle	*Cirsium vulgare*	Corn poppy	*Papaver rhoeas*
Burdock, common	*Arctium minus*	Corn speedwell	*Veronica arvensis*

APPENDIX 1. *(Continued)*

Common name	Scientific name	Common name	Scientific name
Corn spurry	*Spergula arvensis*	Garlic, wild	*Allium vineale*
Cow cockle	*Vaccaria segetalis*	Giant foxtail	*Setaria faberi*
Crabgrass, large	*Digitaria sanguinalis*	Giant ragweed	*Ambrosia trifida*
smooth	*Digitaria ishaemum*	Goatgrass, jointed	*Aegilops cylindrica*
Creeping buttercup	*Ranunculus repens*	Goldenrod spp.	*Solidago* spp.
Cress, corn	*Lepidium compestre*	Goosegrass	*Eleusine indica*
mouse-ear	*Arabidopsis thaliana*	Gourd, Texas	*Cucurbita texana*
rock	*Arabis hirsuta*	Green foxtail	*Setaria viridis*
Crested wheatgrass	*Agropyron cristatum*	Greenflower pepperweed	*Lepidium densiflorum*
Crunchweed	*Sinapsis arvensis*	Groundsel	*Senecio vulgaris*
Crupina, common	*Crupina vulgaris*	Guineagrass	*Panicum maximum*
Curly dock	*Rumex crispus*	Hairy beggarticks	*Bidens pilosa*
Cutgrass, rice	*Leersia oryzoides*	Hawkbit	*Leontodon* spp.
Daisy, ox-eye	*Chrysanthemum leucan-*	Healall	*Prunella vulgaris*
	themum	Hedge mustard	*Sisymbrium officinale*
Dallisgrass	*Paspalum dilatatum*	Hemp dogbane	*Apocynum cannabinum*
Dandelion	*Taraxacum officinale*	Hemp sesbania	*Sesbania exaltata*
Dock, curly	*Rumex crispus*	Hempnettle	*Galeopsis tetrahit*
Dodder, field	*Cuscuta campestris*	Hoary cress	*Cardaria draba*
Dogbane, hemp	*Apocynum cannabinum*	Honeysuckle, Japanese	*Lonicera japonica*
Downy, brome	*Bromus tectorum*	Honeyvine milkweed	*Ampelamus albidus*
Dropseed, sand	*Sporobolus cryptandrus*	Horseweed	*Conyza canadensis*
Ducksalad	*Heteranthera limosa*	Hydrilla	*Hydrilla verticillata*
Dwarf mallow	*Malva rotundifolia*	Ironweed, western	*Vernonia baldwini*
Eastern black nightshade	*Solanum ptycanthum*	Itchgrass	*Rottboellia exaltata*
Eulalia	*Miscanthus sinensis*	Ivyleaf morning-glory	*Ipomoea hederacea*
Fall panicum	*Panicum dichoto-*	Japanese honeysuckle	*Lonicera japonica*
	miflorum	Japanese knotweed	*Polygonum cuspidatum*
False pimpernel	*Lindernia anagallidea*	Jimsonweed	*Datura stramonium*
Falseflax, largeseed	*Camelina sativa*	Johnsongrass	*Sorghum halepense*
smallseed	*Camelina microcarpa*	Jointed goatgrass	*Aegilops cylindrica*
flatseed	*Camelina dentata*	Jointvetch	*Aeschynomene virginica*
Field bindweed	*Convolvulus arvense*	Kikuyu grass	*Pennisetum clandes-*
Field dodder	*Cuscuta compestris*		*tinum*
Flatseed falseflax	*Camelina dentata*	Klamath weed	*Hypericum perforatum*
Fleabane	*Erigeron* spp.	Knapweed, black	*Centaurea nigra*
Flixweed	*Descurainia sophia*	Russian	*Centaurea repens*
Florida beggarweed	*Desmodium tortuosum*	spotted	*Centaurea maculosa*
Florida pusley	*Richardia scabra*	Knotweed, Japanese	*Polygonum cuspidatum*
Foxtail barley	*Hordeum jubatum*	prostrate	*Polygonum neglectum*
Foxtail, giant	*Setaria faberi*	Kochia	*Kochia scoparia*
green	*Setaria viridis*	Ladysthumb	*Polygonum persicaria*
yellow	*Setaria glauca*	Lambsquarters	*Chenopodium album*
Foxtail millet	*Setaria italica*	Large crabgrass	*Digitaria sangvinalis*
bristly	*Setaria verticillata*	Largeseed falseflax	*Camelina sativa*
Fumitory	*Fumaria officinalis*	Leafy spurge	*Euphorbia esula*
Galinsoga, smallflower	*Galinsoga parviflora*	Little barley	*Hordeum pusillum*
Garden orach	*Atriplex hortensis*	Loosestrife, purple	*Lythrum salicaria*

Common name	Scientific name	Common name	Scientific name
Lovegrass	*Eragrostis* spp.	Pennsylvania smart-	*Polygonum pennsylvan-*
Mallow, dwarf	*Malva rotundifolia*	weed	*icum*
round-leaved	*Malva pusilla*	Pennycress	*Thlaspi arvense*
Marigold, wild	*Tagetes minuta*	Pepperweed	*Lepidium campestre*
Medic, black	*Medicago lupulina*	greenflower	*Lepidium densiflorum*
Mediterranean sage	*Salvia aethiops*	Virginia	*Lepidium virginicum*
Medusahead	*Taeniatherum asperum*	Perennial sowthistle	*Sonchus arvensis*
Mesquite	*Prosopis juliflora*	Pigweed, redroot	*Amaranthus retroflexus*
Mexican dock	*Rumex mexicanus*	prostrate	*Amaranthus blitoides*
Milkweed, common	*Asclepias syriaea*	Russian	*Axyris amaranthoides*
honeyvine	*Ampelamus albidus*	tumbling	*Amaranthus albus*
Milkweed vine	*Morrenia odorata*	Pimpernell, scarlet	*Anagallis arvensis*
Moonflower, purple	*Ipomoea turbinata*	Pitted morning-glory	*Ipomoea lacunosa*
Morning-glory, annual	*Ipomoea* spp.	Plantain, broadleaf	*Plantago major*
ivyleaf	*Ipomoea hederacea*	Plantain, buckhorn	*Plantago lanceolata*
pitted	*Ipomoea lacunosa*	Poison ivy	*Toxicodendron radicans*
tall	*Ipomoea purpurea*	Poppy, corn	*Papaver rhoeas*
Moth mullein	*Verbascum blattaria*	Povertyweed	*Monolepis nuttaliana*
Mullein, common	*Verbascum thapsus*	Prickly lettuce	*Lactuca serriola*
Mullein, moth	*Verbascum blattaria*	Prickly pear cactus	*Opuntia* spp.
Multiflora rose	*Rosa multiflora*	Prickly sida	*Sida spinosa*
Musk thistle	*Carduus* spp.	Primrose, evening	*Oenothera biennis*
Mustard, black	*Brassica nigra*	Prostrate spurge	*Euphorbia supina*
dog	*Erucastrum gallicum*	Puncturevine	*Tribulus terrestris*
hare's ear	*Conringia orientalis*	Purple loosestrife	*Lythrum salicaria*
hedge	*Sisymbrium officinale*	Purple moonflower	*Ipomoea turbinata*
Indian	*Brassica juncea*	Purple nutsedge	*Cyperus rotundus*
tumble	*Sisymbrium altissimum*	Purslane, common	*Portulaca oleracea*
wild	*Brassica kaber*	Pusley, Florida	*Richardia scabra*
wormseed	*Erysimum cheiran-*	Quackgrass	*Elytrigia repens*
	thoides	Ragweed, common	*Ambrosia artemisiifolia*
Nightflowering catchfly	*Silene noctiflora*	giant	*Ambrosia trifida*
Nightshade, black	*Solanum nigrum*	western	*Ambrosia psilostachya*
eastern black	*Solanum ptycanthum*	Ragwort, tansy	*Senecio jacobaea*
silverleaf	*Solanum elaegnifolium*	Red alder	*Alnus rubra*
Nutsedge, purple	*Cyperus rotundus*	Red rice	*Oryza sativa*
yellow	*Cyperus esculentus*	Redroot pigweed	*Amaranthus retroflexus*
Oat, wild	*Avena fatua*	Redvine	*Brunnichia cirrhosa*
winter wild	*Avena ludoviciana*	Rhodesgrass	*Chloris gayana*
Oat-grass	*Arrhenatherum elatius*	Rice cutgrass	*Leersia oryzoides*
Oxalis	*Oxalis* spp.	Rocket, yellow	*Barbarea vulgaris*
Ox-eye daisy	*Chrysanthemum leucan-*	Round-leaved mallow	*Malva pusilla*
	themum	Russian knapweed	*Centaurea repens*
Palmer amaranth	*Amaranthus palmeri*	Russian thistle	*Salsola kali*
Panicum, fall	*Panicum dichotomiflo-*	Ryegrass, rigid	*Lolium rigidum*
	rum	Sage, Mediterranean	*Salvia aethiops*
Texas	*Panicum texanum*	Salsify, western	*Tragopogon dubius*
Parsley-piert	*Aphanes arvensis*	Sand dropseed	*Sporobolus cryptandrus*

APPENDIX 1. *(Continued)*

Common name	Scientific name	Common name	Scientific name
Sandbur	*Cenchrus incertus*	musk	*Carduus* spp.
Scarlet pimpernel	*Anagallis arvensis*	Russian	*Salsola kali*
Sesbania, hemp	*Sesbania exaltata*	Tickseed	*Corispermum* spp.
Shepherdspurse	*Capsella bursa-pastoris*	Timothy	*Phleum pratense*
Sicklepod	*Senna obtusifolia*	Toadflax, yellow	*Linaria vulgaris*
Sida, prickly	*Sida spinosa*	Triple-awned grass	*Aristida oligantha*
Signalgrass, broadleaf	*Brachiaria platyphylla*	Tumble mustard	*Sisymbrium altissimum*
Silverleaf nightshade	*Solanum elaegnifolium*	Tumblegrass	*Schedonnardus panicu-*
Skeleton weed	*Lygodesmia juncea*		*latus*
Slender naiad	*Najas flexilis*	Tumbling pigweed	*Amaranthus albus*
Smallflower galinsoga	*Galinsoga parviflora*	Tussilago farfara	*Tursilago farfara*
Smallseed falseflax	*Camelina microcarpa*	Velvetleaf	*Abutilon theophrasti*
Smartweed, Pennsyl-	*Polygonum pennsylvan-*	Virginia pepperweed	*Lepidium virginicum*
vania	*icum*	Waterhyacinth	*Eichornia crassipes*
Smooth bromegrass	*Bromus inermis*	Waterprimrose, winged	*Jussiaea decurrens*
Smooth crabgrass	*Digitaria ischaemum*	Western ironweed	*Vernonia baldwinii*
Soapwort	*Saponaria officinalis*	Western salsify	*Tragopogon dubius*
Sorrel	*Rumex acetosa*	Wild buckwheat	*Polygonum convolvulus*
Sowthistle, perennial	*Sonchus arvensis*	carrot	*Daucus carota*
Speedwell, birdseye	*Veronica persica*	chamomile	*Matricaria chamomilla*
corn	*Veronica arvensis*	garlic	*Allium vineale*
Spurge, leafy	*Euphorbia esula*	marigold	*Tagetes minuta*
prostrate	*Euphorbia supina*	marjoram	*Origanum vulgare*
Spurred anoda	*Anoda cristata*	mustard	*Brassica kaber*
Spurry, corn	*Spergula arvensis*	oat	*Avena fatua*
St. Johnswort	*Hypericum perforatum*	parsnip	*Pastinaca sativa*
(Klamath weed)		sunflower	*Helianthus* spp.
Stinkgrass	*Eragrostis cilianensis*	Winged waterprimrose	*Jussiaea decurrens*
Stinking clover	*Cleome serrulata*	Winter wild oat	*Avena sterilis*
Strangler vine	*Morrenia odorata*	Witchweed	*Striga lutea*
Sunflower, wild	*Helianthus* spp.	Wormseed mustard	*Erysimum cheiran-*
Tall morning-glory	*Ipomoea purpurea*		*thoides*
Tansy ragwort	*Senecio jacobaea*	Yarrow, common	*Achillea milliefolium*
Texas gourd	*Cucurbita texana*	Yellow bedstraw	*Gallium verum*
Texas panicum	*Panicum texanum*	Yellow foxtail	*Setaria glauca*
Thistle, bull	*Cirsium vulgare*	nutsedge	*Cyperus esculentus*
Canada	*Cirsium arvense*	rocket	*Barbarea vulgaris*

APPENDIX 2. COMMON AND CHEMICAL NAMES OF HERBICIDES

Common name[1]	Chemical name
Alachlor	2-chloro-*N*-(2,6-diethylphenyl)-*N*-(methoxymethyl)acetamide
Ametryn	*N*-ethyl-*N'*-(1-methylethyl)-6-(methylthio)-1,3,5-triazine-2,4-diamine
Amiben	(See *Chloramben*)
Amitrole	3-amino-*s*-triazole
Atrazine	6-chloro-*N*-ethyl-*N'*-(1-methylethyl)-1,3,5-triazine-2,4-diamine
Barban	4-chloro-2-butynyl *m*-chlorocarbanilate
Bentazon	3-(1-methylethyl)-(1*H*)-2,1,3-benzothiadiazin-4(3*H*)-one 2,2-dioxide
Bromoxynil	3,5-dibromo-4-hydroxybenzonitrile
Butylate	*S*-ethyl bis(2-methylpropyl)carbamothioate
CDAA	2-chloro-*N*-*N*-diallyl-2-chloroacetamide
Chloramben	3-amino-2,5-dichlorobenzoic acid
Chlorimuron	2-[[[[(4-chloro-6-methoxy-2-pyrimidinyl)amine] carbonyl]amino)sulfonyl]benzoic acid
Chloropropham	isopropyl *m*-chlorocarbanilate
Chlorsulfuron	2-chloro-*N*-[[4-methoxy-6-methyl-1,3,5-triazin-2-yl]amino]carbonyl]benzenesulfonamide
Cyanazine	2-[[4-chloro-6-(ethylamino)-1,3,5-triazin-2-yl]amino]-2-methylpropanenitrile
Cycloate	*S*-ethyl cyclohexylethylcarbamothioate
Dalapon	2,2-dichloropropionic acid
DCPA	dimethyl 2,3,5,6-tetrachloro-1,4-benzenedicarboxylate
Diallate	*S*-(2,3-dichloroallyl) diisopropylthiocarbamate
Dicamba	3,6-dichloro-2-methoxybenzoic acid
Diclofop	2-[4-(2,4-dichlorophenoxy)phenoxy]propanoic acid
Dinoseb	2-*sec*-butyl-4,6-dinitrophenol
Diphenamid	*N*,*N*-dimethyl-2,2-diphenylacetamide
Diquat	6,7-dihydrodipyridol[1,2-α:2',1'-*c*]-pyrazinedinium ion
Diuron	*N'*-(3,4-dichlorophenyl)-*N*,*N*-dimethylurea
EPTC	*S*-ethyl dipropyl carbamothioate
Fenac	(2,3,6-trichlorophenyl)acetic acid
Fenuron	1,1-dimethyl-3-phenylurea
Glyphosate	*N*-(phosphonomethyl)glycine
Imazaquin	2-[4,5-dihydro-4-methyl-4-(1-methylethyl)-5-oxo-1*H*-imidazol-2-yl]-3-quinolinecarboxylic acid
Imazethapyr	2-[4,5-dihydro-4-methyl-4-(1-methylethyl)-5-oxo-1*H*-imidazol-2-yl]-5-ethyl-3-pyridinecarboxylic acid
Linuron	3-(3,4-dichlorophenyl)-1-methoxy-1-methylurea
Maleic Hydrazide	1,2-dihydro-3,6-pyridazinedione
MCPA	[(4-chloro-*o*-tolyl)oxy]acetic acid
Metolachlor	2-chloro-*N*-(2-ethyl-6-methylphenyl)-*N*-(2-methoxy-1-methylethyl)acetamide
Metribuzin	4-amino-6-(1,1-dimethylethyl)-3-(methylthio)-1,2,4-triazin-5(4*H*)-one
Monuron	3-(*p*-chlorophenyl)-1,1-dimethylurea
Oryzalin	4-(dipropylamino)-3,5-dinitrobenzenesulfonamide

---

[1]Names accepted by the Weed Science Society of America

APPENDIX 2. (*Continued*)

Common name	Chemical name
Paraquat	1,1-dimethyl-4,4-bipyridinium ion
Pendimethalin	*N*-(1-ethylpropyl)-3,4-dimethyl-2,6-dinitrobenzenamine
Picloram	4-amino-3,5,6-trichloro-2-pyridinecarboxylic acid
Prometon	2,4-bis(isopropylamino)-6-methoxy-*s*-triazine
Prometryn	2,4-bis(isopropylamino)-6-(methylthio)-*s*-triazine
Propachlor	2-chloro-*N*-(1-methylethyl)-*N*-phenylacetamide
Pyrazon	5-amino-4-chloro-2-phenyl-3-(2*H*)-pyridazinone
Sethoxydim	2-[1-(ethoxyimino)butyl]-5-[2-(ethylthio)propyl]-3 hydroxy-2-cyclohexen-1-one
Silvex	2-(2,4,5-trichlorophenoxy)propionic acid
Simazine	6-chloro-*N'*,*N'*-diethyl-1,3,5-triazine-2,4-diamine
Sodium Azide	N_3Na
Sulfometuron	2-[[[[(4,6-dimethyl-2-pyrimidinyl)amino]carbonyl] amino]sulfonyl]benzoic acid
Terbacil	5-chloro-3-(1,1-dimethylethyl)-6-methyl-2,4-(1*H*,3*H*)-pyrimidinedione
Triallate	*S*-(2,3,3-trichloro-2-propenyl) bis(1-methylethyl)carbamothioate
Trifluralin	2,6-dinitro-*N*,*N*-dipropyl-4-(trifluoromethyl)benzeneamine
2,4-D	(2,4-dichlorophenoxy)acetic acid
2,4-DB	4-(2,4-dichlorophenoxy)butyric acid
2,4,5-T	(2,4,5-trichlorophenoxy)acetic acid

——GLOSSARY——

Abiotic: The nonliving portion of the environment.

Abscisic acid: A naturally occurring, plant growth inhibitor; commonly considered to be a key factor in stomatal control, leaf senescence, and bud and seed dormancy.

Absorption: Penetration of a substance into the body of an organism or particle.

Achene: A dry, single-seeded, indehiscent fruit whose pericarp and seed coat are not fused.

Acropetal: Plant structures produced in succession toward the apex; also, translocation toward the apex.

Additive design: With respect to studies of effects of weed infestation level on crops, experimental designs in which the total number of plants on a given area (weeds + crops) is either increased or decreased as their proportion changes. See *replacement design.*

Adsorption: Retention of a substance on a surface.

Adventitious (bud or root): One originating from mature rather than from meristematic tissue.

Agroecosystem: The agricultural species and production practices functioning together in a production system.

Allelochemicals: Active chemicals responsible for allelopathic effect.

Allelopathy: Any harmful effect of one plant on another from the production of chemical compounds that escape into the environment.

Allopatry: Speciation resulting from the compensation for environmental conditions of geographically separated segments of a species. See *sympatry.*

Amphoteric: A chemical capable of reacting either as an acid or as a base.

Annidation: The complementary use of resources by two or more plant species occupying a given area.

Annual: A plant that completes its life cycle in 1 year or less. A summer annual germinates in the spring or summer, flowers, produces seed, and dies that growing season. A winter annual germinates in late summer or fall, overwinters, then flowers, produces seed, and dies the following spring.

Antidote (herbicidal): A chemical used to protect a desired plant from a herbicide. See *safener.*

Apical dominance: Inhibition by the apical meristem of growth of buds on a rhizome or along the meristem.

Apoplast: The translocation pathway of a plant that involves nonliving cells.

Assimilation: The incorporation or conversion of absorbed substances into living matter.

Augmentation: Modifying the environment or supplementing the population of a native biotic agent to increase its effectiveness in biological control.

Auxin: See *plant growth regulator.*

Axillary bud: One originating at the angle between the stem and a branch or leaf petiole.

Basipetal: Structures produced in succession toward the base; also, translocation toward the base.

Biennial: A plant that requires more than 1 but less than 2 years to complete its life cycle.

Binomial: With reference to taxonomy, the two-part Latin name used in the scientific identification of plants (e.g., *Chenopodium album* = lambsquarters).

Biochemical site of action: The single reaction affected at a concentration lower than any other reaction, or first affected at a given low concentration.

Biodegradation: Degradation of a compound mediated by living organisms, primarily microorganisms.

Bioeconomic weed management: A system that integrates weed threshold information and economic information in deciding on action to be taken.

Bioherbicide (mass exposure): A biological agent effective in controlling a weed at the applied concentration.

Biological control: Intentional use of living organisms to reduce vigor, reproductive capacity, density, or impact of weeds.

Biomass: The total quantity of plant tops and roots produced on a given area.

Biosystem (weed–crop): Regularly interacting and interdependent weeds and crops in a plant community.

Biotic: The living portion of the environment.

Biotic agent: A biological organism used to suppress or control weeds.

Bulb: A specialized underground organ consisting of a short, fleshy, usually vertical stem axis (basal plate) bearing at its apex a growing point or a flower primordium that is enclosed by thick fleshy leaves.

Canopy: The cover of leaves and stems formed by the tops of plants as viewed from above.

Carrying capacity: The maximum biomass that can be maintained on an area over time.

Casparian strip: A continuous, impermeable, waxy band in cell walls of the root endodermis that serves as a barrier to free passage of water and solutes to vascular tissue.

Chlorosis: Lack of green color in foliage caused either by loss of chlorophyll or by its failure to develop.

Climax: A stabilized ecosystem in ecological succession.

Clone: A group of organisms descended by asexual reproduction from a common ancestor.

Coexistence: Persistence on a given area over time of two or more plant species sharing common resources.

Coleoptile: Sheath surrounding the plumule in a grass seedling.

Coma: A tuft of hairs attached to a seed.

Cometabolism: Biodegradation of a compound that is not used as an energy source by the degrading microorganism.

Community: The assemblage of plant populations on a given area.

Competition: Relationship between two or more plants in which the supply of a growth factor falls below their combined demands.

Competitive exclusion principle: An ecological separation of closely related and similar species in response to competition.

Conservation tillage: Any tillage practice that reduces soil or water loss compared to moldboard plowing. Conservation tillage commonly leaves at least 30% of plant residues on the soil surface.

Contact herbicide: A chemical that kills plants mainly by contact with tissue rather than as a result of translocation to another site.

Control: That part of weed management that focuses on the reduction of a given weed infestation for a specified period of time, usually part of a crop growing season.

Conventional tillage: Any seedbed preparation practice that buries most of the plant residues. Commonly begins with moldboard plowing followed by disk or spring-tooth harrowing.

Corm: The swollen base of a stem axis enclosed by dry, scale-like leaves. Distinguished from a bulb by its solid stem structure with distinct nodes and internodes.

Correlative inhibition: The inhibiting effect of one bud on the growth of another as on rhizomes of a perennial weed.

Cover crop: A close-growing crop grown primarily to protect soil between periods of regular crop production.

Creeping rootstock: Laterally growing roots whose tissue commonly stores carbohydrates and may form adventitious buds for regeneration of growth, such as in bindweed.

Culm: The jointed, usually hollow, stem of various grasses.

Cultivation: The mechanical loosening or tilling of soil around growing plants.

Day length: See *photoperiod.*

Degradation: Breakdown of organic compounds to simpler components.

Deleterious rhizobacteria: Nonparasitic bacteria colonizing plant roots able to suppress plant growth without invading root tissues.

Density: Number of plants on a specified area.

Depletion zone: The volume of soil in which the supply of a growth factor is reduced by plant roots.

Dermal tissue: The outer system of cells of a plant.

Dicot (dicotyledonous): Those plants having two cotyledons, as in broadleaf weeds. See *monocot.*

Diffusion: With respect to a soil-supplied nutrient, its movement in response to uniform mixing (equalizing concentrations) in the soil solution.

Directed application: Application of a herbicide to a specific area, or plant part, commonly the base of plants.

Dispersal: The movement of a reproductive unit (seed or perennating part) from its place of production.

Diurnal: A daily event or process usually associated with changes from day to night.

Dockage: Reduction in the price paid for a farm product as a result of contamination with weeds or weed parts.

Dominants: Species in a community that exert the major controlling influence on energy flow and on the environment of other species in the ecosystem.

Dormancy: State in which growth of a specific plant part is not resumed even though the environment supports germination, seedling growth, or development of other, apparently identical, tissues of the same species or plant. See *enforced dormancy, induced dormancy,* and *innate dormancy.*

Ecological (biological) niche: See *niche.*

Ecological succession: See *succession.*

Ecology: The study of the relationship between living organisms and their environment.

Economic threshold: That point in weed density where cost of weed control equals value of crop yield saved.

Ecosystem: Any unit that includes the living and nonliving environment regularly interacting to form a unified whole with clearly defined trophic structure, biotic diversity, and materials cycles.

Ecotone: The juncture zone of two or more ecological communities.

Ecotype: A locally adapted population of a species.

Edge effect: Tendency for greater species diversity at the juncture of ecological communities.

Emergence: The stage in plant development when the seedling is first visible above the soil surface.

Enforced dormancy: Failure to germinate and grow because conditions necessary to support growth are lacking.

Epinasty: The twisting or curling of leaves and stems caused by uneven growth of cells.

Epithet: With reference to the binomial nomenclature, the second part of the species' Latin name (e.g., *album* in *Chenopodium album*).

Eradication: The complete elimination of all live plants, perennating parts, and seeds of a weed from a given area.

Escapes: Individual specimens or species of plants not controlled by a specific control practice.

Factor compensation: Modification of plant growth form and of the physical environment by a plant that serves to limit the effect of the physical conditions to which it is exposed.

Fallowing: Allowing cropland to lie idle, either tilled or untilled, for the entire or most of the growing season.

Fibrous root system: A root system having a large number of small, finely branched, spreading roots but no large individual roots or central root.

Fundamental tissue: The tissue within which the vascular tissue is imbedded and from which the various plant parts originate.

General-purpose genotype: Genotypes that survive over a wide range of climatic conditions while still maintaining the ability to evolve new forms through genetic recombinations.

Genus: With reference to the binomial nomenclature, the first part of a species' Latin name; it provides generic identity to the taxonomic group between family and species within the plant kingdom.

Germination: Resumption of growth of a seed or of a vegetative perennating part.

Gibberellins: A specific, naturally occurring group of plant growth regulators that stimulate growth.

Growth factor: Any one of the five factors—light, carbon dioxide, water, nutrients, and oxygen—required for plant growth.

Growth form: A plant's aboveground general shape, including its height, leaf type, leaf arrangement, and attitude toward light.

Growth substance: See *plant growth regulator.*

Gymnosperm: A plant that does not have flowers in the ordinary sense, but is naked seeded, such as ground hemlock (*Taxus canadensis*).

Habitat: Place where a plant lives.

Herbaceous perennial: A vascular plant that lives for more than 2 years and does not develop woody tissue.

Homeostasis: Tendency of a biological system to resist change and remain in a state of equilibrium.

Hybrid: A plant resulting from a cross between parents of different species, subspecies, or ecotypes.

Hydrophilic: With respect to surfactants, that portion of the molecule soluble in water.

Hydrophobic: See *lipophilic.*

Incorporation: The mechanical mixing of a herbicide in the surface 2.5 cm to 7.5 cm of soil.

Induced (secondary) dormancy: Creation of the dormant state as a result of conditions to which the reproductive part is exposed after separation from the parent.

Inflorescence: The flowering part of a plant.

Innate (primary) dormancy: Presence of the dormant state in the reproductive part when released from the parent.

Inoculation: Release of a biotic agent to build to levels sufficient to hold a weed infestation below an economic threshold level.

Interference: The deleterious effect of one plant on another.

Juvenile period: The stage in the early development of a plant before initiation of reproduction (of seed or perennating parts).

K-strategy: With respect to survival, a weed that depends upon strong competitive ability, such as is found with many perennial species. See *r-strategy*.

Land equivalent ratio: The land area required to produce a given total yield in monoculture compared to the area needed to produce the same total yield in intercrops.

Leaching: The movement, most commonly downward, of a herbicide in soil with percolating water; also, the introduction of an allelochemical into the plant environment as a result of dissolution from living or dead plant parts.

Leaf area index: The blade area of a leaf, or leaves, relative to the given soil surface area it covers.

Life cycle: The life span of a plant from germination of the seed, to growth, maturation, reproduction, senescence, and death.

Limits of tolerance: Range of supply of a growth factor or environmental attribute within which an organism will exist.

Lipophilic: With respect to surfactants, that portion of the molecule soluble in nonaqueous substances.

Longevity: The time, after being produced, a seed or vegetative perennating part retains its ability to resume growth.

Manipulative method: With respect to biological control, steps taken to conserve or augment the number of individuals of a biotic control agent present so as to attain the desired weed control level.

Mass exposure: See *bioherbicide*.

Mass flow: Movement of a soil-supplied nutrient or herbicide by movement of the soil water.

Mechanism of action: The biochemical and biophysical responses of a plant to a herbicide.

Mode of action: The total phytotoxic effects and fate of a herbicide on or in a plant.

Moisture extraction profile: The cross-section profile of lateral and vertical extraction of water from soil by plant roots.

Monocarpy: The production of only one seed per carpel (seed-bearing structure).

Monocot (monocotyledon): Any seed plant having only one seed leaf.

Mutually exclusive: With respect to weed control in a crop with herbicides, the separation of the herbicidal effect, if any, from the effect of interspecific competition.

Mycoherbicide: A bioherbicide composed of a fungal pathogen.

Mycorrhizae: Fungal association with the roots of higher plants, which may be a factor in nutrient or water uptake in weed–crop situations.

Niche: Physical space, functional role in the community, and position in environmental gradients of temperature, moisture, pH, soils, and other conditions of existence occupied by an organism (what an organism does).

Ontogeny: The developmental history of an individual plant.

Organism: See *species*.

Pappus: The modified calyx limb forming a crown at the summit of the achene in Compositae and other plants.

Patch: The concentration of individuals of a single weed species, commonly perennials, to a specific site or portion of a crop field.

Penumbra: With respect to reception of light by a given plant, the extent of shading of one leaf by another over the total photoperiod.

Perennating part: A specialized structure for vegetative regrowth of perennial plants.

Perennial: A vascular plant that lives for more than 2 years. See *herbaceous perennial* and *woody perennial.*

Periodicity: The tendency for seeds or perennating parts of an individual species to have a flush of resumption of growth at a certain time in the growing season.

Persistence: The duration of toxic levels of a herbicide in soil.

Phloem: The food- (photosynthate-) conducting tissue of vascular plants.

Photodegradation: Degradation of compounds mediated by radiation.

Photoperiod (day length): Period of photosynthetically active radiation.

Photosynthate: The primary product of photosynthesis, mainly carbohydrates.

Phytochrome: A light-absorbing pigment that affects plant morphogenesis; especially important in weed seed germination.

Phytotoxin: Metabolite produced by microorganisms that has a toxic effect on plants.

Plagiotropic: With respect to light reception of an individual leaf, change in inclination toward the sun to maximize light capture throughout the photoperiod.

Plant growth regulator: Any synthetic or naturally occurring organic compound that in very low concentrations affects plant growth and development.

Plant hormone: See *plant growth regulator.*

Plasmodesmata: Minute tubules through pores in plant cell walls.

Plasticity: The extent to which any aspect of plant growth changes in response to changes in the environment.

Polycarpy: The production of two or more seeds per carpel (seed-bearing structure).

Polymorphism (seed): Production by an individual plant of seeds differing in color, form, dormancy, or other characteristics.

Polyploidy: Having more than 2 times the basic chromosome number.

Population: Collective group of organisms of the same species.

Preemergence: Application of a herbicide prior to emergence of specified weed or seeded crop.

Prevention: The part of weed management that focuses on preventing influx and reproduction of weeds and on practices to minimize their competitive effects. See *control.*

Propagule: An individual unit of reproduction (seed or perennating part).

Protectant: See *safener.*

Ramet: An individual reproductive unit of clonal growth that may develop independently if severed from the parent.

Recirculating sprayer: A type of application equipment designed to apply a spray horizontal to the ground and to catch that spray not intercepted by the plants in a receptacle for recirculation.

Reduced herbicide rate: Rate below that on the herbicide label.

Relative yield total: The sum of the yields of each species in a mixture divided by its yield in pure stand.

Replacement design: With respect to studies of effects of weed infestation level on crops, experimental designs in which the total number of plants (crops + weeds) is kept constant as their proportion is changed. See *additive design.*

Resources allocation: The portion of photosynthate or minerals utilized in the production of distinct plant parts. Commonly applied to partitioning of photosynthate or minerals to production of reproductive versus vegetative growth.

Rhizome: A specialized horizontal stem that grows belowground or just at the soil surface.

Ridge-tillage: A production system that involves movement of soil to and from the crop row during the growing season.

Root depletion zone: The soil area surrounding a root or root system in which the quantity of a soil-supplied growth factor has been reduced through root uptake.

Rootstock: Root tissue that commonly stores carbohydrates and may form adventitious buds for regeneration of growth. See *creeping rootstock* and *taproot.*

Rope-wick: An applicator for herbicides in which the herbicide in solution moves by capillary action to an applying surface (commonly a rope or fabric) where it is wiped onto the weed leaves.

r-strategy: With respect to survival, a weed that depends upon large numbers of reproductive units, such as is found with most annual species. See *K-strategy.*

Safener: A chemical used to protect a desired plant from a herbicide.

Secondary plant compounds: Plant metabolites that do not function in primary activities that support growth, development, or reproduction.

Seed: A fertilized, mature ovule having an embryonic plant, stored food material (rarely missing), and a protective coat or coats.

Seedbank: The reservoir of viable weed seeds in soil.

Seedling establishment: The stage in the life cycle of a plant when the newly emerged plant becomes independent of the parent or seed.

Selectivity: The extent of tolerance of desired plants to the amount of herbicide needed to control specified weeds.

Speciation: The product of natural selection and genetic mutation resulting in a new gene pool.

Species: A biological unit that shares a common gene pool.

Stability index: A number that indicates the extent of changes in weed numbers of a given cropping system. It can be found by summing percentage changes in numbers of individual weed species at any specified time following initiation.

Stale seedbed: A weed management practice that involves seedbed preparation enough in advance of crop planting to ensure germination of many weed seeds that are then destroyed by herbicides.

Stolon: A general term for any of several specialized horizontal stems that take root at the nodes (e.g., tubers and rhizomes). In weed science, it is best to restrict the term to aboveground structures to distinguish them from rhizomes.

Succession: Orderly process of community development involving changes in species with time.

Surfactant: A material that improves the emulsifying, dispersing, spreading, wetting, and other surface-modifying properties of a herbicide formulation.

Survival strategy: See *r-strategy* and *K-strategy*.

Sympatry: Speciation under very local conditions, such as a result of polyploidy, hybridization, self-fertilization, and asexual reproduction.

Symplast: The part of a plant's transport pathway consisting of phloem and living matter of the fundamental tissue.

Synergistic: Complementary effect of one plant on another so that the collective growth exceeds either one alone. Also, complementary action of two or more herbicides (chemicals) resulting in a greater combined effect than the sum of the independent effects.

Systemic: A herbicide readily translocated within the plant that commonly has its effect at a site other than the point of entry.

Taproot: A central, dominant root that normally grows vertically and from which most or all of the smaller roots spread out laterally; also, root tissue that commonly stores carbohydrates and may form adventitious buds for regeneration of growth, such as in dandelion.

Threshold level (value): That point in infestation by weeds (numbers or weight) at which crop yield begins to be reduced.

Tillage: The mechanical manipulation of soil, with *conventional* being the combined primary and secondary tillage commonly performed in preparing a seedbed for a given crop and area; *minimum* being the least amount of soil manipulation necessary for crop production or to meet tillage requirements under the existing soil and climatic conditions; *primary* being the initial major soil-working operation commonly designed to loosen or reduce soil strength; *reduced* being any combination of tillage operations designed to lessen the total amount of tillage; and *secondary* being any soil-working operation following primary tillage, commonly designed to refine the seedbed for crop planting.

Tiller: An erect or semierect branch arising from a bud in the axils of leaves at the base of a plant, as in johnsongrass.

Tissue system: One of the three broad types of tissues common to vascular plants (e.g., dermal, vascular, and fundamental).

Tuber: A relatively short, thickened stem structure that develops belowground as a consequence of the swelling of the subapical portion of a rhizome and subsequent accumulation of reserve materials.

Vascular tissue: The conducting tissue of vascular plants.

Viability: The extent to which seed or perennating parts retain the capability to resume growth when provided conditions favorable to growth.

Weed: A plant that originated under a natural environment and, in response to imposed and natural environments, evolved, and continues to do so as an interfering associate with our desired crops and activities.

Weed composition: The makeup or the proportion of weed species in a weed community.

Weed–crop ecology: The study of the interrelationships between crops and weeds and their environment.

Weediness: The extent of weed abundance in a field, lawn, garden, and other areas used by humans.

Weed management: The approach to weeds in which prevention and control have companion roles.

Woody perennial: A vascular plant that lives for more than 2 years and develops woody tissue.

Xylem: The principal conducting tissue for water and mineral nutrients in plants.

INDEX

Abiotic components, of biosystems, 37-38
Abiotic degradation, of herbicides, 321-22
Abscisic acid, 196, 199
   in axillary branch production, 188-89
   in dormancy, 138, 140, 157
Absorptive barriers, for herbicides, 299-300
Actinomycetes
   herbicide degradation by, 327-28
   natural herbicides from, 247
Activated charcoal, 299-300
Active uptake, 290
Adsorption of herbicides, 304-8
   leaching and, 309, 310, 312
   persistence and, 326
   volatilization and, 314-15, 317
Adventitious buds, on roots, 107-9
   distribution in soil, 110, 111
   environmental factors and, 118, 162
Adventitious roots, of monocots, 290-91
Aerial application, of herbicides, 274. *See also*
   Postemergence herbicide use
Agamospermy, 49-50
Agroecosystems. *See* Ecosystem, weed–desired
   plant
Albuminous cells, 279
Alien weeds. *See* Introduced weeds
Allelochemicals, 203-4, 216-23
   antimicrobial, 209
   chemical isolation of, 206
   chemical nature of, 217-18
   locations in plants, 218
   modes of entry into environment, 220-23
   plant processes affected by, 223
   quantities produced, 218-20
Allelopathy
   competition accentuated by, 219-20
   competition distinguished from, 170, 204, 206,
      211, 215-16
   definition of, 203
   experimental proof of, 205-6
   historical background, 204-5
   interference with desired plants
      germination effects, 209-11
      growth effects, 211-16
   of perennial weeds, 228-29
   weed composition and, 207-9
   in weed management

integrated with biological control, 253, 254-
   55
for minimizing competition, 225-27, 414-16
for reducing emergence, 410-11
for reducing propagules, 405
Allopatric speciation, 46-47
Ammonium, fertilization with, 375
Amphoteric herbicides, 310
Animals
   predation on seeds by, 79
   seed dispersal by, 70, 364
   weed control by, 246
Annidation, 174-75, 176
Annual weeds, 12, 13
   biological control of, 257
   in ecological succession, 59, 61
   ploidy of, 50-51
   relative freedom from pests, 68
   short juvenile period of, 73-74
   survival strategies of, 53
   tillage practices and, 377-78
Antagonists, herbicide, 304
Anthracnose fungus, weed control by, 241-42,
   255-56
Antibacterial agents, in weed seeds, 209
Antidotes, herbicide, 304
Antifungal agents, in weed seeds, 209
Apical dominance, 153-56, 188
Apomixis, 49-50
Apoplast, 278. *See also* Xylem
Aquatic weeds, biological control of, 246-47
Aqueous phase, of cuticle, 282
Asexual reproduction. *See* Vegetative reproduction
Augmentation, of biological control, 234
Auxins, 196-98
Avena species, annidation with, 175
Axillary buds, apical dominance over, 153-56, 188

Bacteria. *See also* Microorganisms
   herbicide degradation by, 319, 321, 327-28
   herbicides derived from, 247
   weed control by, 242-44, 248
      combined with cultural practices, 253
      combined with insects, 252
      combined with mycoherbicides, 252
      for preventing emergence, 411

443

PERSHORE & HINDLIP COLLEGE
LIBRARY.

7 DAY BOOK

Return on or before the last date stamped below.

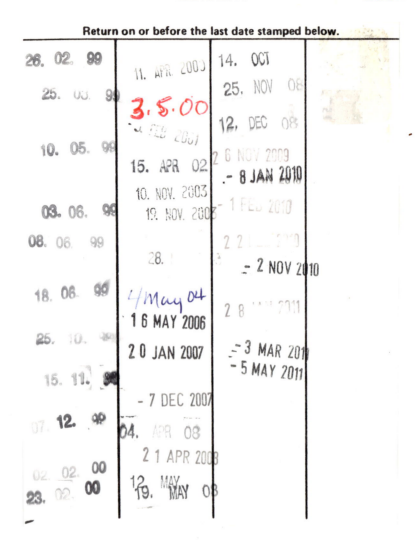

26. 02. 99

25. 03. 99

10. 05. 99

03. 06. 99

08. 06. 99

18. 06. 99

25. 10. 99

15. 11.

07. 12. 99

02. 02. 00

23. 02. 00

11. APR 2000

3.5.00

8 FEB 2001

15. APR 02

10. NOV. 2003

19. NOV. 2003

28.

4/May 04

1 6 MAY 2006

2 0 JAN 2007

- 7 DEC 2007

04. APR 08

2 1 APR 2008

19. MAY 08

14. OCT

25. NOV 08

12. DEC 08

2 6 NOV 2009

- 8 JAN 2010

- 1 FEB 2010

2 2 ... 2010

- 2 NOV 2010

2 8 ... 2011

- 3 MAR 2011

- 5 MAY 2011